The Integrated Nervous System

A Systematic Diagnostic Approach

Walter J. Hendelman
Peter Humphreys
Christopher Skinner

CRC Press
Taylor & Francis Group
Boca Raton London New York

CRC Press is an imprint of the
Taylor & Francis Group, an **informa** business

CRC Press
Taylor & Francis Group
6000 Broken Sound Parkway NW, Suite 300
Boca Raton, FL 33487-2742

Printed in the United States of America on acid-free paper
10 9 8 7 6 5 4 3 2 1

International Standard Book Number: 978-1-4200-4597-0 (Hardback)

Library of Congress Cataloging-in-Publication Data

Hendelman, Walter.
The integrated nervous system : a systematic diagnostic approach / authors, Walter Hendelman, Peter Humphreys, Christopher R. Skinner.
p. ; cm.
Includes bibliographical references and index.
ISBN 978-1-4200-4597-0 (hardcover : alk. paper)
1. Nervous system--Diseases--Diagnosis. 2. Neuroanatomy. 3. Neurosciences. I. Humphreys, Peter, 1942- II. Skinner, Christopher R. III. Title.
[DNLM: 1. Nervous System Diseases--diagnosis. 2. Nervous System Physiological Phenomena. 3. Nervous System Diseases--therapy. 4. Neurologic Examination. WL 141 H495i 2010]

RC348.H3975 2010
616.8'0475--dc22 2009029622

Visit the Taylor & Francis Web site at
http://www.taylorandfrancis.com

and the CRC Press Web site at
http://www.crcpress.com

Dedication

We wish to dedicate this book to our families

and

To the many teachers, mentors and colleagues

we have encountered in our careers

and

To all the students who have inspired us in this

learning partnership

Contents

SECTION II Applying the Basics to Clinical Cases

SECTION III Supplementary Considerations: Rehabilitation and Ethics

List of Illustrations

List of Tables

Chapter 12

Chapter 13

Preface

This book is about integration—how the nervous system itself integrates information from different sensory modalities, from the past and the present, and how you, as a non-neurologist, can integrate your knowledge of neuroanatomy with the art and science of clinical neurology.

The coauthors of this book include a neuroanatomist (WH), an adult neurologist (CS), and a pediatric neurologist (PH), all of whom have been involved in teaching a combined neuroscience and neurology course to second-year medical students. This is a six-week course, problem-based with lectures, organized on the principle of adult learning in small groups, with expert tutors. Clinical disease entities (e.g., multiple sclerosis) are used as the "problems" in the course. One of the real challenges in a course that includes both neuroscience and neurology is the enormous scope of the subject matter; this leads to significant information overload when the process is compressed into a short period of block learning. A by-product of this compressed learning process is the observation by both clinicians that most of their students—members of their tutor groups who knew their material and passed the course, often with a good grade—when returning as clinical clerks or while taking a neurology elective, were unable to use the knowledge they once had mastered to solve clinical problems at the bedside.

This lack of integration of basic science and clinical information has been postulated as the basis of a syndrome called *neurophobia*, which apparently can affect one of every two medical students. This lack of ability to reason through clinical problems results in anxiety and dislike of the subject matter, and eventually negative sentiments about and even fear of neurology (Jozefowicz, 1994). This book has been created to bridge these two worlds and to overcome this pedagogical deficit by use of a problem-based approach with clinical disease entities. Our objective is to bridge the gap between the book and the bedside, or in other words, between the classroom and the clinic.

To help you understand the neurological disorders that you will encounter in this text, and the challenges that patients and relatives face when such disorders appear, we have created a fictional family (the Family Tree can be found preceding the Introduction) whose various members develop different clinical symptoms related to nervous system dysfunction. We meet the central figure of this family in the Introduction, which lays out the complexity of the nervous system and its capacity to multitask. The Introduction exposes the reader to neuroanatomical pathways which, by the end of the book, should be quite familiar in terms of both their function and their importance in neurology.

The first chapter deals with the basics of nervous tissue and presents an overview of the nervous system, enough to set the stage in terms of the basic knowledge of the nervous system needed for this book. The second chapter is devoted to the neurological history and physical examination, and the integration of the information garnered from these activities with respect to the functioning nervous system, for example, the assessment of reflex activity. Patients with neurological problems present great challenges to their physicians (as well as to their relatives and caregivers) to accurately and completely gather the information required to make a working diagnosis and plan of investigation based on a single clinic visit.

Chapter 3 introduces the student to *neurological clinical reasoning* for the purpose of localizing the disease or lesion within the nervous system and determining the etiology, the pathophysiological mechanism of disease. The approach used in this chapter and its accompanying worksheets is designed to provide students and non-neurologist clinicians with practical guidelines and tools with respect to diagnosis for the full range of neurological problems seen by a neurological generalist. This approach is applied throughout the book.

The succeeding nine chapters deal with important specific clinical diseases or syndromes that have afflicted our fictional family and their friends, each with the focus on a different component of the nervous system (e.g., peripheral nervous system, brainstem, etc.). For each of these cases, the history is presented followed by the findings of the neurological examination. Additional neuroanatomical, neurophysiological, and neurochemical information is added, where required, as it pertains to the clinical condition discussed in the chapter. Notwithstanding the information given, it is suggested that students review their knowledge of neuroanatomy, neurophysiology, neuropharmacology, and neuropathology, using other resource books (given in the suggested readings and references sections at the end of each chapter and in the annotated bibliography at the end of the book). In each of these chapters, there is an application of the process of neurological reasoning to narrow the possibilities of *where* the lesion is located. This is followed by a systematic analysis in order to determine *what* disease (or diseases) should be considered. Relevant selected investigations are then presented and the results discussed. Finally, the diagnosis is made, with its prognosis, and an outline of the appropriate management is given, ending with the outcome of the case.

The text illustrations have been prepared with an emphasis on the functioning nervous system. In addition to neuroanatomical drawings related to the cases and tables with relevant clinical data, there are figures illustrating neurophysiological concepts, clinical findings (such as radiographic images and EEGs) and microscopic neuropathological images. Again, the information is described in the context of the disease presented in that chapter. The glossary of terms also emphasizes clinical terminology.

The final two chapters were written by guest authors: a physiatrist (a specialist in rehabilitation medicine) and a senior neurologist with expertise in ethics. Both discuss other important dimensions of neurological problems: rehabilitation and ethics. Currently, *rehabilitation* has much to offer for those afflicted by disease or injury of the nervous system; there is now a certain air of hopefulness that there can be recovery of function following an insult to the nervous system, in adults as well as in children. The ethical principles and reasoning on the basis of which decisions (sometimes quite unique) are taken in neurological cases are presented in the context of an inherited disease of the nervous system.

There is a DVD included with this textbook. It contains the worksheets that have been developed to apply the clinical reasoning approach to neurological problem solving. It is highly recommended that you, the non-neurologist student/learner, apply this approach when confronted with a neurological patient. The authors want you to learn how to think like a neurologist and propose that you adopt a contract for this purpose. Please read The Learning Contract; hopefully, you will agree to fulfill its obligations. In addition, the DVD has all the illustrations found in the book, with animation added to assist in the understanding of various pathways and reflect circuits. It also includes the glossary. The DVD also has a learning module to assist you, as a nonexpert, in understanding neuroimaging, how the various modes of CT and MRI assist in localizing a lesion and defining the likely etiology.

Now for the final value-added feature—the additional cases for each of the clinical case chapters, referred to in the text as **e-cases**. This book is not intended to be a detailed compendium of the basic sciences related to neurology, nor is it intended to contain all of the most frequent neurological disease entities. The e-cases enlarge the scope of the book by adding other commonly seen neurological diseases for each level of the nervous system. These are presented in a more straightforward fashion, although once having learned the analytic approach it is hoped (expected) that you, the learner, will work through each clinical case on your own, using the worksheets, before reading the case evolution, investigations, and resolution. The e-cases are found on the accompanying DVD and are listed as *case studies* for each of the clinical chapters; they will also be found on the text Web site (http://www.integratednervoussystem.com). The Web site will also be utilized to provide updates on the cases presented as well as new cases, so it may be wise to check it periodically. We would appreciate your honesty in not copying the DVD for others and in not sharing the URL.

We believe that this book, the DVD, and the associated Web site will be of practical value to all the professionals who deal with people who have neurological conditions, not only medical students and residents. This includes physiatrists, physiotherapists, occupational therapists and speech therapists, and nurses who specialize in the care of neurological patients. We think that this text will also be of value for family physicians and specialists in internal medicine and pediatrics, all of whom must differentiate between organic pathology of the nervous system and other conditions.

The aim of this book is to enable you, the learner, to use your knowledge of the nervous system combined with a neurologically based, problem-solving clinical reasoning approach to neurology to help in the diagnosis and treatment, in the broadest sense, of those who suffer from a neurological disease or injury. We hope that this approach meets with success, insofar as it leads to an improvement in the diagnosis and care of persons afflicted by neurological problems.

Dr. Walter J. Hendelman,
Dr. Peter Humphreys, and
Dr. Christopher Skinner
Faculty of Medicine
University of Ottawa
Ottawa, Canada

Reference

Jozefowicz, R.F. Neurophobia: The fear of neurology among medical students. *Arch. Neurol.* 51 (1994): 328–329.

Authors

Dr. Walter J. Hendelman

Walter Hendelman, a Canadian born and raised in Montreal, did his undergraduate studies at McGill University in science with honors in psychology. He then completed medical studies at McGill, received his M.D.,C.M., and subsequently did a year of internship and a year of pediatric medicine, both in Montreal.

Having chosen the brain as his lifelong field of study and work, Dr. Hendelman chose the path of brain research. Postgraduate studies followed in the emerging field of developmental neuroscience, using the "new" techniques of nerve tissue culture and electron microscopy at Columbia University Medical Center in New York City.

Dr. Hendelman then returned to Canada and made Ottawa his home for his academic career at the Faculty of Medicine at the University of Ottawa. He began his teaching in gross anatomy and neuroanatomy, and in recent years he has concentrated on the latter, first assuming the responsibility of coordinator for the course and then becoming co-chair for the teaching unit on the nervous system in the new curriculum. His research has focused on the examination of the development of the cerebellum and the cerebral cortex.

As a teacher, Dr. Hendelman is dedicated to assisting those who wish to learn functional neuroanatomy. He began by producing teaching videotapes on the brain based upon laboratory demonstrations. During the 1990s, when digital technology became available, Dr. Hendelman recognized its potential to assist student learning, particularly in the anatomical subjects. He has collaborated in the creation of two computer-based learning modules (one on the spinal cord based upon the disease syringomyelia and the other on voluntary motor pathways). He is the author of a teaching atlas of neuroanatomy, the *Atlas of Functional Neuroanatomy* (accompanied by a CD-ROM), published by CRC Press and presently in its second edition.

In 2002, Dr. Hendelman received a master's degree in education from the Ontario Institute of Studies in Education, affiliated with the University of Toronto.

He welcomes feedback on this book and can be reached at whendelm@uottawa.ca.

Dr. Peter Humphreys

A graduate of the McGill University Faculty of Medicine in 1966, Dr. Humphreys trained in pediatrics at Boston Children's Hospital and at St. Mary's Hospital in London, followed by training in neurology at the Montreal Neurological Institute. After a six-year stint on the neurology staff of the Montreal Children's Hospital, he became the founding head of the Neurology Division at the Children's Hospital of Eastern Ontario, Ottawa, a position he held for twenty-three years. Currently a full professor in the Department of Pediatrics at the University of Ottawa, Dr. Humphreys combines a busy academic neurology practice with an active teaching schedule. His principal area of interest is in disorders of brain development.

As a teacher, Dr. Humphreys has been active at all levels of the medical curriculum. For many years he was a tutor in small-group learning sessions for second-year medical students doing a problem-based introductory course on the nervous system. During the same course he conducts a full-class lecture on brain developmental disorders as well as a live patient demonstration devoted to the pediatric neurological examination. For senior medical students he conducts small-group case-based seminars and supervises them in clinics and on the wards. He also does bedside and clinic instruction for residents in neurology and pediatrics. Finally, he participates in a teaching role in refresher courses for the University of Ottawa, the Canadian Paediatric Society, and the Canadian Neurosciences Federation.

Dr. Christopher Skinner

Dr. Skinner received his B.Eng. (Electrical) from the Royal Military College in 1970. He worked as a systems engineer with the Department of National Defence, implementing nationwide information systems until 1975. He received his medical degree from Queen's University in 1979. He received his specialist certification in general internal medicine in 1986, in neurology in 1987, and qualified as a Diplomat of the American Board of Sleep Medicine in 2005. He was Chief Information Officer at the Ottawa Hospital from 1996 to 1998.

He has been a clinical teacher and lecturer in the Faculty of Medicine at the University of Ottawa since 1993. He has taught clinical neurology, occupational neurology, and sleep medicine to all levels of study, including medical students, residents, physician assistants, and military flight surgeons. He was also involved in the design and implementation of the problem-based digital curriculum portal used for the teaching of medical students.

Dr. Skinner currently practices general neurology and sleep medicine at the Ottawa Hospital.

Acknowledgments

The authors wish to express their appreciation and respect for the two illustrators who have helped shape this book, Perry Ng and Dr. Tim Willett. Without their creative and conscientious efforts, we could not have achieved what is necessary to convey our message to you, the learner.

The authors would also like to acknowledge the work of David Skinner, who diligently crafted the DVD and Web site to emulate the problem-solving methodology of the text.

A special note of gratitude is extended to our chapter contributors, Dr. R. Nelson (Neurology, The Ottawa Hospital) and Dr. A. McCormick (Physiatry, The Children's Hospital of Eastern Ontario). Dr. Nelson is a neurologist's neurologist, highly regarded by his colleagues, with a special interest in ethical issues. Dr. McCormick, who carries with her an air of enthusiasm and hope, has successfully championed the cause of pediatric rehabilitation and is one of the few people in her field who actively treats both children and adults.

Many colleagues have contributed illustrations and clinical material collegially and willingly to this book. We are particularly grateful not only for their particular and unique contribution but also for the spirit in which it has been donated. In many cases their staff helped with the preparation of this material and we thank them as well.

Dr. D. Grimes: Divison of Neurology, The Ottawa Hospital
Dr. R. Grover: Neuroradiology, The Ottawa Hospital
Dr. M. Kingstone: Neuroradiology, The Ottawa Hospital
Dr. J. Marsan: Otolaryngology, The Ottawa Hospital
Dr. J. Michaud: Neuropathology, The Ottawa Hospital
Dr. M. O'Connor: Ophthalmology, The Children's Hospital of Eastern Ontario
Dr. C. Torres: Neuroradiology, The Ottawa Hospital
Dr. S. Whiting: Neurology, The Children's Hospital of Eastern Ontario, with S. Bulusu (Chief Technologist, Clinical Neurophysiology Laboratory, CHEO)
Dr. J. Woulfe: Neuropathology, The Ottawa Hospital

The Health Sciences library staff is thanked for assistance in creating the annotated bibliography and checking some of the references.

We also wish to thank the secretarial staff that we work with in our various offices, but particularly Orma Lester, who has carried the major load in assisting with the various drafts of the textbook in addition to her regular duties in a busy hospital office.

Last, but not least, the authors gratefully acknowledge the assistance and cooperation of the production team of CRC Press, particularly Tara Nieuwesteeg (production editor), Pat Roberson (production coordinator), and Barbara Norwitz, our executive editor.

The Learning Contract

The basis of this book is to teach you—whether medical student, resident, physiotherapist, general practitioner, or other health professional—how to apply your knowledge of the nervous system in solving clinical neurological problems, thus helping those afflicted with neurological diseases.

The approach presented in Chapter 3 is the one used throughout the nine chapters that deal with important clinical diseases or syndromes. The worksheets found in that chapter for analyzing a neurological problem—to determine the localization and the etiology of the disease process or clinical disorder—are also on the DVD accompanying the book. The student/learner should engage in the process of using this approach to work through the case presentations. The goal is to utilize this approach so that it becomes internalized, along with the formal knowledge of the neuroanatomy and pathology of the disease itself. The worksheets should be printed out before tackling the clinical problem, either in the text or with the e-cases on the DVD and Web site.

We suggest that you sign a learning contract with yourself. This is not unlike the "contract" between patient and physician, whereby the patient agrees to divulge full information to the doctor and allows him or her to probe his or her body. At the same time, the physician agrees to certain rules (including respect and confidentiality) and commits to using his or her knowledge and expertise to help the patient.

The learning contract that needs to be agreed upon for this approach to work is that, rather than skipping to the last page of each chapter to find the correct diagnosis, you—the learner—will undertake to follow the prescribed clinical reasoning approach, as presented in Chapter 3 and on the DVD and Web site.

We are confident that the effort taken to apply this process of neurological reasoning will reward you in the long run and free you from memorizing innumerable tables with names of neurological diseases, often times with long names or conditions named after persons who first described the disease entity.

It might be too much to ask, but perhaps you would like to put the following contract in your own words!

My Learning Contract

Family Tree Illustration

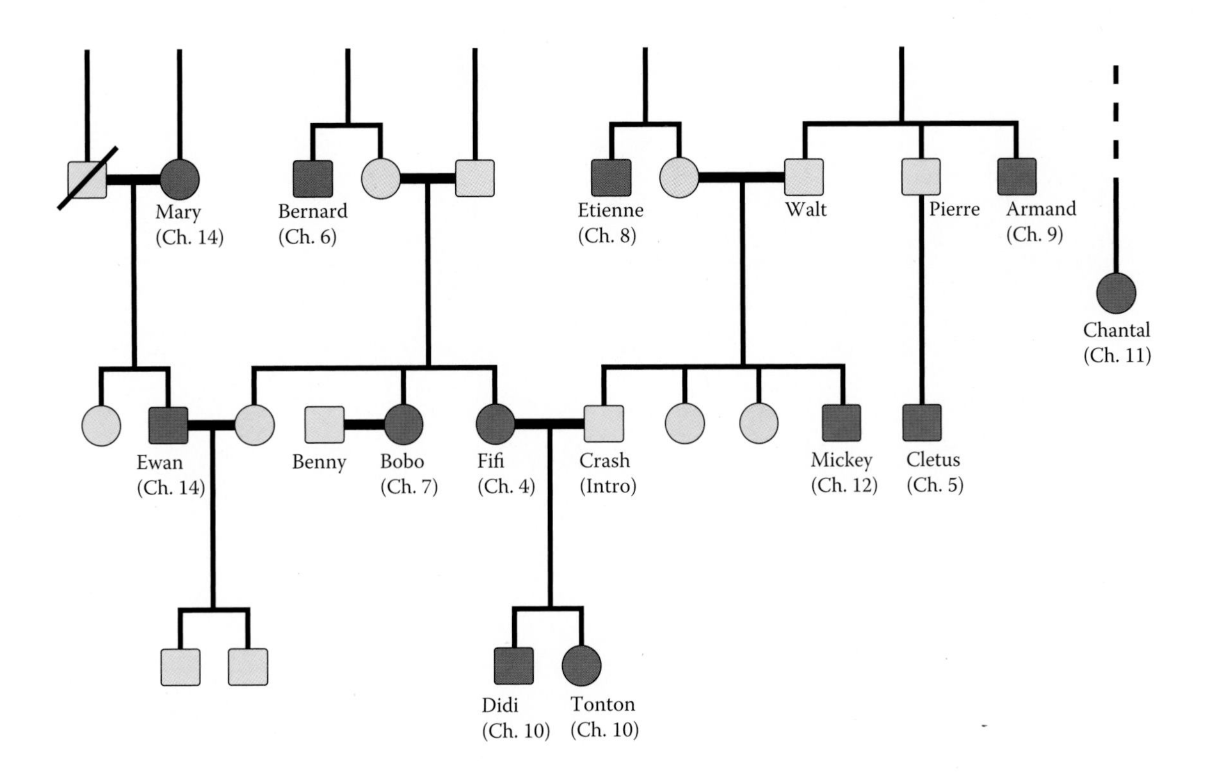

The McCool family tree.

Introduction

What do you think would be one of the most difficult challenges that the human nervous system could face in modern times?

Take a ride with Crash McCool, an experienced ex-military airline pilot, and see what he has to do on a day full of surprises. Add on the challenges of multiprocessing in a complex air transport environment, the G forces, heat stress, and basic activities of self-preservation.

The human nervous system has indeed been designed to deal with many concurrent tasks: visual, auditory, tactile, proprioceptive, vestibular, emotional, and executive.

In this introduction, we will show you how Crash's normal nervous system manages multiple complex tasks concurrently. We will then teach you how to troubleshoot the nervous system to deal with the disease processes that degrade it.

Chris McCool (or "Crash" as he is known to his buddies) is an experienced pilot with the new low-budget Canadian airline called DuckAir. He is the captain of a shiny new Airbus A380 with pilot Flash Gordon as his first officer; they are presiding over an inaugural flight from Halifax to Vancouver, six hours non-stop with all of the important Canadian press aboard.

On takeoff, as he holds the control stick of his A380 with his right hand, Crash makes fine adjustments to keep his distance from the ground while rotating the aircraft during its ascent. Little does he know that his hand is sensing the position and tension on the stick through the proprioceptive receptors in the joints of his hands; this information is passed upward via the median, radial, and ulnar nerves through the brachial plexus to the spinal roots of the cervical spinal cord. After entering into the dorsal root ganglia, the proprioceptive nerve fibers enter the spinal cord in the sensory processing area of the spinal cord. These fast information fibers head backward to the posterior columns of the spinal cord to ascend to the medulla in the brainstem.

The proprioceptive fibers end in the lateral of two posterior column nuclei; the next sensory fibers in the chain cross over into a fiber bundle called the medial lemniscus on the opposite side of the brainstem to travel with other sensory information through the pons, then the midbrain and upward to the thalamus.

The fibers from the posterior column nucleus end on the ventral posterolateral nucleus (VPL) in the thalamus where they converge with sensory data from other areas of the right hand and arm giving position and force information; all of this information is filtered and modified by cells in the thalamus. The processed information is then sent via a third nerve fiber bundle to the sensory area of the cerebral cortex, which keeps a holographic neural image of the right hand and arm.

Messages from the visual cortex tell him that he needs a bit more speed and must delay pulling back on the stick to rotate the aircraft during takeoff. His motor cortex sends a message to the putamen of the basal ganglia, which forwards the signal to the globus pallidus; in collaboration with other elements of the basal ganglia, these structures access a pre-programmed sequence of muscle actions involving alternating the flexors and extensors of his right hand to make the adjustments. This sequence is sent to the thalamus, which makes a last minute comparison of the planned sequence with the actual position of the hand and sends the output back the motor cortex. The motor cortex then sends the final commands via the corticospinal tract, a collection of upper motor neuron fibers projecting directly to the spinal cord, as well as to nuclei in the brainstem such as the reticular formation in the upper medulla. The cerebellum applies fine braking actions to smooth out the movements.

The signals in the corticospinal tract terminate in the gray matter of the spinal cord on the lower motor neurons of the ventral horn of the spinal cord. The message is then transmitted to the lower motor neuron, which carries the signals in the form of motor units. Motor units consist of the lower motor neuron connected to an axon that terminates on a collection of single muscle cells that receive the same message to contract simultaneously. Multiple motor units are activated for one single movement of the hand. The stimulation of the motor units consists of tightly coordinated flexor/extensor sequences to maintain fine control and appropriate tone to the limb.

Two and a half hours into the six-hour flight, while over Northern Quebec, McCool's copilot casually says: "Hey, boss, I think that we are running out of gas." "Not again—not another of those liter-to-gallon mix-ups!" Crash says to himself as he hears each of the four engines become silent in sequence. The words of his copilot and the sounds of the engines enter through the inner ear and are transduced through the cochlea in the inner ear into impulses on the acoustic or eighth cranial nerve, which terminates in the cochlear nucleus in the ponto-medullary junction on both sides. The information is modulated and forwarded to the medial geniculate body of the thalamus through a fiber bundle called the lateral lemniscus. The information is filtered and transformed, then forwarded through white matter bundles to the infrasylvian cortex of the temporal lobe. The auditory information is again stored in a real time neural network in this area, which is connected to the association cortex of the superior gyrus of the temporal lobe. This structure interprets the speech information in the message by recalling language objects that have been remembered from past experience. McCool interprets the sound objects and assesses the situation immediately as the processed speech information is passed forward to the prefrontal cortex through the arcuate fasciculus.

The decoded language information, along with the position and height of the aircraft and other associated data, are passed forward to the dorsolateral frontal cortex where an executive decision is made to take action. He simultaneously puts the aircraft into a glide, puts the landing gear down and says to his copilot: "We are over southern James Bay; there is a small airfield at Moosonee where I originally got my wings. Call the airport manager there and have him light up the runway with some burning barrels. This guy also runs the taxi company; if we make it in one piece, we will give him lots of business tonight. Also, tell the cabin crew to get those paparazzi in the back to put away their caviar and champagne and put their heads between their knees for a bumpy landing."

The speech is generated with preprocessing in the frontal lobe; this information is then forwarded to the opercular area that actually produces speech (Broca's area). Broca's area generates the speech sequence, which descends through the corticobulbar tract to the motor nucleus of the vagus nerve. The lower motor neuron to the vocal cords originates in the motor nucleus of the vagus nerve in the dorsum of the medulla and descends into the thorax via the vagus nerve, then comes back up via its recurrent laryngeal branch to the muscles of the larynx.

At the same time, other axons project from the speech cortex to neurons in the pons and medulla that travel to the tongue and face muscles. Still others project to the upper spinal cord to connect to neurons controlling the movement of the diaphragm and intercostal muscles for the generation of respiratory movement. Subsequent diaphragmatic and intercostal contractions cause air to be expelled from the lungs into the trachea and larynx where the air flow causes vibration of the vocal cords. This sound is modulated by the length and tension of the vocal cords in the larynx. The sound is further modified by muscles of the face and upper airway to create the words.

McCool has practiced the maneuver of landing an unpowered passenger aircraft in the flight simulator many times, as it is required training for Canadian airline pilots. He doesn't even have to think about how to perform the task. The training sessions for this sequence have created information pathways involving multiple structures such as the cerebellum, which provides constant complex feedback information to the preprogrammed maneuver, which is executed through the cortex, basal ganglia, thalamus and corticospinal tract.

During all of this sequence, McCool does not notice that his left ankle is causing him pain after he had twisted it playing volleyball the night before. The pain receptors in the ankle joint send impulses through the tibial nerve, which joins the peroneal nerve to form the sciatic nerve in the back of the leg. The sciatic nerve enters the pelvis through the sciatic notch where it divides into the lumbosacral plexus before reforming into spinal roots of the lumbar and sacral segments of the lower spinal cord. The pain information enters the spinal cord through the dorsal roots and further into the dorsal gray matter of the cord. This first sensory neuron terminates here and is connected to interneurons, which modulate the output; the next pain neurons in the chain form a white matter tract which crosses, through the anterior commissure of the spinal cord, to the opposite side to form the lateral spinothalamic tract. This tract system ascends on the opposite side of the spinal cord up through the brainstem to join the thalamus in the VPL nucleus.

Similar to the proprioceptive preprocessing in the thalamus, the pain information from the right ankle is filtered and modulated in the thalamus and forwarded to the parietal sensory cortex. The high level of vigilance required during these complex flying maneuvers has caused the thalamus to suppress the pain impulses to the cortex so that he could attend to the essential tasks of flying his aircraft safely.

The visual system, the most complex of all, starts with the images of the surroundings entering his eyes through the cornea, lens and vitreous to focus a full color image on the macula of the retina. The signal is transduced by the rods and cones, the photosensitive cells of the retina. The signal is then modulated by the other cell layers and outputs through the optic nerve, which exits each eye through the optic disc. The two optic nerves join together in the optic chiasm where they separate from being eye-specific to visual field specific. The two visual fields separate into the optic tracts, which take the right visual field to the left side of the brain and left visual field to the right side of the brain.

The two optic tracts terminate in the thalamus in the lateral geniculate bodies (LGB) on each side. The majority of the nerve cells in the LGB send axons to the occipital cortex via the optic radiation. This is a densely packed tract of white matter that passes through the temporal and parietal lobes before terminating in the occipital lobes. The visual images of the surroundings are stereotopically represented in the occipital lobes. The images are then exported to association areas in the parieto-occipital cortex that process color and recognize faces and other objects in a similar fashion to what the temporal cortex does for language.

Crash sees the airfield in the distance; it is snow covered and brightly lit with flaming fuel drums. He judges the distance from the runway, speed and glide angle all simultaneously. Whoever would have thought that today he would have to land a 450-ton aircraft on a 5000-foot runway with a 20 mph crosswind in blowing snow with no power? All of his faculties are now running at full capacity and attention.

With the help of Flash, Crash successfully lands the aircraft, only blowing a few tires, his brakes glowing cherry red, having 50 feet of runway to spare. Flash says, "Not bad, boss; just think of the paperwork that we are going to have to do."

After evacuating the passengers safely, Crash pulls out his satellite phone while thinking of the smell and taste of some steak and red wine that his wife Fifi was to prepare for him tonight following his arrival in Vancouver. "Fifi, we have been slightly diverted; I will be home a little late tonight; don't wait up," His testosterone-soaked frontal lobes cause some minor modification of the truth.

Flash asks: "So what are we going to call this mess-up?"

Crash scans his hippocampus and amygdala for some light humor. "How about the 'Moosonee Mallard'?!"

Section I

The Basics of Neurological Problem Solving

Chapter 1

Synopsis of the Nervous System

Objectives

- To review the basic histological knowledge of the nervous system from a functional (neurological) perspective
- To organize the nerves, nuclei and tracts of the nervous system into functional systems

In the Introduction, Crash, an aircraft pilot and a member of our textbook family, executes a number of intricate tasks in response to input from several sensory systems: tactile, muscle and joint, visual, vestibular, and auditory. He reacts to all of these stimuli appropriately and performs highly accurate and skilled motor movements. Pathways were sketched for the sensory input and for executing the motor movements.

How does the brain process all this information? Which parts of the nervous system are involved in the exquisite motor control required to fly a jet aircraft? Where in the "brain" are the integration and decision-making functions carried out?

A neurologist's view of the nervous system is one of functionality—are all the components operational in order to receive information, analyze and assess its significance, and produce the appropriate action? If not, the task of the physician is to determine where the problem is occurring and what is its most likely cause—the localization of the lesion and its possible etiology.

In order to determine where, the *localization*, one needs to have knowledge of the anatomy and physiology of the nervous system. This chapter will provide that information from a functional perspective, but the student should expect to consult other resources—details of neuroanatomy, neurophysiology, and neuropathology—to supplement this presentation (see the Annotated Bibliography). One needs this knowledge to understand the significance of the findings of the neurological examination, which is outlined in the next chapter.

Determining the likely cause, the *etiology*, requires knowledge of disease processes. This determination is based initially on the nature of the symptoms and the history of the illness—how long the problem has been occurring (acute, subacute, chronic) and how the symptoms have evolved over time. The task of the practitioner—physician, resident, or student—will be to determine what disease process (e.g. infectious, neoplastic) is occurring and its pathophysiology, and to identify diseases that most likely account for the patient's signs and symptoms.

Laboratory investigations, including blood work, special tests (e.g., disease-specific antibody levels) and particularly neuroimaging, usually provide additional information to help pinpoint the localization of the disease and often limit the possible list of most likely diseases.

Lastly, the neurologist will synthesize the patient's history and the symptoms with the signs found on neurological examination as well as the additional information provided by the investigations to come up with the definitive diagnosis. This diagnosis allows for a therapeutic plan and some idea of the likely outcome: the prognosis. All of this must be communicated sensitively to the patient and family in a way that can be readily understood.

1.1 Neurobasics

The nervous system is designed to receive information, analyze the significance of this input and respond appropriately (the output), usually by performing a movement or by communicating ideas through spoken language. In its simplest form, this process would require a minimum of three neurons, but as we come up through the animal kingdom the complexity of analysis increases incredibly. This evolutionary development culminates in the human central nervous system with all its multifaceted functions.

The nervous system consists of two divisions (Figure 1.1), a peripheral component, called the *peripheral nervous system*, the PNS, and a central set of structures, called the *central nervous system*, the CNS. The PNS consists of sensory neurons and their fibers, which convey messages that originate from the skin, muscles and joints, and from special sensory organs such as the cochlea (hearing); it also carries the motor nerve fibers that activate the muscles. The *autonomic nervous system*, the ANS, is also considered part of the PNS; it is involved with the regulation of the cardiac pacemaker system and of smooth muscle and glands, including control of bowel and bladder functions. The CNS consists of the spinal cord, the brainstem and the brain hemispheres. The CNS adds analytic functionality and varying levels of motor control, culminating in a remarkably intricate capacity for "thinking" forward and backward in time, in consciousness, language

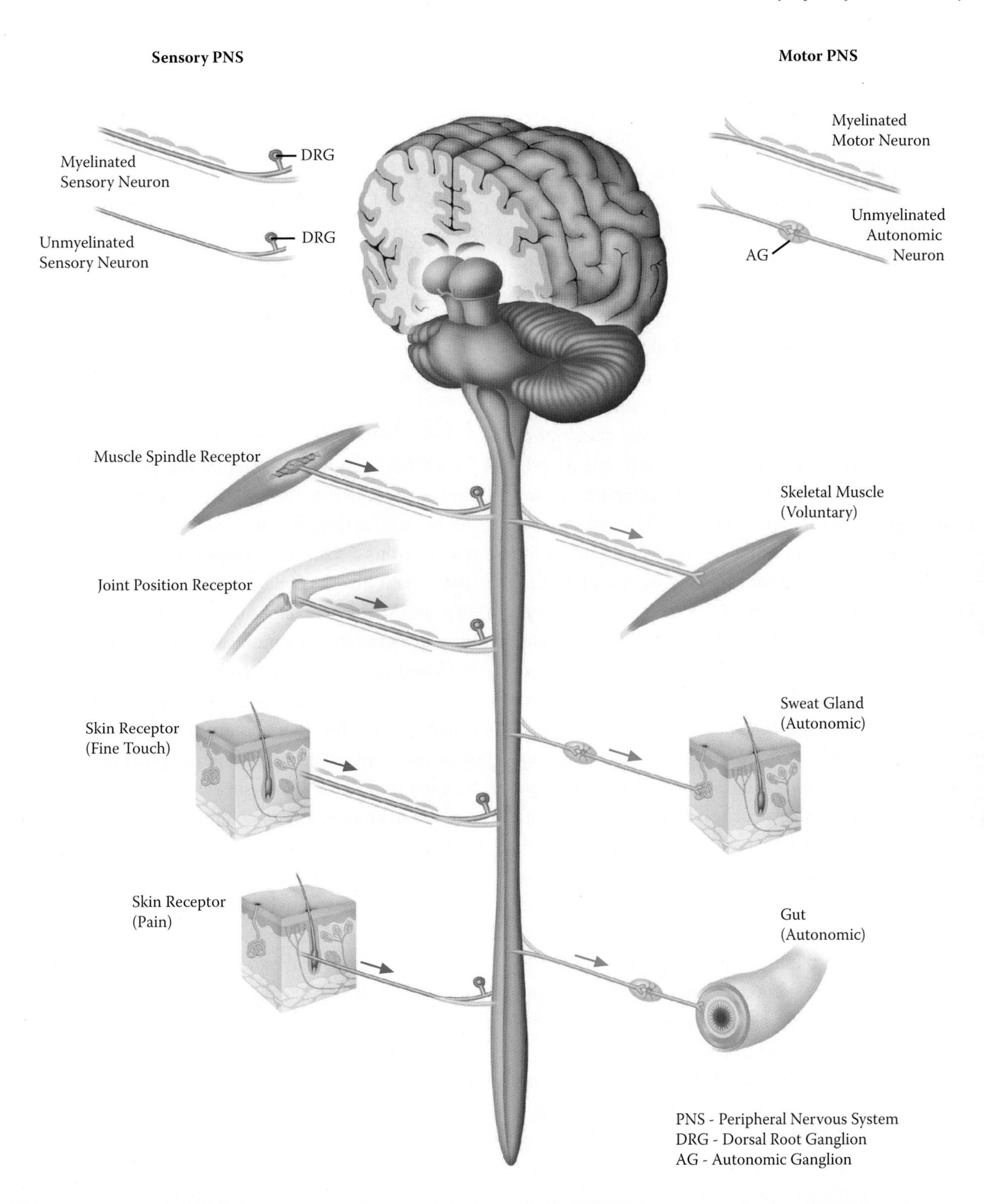

FIGURE 1.1 Overview of the nervous system—CNS, PNS, and ANS—central, peripheral and autonomic nervous systems. Sensory fibers, colored purple, convey information toward the CNS (afferent) from receptors in skin and muscle. Motor fibers, colored green, carry instructions away from the CNS (efferent) to muscle and via autonomic ganglia to glands and viscera.

and executive functions, processes performed in different areas of the cerebral cortex.

1.1.1 The Neuron

A *neuron* is the basic cellular element of the nervous system. In the most simplistic language of today's electronic world, each of the billions of neurons in the human central nervous system is equivalent to a unique microchip, possessing a specific information processing capacity. Like other cells, the neuron has a cell body (the soma or perikaryon) with a nucleus and the cellular machinery to be its nutritive center. Morphologically, it is the cellular processes—dendrites and axon—that distinguish a neuron from other cells. The electrochemical nature of its membrane, whereby the interior of the cell and its processes have a negative charge, is a characteristic feature of the neuron (see Figures 10.7 and 10.8). The synapse, the electrochemical communication between neurons, is the other unique feature of nervous tissue (discussed below).

The typical neuron in the CNS (Figure 1.2A) has *dendrites* that extend from the cell body for several microns. Dendrites receive information from other neurons at specialized receptor areas, the synapses, some of which form small excrescences on the dendrites, called *synaptic spines* (discussed below). More complex neurons have an extensive arborization of dendrites and receive information from perhaps hundreds of other neurons. Neurons of a certain functional type tend to have a typical configuration of their dendrites and group together to form a *nucleus* (somewhat confusing terminology!) in the CNS, or a layer of *cortex* (e.g., the cerebral or cerebellar cortex). Brain tissue is traditionally fixed in formalin for the purpose of study and neuronal areas (e.g., the cortex) become grayish in appearance when the brain is cut; hence, the term *gray matter* is used for areas of neurons, their processes and synapses.

The neurons in the PNS include both sensory neurons and those associated with the ANS. The cell body of a sensory neuron is displaced off to the side of its two processes: a peripheral process (e.g., to the skin) and a central process, which will form synapses within a nucleus in the spinal cord or brainstem (Figure 1.2B). The distal endings of sensory neurons (e.g., in the skin) are sensitized to receive information of a certain type and hence behave functionally as dendrites. Often they are enveloped by specialized receptors (e.g., specialized touch and temperature receptors in the skin). Receptors for pain sensation are "naked" nerve endings in the skin and are located within all tissues (except brain tissue of the CNS). For the special senses such as hearing, there are highly developed receptor cells (e.g., hair cells of the cochlea) which are in contact with the sensory neurons.

The cell bodies of sensory neurons congregate in a specific location forming a peripheral *ganglion* (plural, ganglia), typically located along the dorsal root (see below, the dorsal root ganglia, and Figure 1.1). Neurons of the autonomic nervous system also form ganglia (see Figure 1.1), located alongside the vertebra (sympathetic) and within or closer to the end organ (parasympathetic; see Section 1.2.2).

1.1.2 Axons (Nerve Fibers)

Each CNS neuron has (with rare exception) a single *axon*, also called a nerve fiber, which is the efferent process of the neuron, acting like an electric wire to convey information from a neuron to other neurons, muscle or other tissue, with the possibility of many branches (collaterals) en route. Millions of axons course within the CNS providing extensive intercommunication within and between the CNS neuronal nuclei and cortical areas (see below). The axons of functionally linked sensory and motor neurons usually bundle together and are called tracts or pathways. A neuron within the brainstem or spinal cord that sends its axon (via the PNS) to skeletal muscle fibers is called a motor neuron, also known clinically as the lower motor neuron (discussed below).

A typical (observable, dissectible) nerve in the PNS usually has both motor and sensory nerve fiber bundles, and often autonomic fibers. The postganglionic fibers of the ANS (sympathetic and parasympathetic) are distributed to smooth muscle and glands.

1.1.3 Myelin

Myelin is the biological (lipid-protein) insulation surrounding axons; its function is to increase the speed of axonal conduction (Figure 1.2C). Since axonal conduction velocity increases in proportion to axonal diameter, an alternative to a myelin sheath would be a marked increase in axonal diameter; this, if present in most CNS axons, would cause the nervous system to be extremely bulky. Such an arrangement would require much longer axons that would be easily susceptible to transmission degradation, rendering the nervous system more inefficient. In the human brain, axons may travel for long distances, and the longer the distance

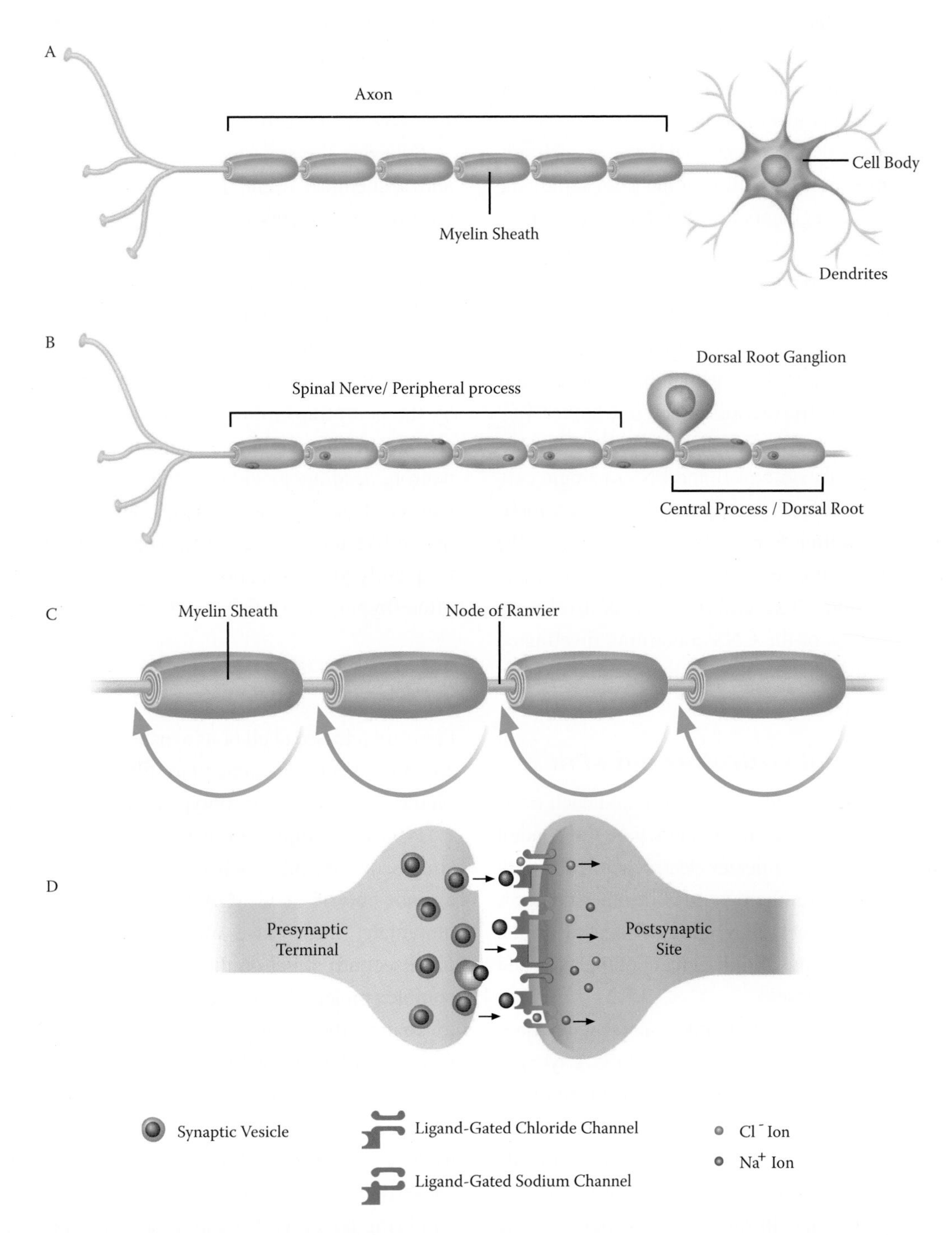

FIGURE 1.2 (A) A "typical" CNS neuron has several dendrites and a single (myelinated) axon; its terminal branches communicate with other neurons. (B) A PNS neuron has both central and peripheral (myelinated) processes. The sensory endings of the peripheral process are located in the skin joints and muscles. The cell body of the PNS neuron is located within the dorsal root ganglion (DRG). The central process enters the CNS via the dorsal root. (C) An axon of a nerve with its myelin sheath: each internode segment is the territory of a single Schwann cell in the PNS or an oligodendrocyte process in the CNS. A node of Ranvier separates each segment and the impulse "jumps" from node to node, a process called saltatory conduction. (D) A "generic" synapse is illustrated, including the presynaptic ending with the synaptic vesicles containing a neurotransmitter, the synaptic "gap" and the postsynaptic site with its Na and Cl ion channels.

the more likely it is that the fibers are myelinated; faster transmission such as that required for certain sensory and motor functions requires axons with larger diameters, and these axons have thicker myelin sheaths.

Myelin is composed of segments and between each segment is a very short "naked" section of the axon, called the node of Ranvier; the segments are therefore called internodes. At the nodes there is a concentration of sodium ion channels within the axonal membrane. Impulse conduction along a myelinated axon "jumps" from node to node, a process called saltatory conduction, which thus speeds the transmission of the impulse; the membrane potential is recharged at each node (further discussed in Chapter 4).

Myelin is formed and maintained by glia, supporting cells of the nervous system. In the PNS, a single cell, known as the Schwann cell, is responsible for each internode segment of myelin (see Figures 1.2C and 4.4). In the CNS, the equivalent glial cell is the oligodendrocyte, and each cell is responsible for several segments of myelin, on different axons. Areas of the CNS containing myelinated tracts have a whitish appearance with formalin fixation and are hence called the *white matter.*

1.1.4 Synapse and Neurotransmission

In the CNS, the terminal end of each axon and each of its collaterals is a *synapse*, a specialized junction, the conduit by which one neuron communicates electrochemically with another. Synapses abut on the dendrites of other neurons, typically at synaptic spines; they are also located on the cell body of neurons, on the initial segment of the axon and sometimes on other synapses.

Synapses can be seen with light microscopic techniques but are best visualized with electron microscopy. A synapse (Figure 1.2D) consists of an enlargement of the terminal end of the presynaptic axon containing small (synaptic) vesicles, the presynaptic membrane, and a postsynaptic receptor site (e.g., a dendritic spine) where the membrane is specialized for neurotransmission; in some cases the postsynaptic membrane is thickened. In between is a space or cleft, the synaptic gap, which is sometimes widened compared to the usual space between adjacent cells in the CNS.

Biologic agents that can alter the membrane properties of neurons at the postsynaptic site are called *neurotransmitters*. These are synthesized in the cell body, transported down the axon and stored in the synaptic ending in packets, the synaptic vesicles. In some cases, the neurotransmitter may be synthesized within the nerve terminal. These endings are activated when the axonal electrical impulse, the action potential, invades the terminal, setting off a process whereby the transmitter is released into the synaptic cleft.

Synaptic transmission is therefore both an electrical and a chemical event. In some instances synapses have built-in mechanisms for recapturing the neurotransmitter (recycling); alternatively, enzymes in the synaptic cleft may destroy the active neurotransmitter. Glial cells (astrocytes) may be involved in this process, for example by removing the neurotransmitter from the synaptic site.

Receptors on the postsynaptic neuron are activated by the neurotransmitter, causing a shift of ions and a net change in the membrane potential of the postsynaptic neuron, leading to either depolarization or hyperpolarization of the membrane. This contributes to an increase in the likelihood of the neuron either to discharge more frequently (depolarization, excitatory) or to discharge less often (hyperpolarization, inhibitory).

1.1.5 Neurotransmitters

A neuron may synthesize one or more neurotransmitters and these are released at all of its synaptic endings. The action of any single neurotransmitter may differ at each site depending on the receptor type or subtype in the postsynaptic neuron.

Typical examples of neurotransmitter chemicals are simple amino acids, such as the inhibitory-acting gamma-amino butyric acid (GABA) or the excitatory-acting glutamate; these have an immediate but short-lasting (millisecond) effect on the postsynaptic membrane. More complex molecules may act long term (seconds or minutes) to change the nature of the response (an effect called neuromodulation), or to alter the properties of the membrane of the postsynaptic neuron, thereby strengthening or diminishing the synaptic relationship, a process known as long-term potentiation. Other neurotransmitters may cause the release of messengers which enter the nucleus and bring about the activation or deactivation of genes in the nucleus, thereby producing a long-lasting effect on the cell and its synaptic relationships. (Neurotransmitters are further discussed in Chapter 10.)

1.1.6 Muscle

Although voluntary skeletal muscle is not part of the nervous system, the output of the nervous system most often includes some form of muscular activity. The neuromuscular junction between the motor neuron and its associated muscle fibers is an essential link in the chain, part of the PNS. Neurologists must therefore assess patients for muscle diseases (such as muscular dystrophy) and need to distinguish these entities from diseases that affect the synapse at the neuromuscular junction (e.g., myasthenia gravis), from diseases of the peripheral nerves, or from lesions of the spinal cord.

1.2 Nervous System Overview

The perspective of this book is an understanding of the functioning nervous system as it goes about achieving its three essential tasks:

- To detect what is happening in the external environment (e.g., vision, hearing) or internally (e.g., within muscles and joints)
- To integrate this information with ongoing brain activity and, if possible, relate it to previous experience
- To act or react in an appropriate fashion, in order to accommodate to the new situation or perhaps to alter it

When disease or injury affects the nervous system, there is a disruption of function. It is the physician's (neurologist's) task to use his or her knowledge to diagnose where the nervous system is malfunctioning and the nature of the problem.

We detect changes in the external environment via the PNS, including particularly the special senses. The CNS is the integrative center for analyzing the incoming information and organizing the output. The CNS consists of several distinct areas each contributing a piece to this operation; all parts must function harmoniously in order to carry out complex tasks. Most of our responses include movements to adapt to these changes via the nerve fibers (of the PNS) that activate the skeletomuscular system.

One can discuss the nervous system as consisting of a set of *modules*:

- The periphery—sensory and motor nerves; neuromuscular junctions and muscles
- The spinal cord—the location of the lower motor neurons and the site of sensorimotor reflex activity; pathways ascending and descending
- The brainstem—three divisions each with cranial nerve nuclei; reticular formation; pathways ascending and descending
- The diencephalon—hypothalamus for vegetative functions, and the thalamus for amalgamation with the cerebral cortex
- The cerebellum—a major modulator of the motor systems
- The basal ganglia—several nuclei with both motor and nonmotor functions
- The cerebral cortex—integration, visuospatial orientation, language, memory, and executive function
- The limbic system—involved in the development and expression of emotional reactions

It is the characteristic contribution of each part that permits the physician to determine the localization of any damage or lesion.

In order for the CNS to function collaboratively, pathways (tracts) are needed to carry information from the special senses, skin, muscle, joints, and viscera to higher "centers" in the brain, including the cerebral cortex, as well as from these coordinating areas back down to the (lower) motor neurons producing actual movements. At the same time, there is a need for the various CNS modules to exchange information about the task that each is performing and what it is accomplishing. In fact, much of the substance of the hemispheres consists of nerve fibers interconnecting various parts of the brain and the two hemispheres with each other; these nerve fibers constitute the CNS white matter.

1.2.1 The Periphery/PNS

Strictly speaking the peripheral nervous system, the PNS, is the nervous system outside the brainstem and spinal cord. It includes the peripheral nerves, both sensory and motor, as well as the neuromuscular junctions. Muscle diseases are within the sphere of neurology; examples of muscle disease will be introduced in the Chapter 4 e-cases.

Information from the skin and from receptors in muscles and joints is constantly needed for adaptation to a changing environment. Sensory information from the skin is detected by nonspecialized and specialized receptors for two main categories of sensation, called modalities:

1. Highly discriminative information such as fine touch and texture. Discriminative touch sensation is carried to the spinal cord by larger fibers with thicker myelin; therefore the information is conveyed more rapidly (Figure 1.1; discussed further below).
2. Pain and temperature. Fibers are mostly smaller and tend to be thinly myelinated or unmyelinated; impulse conduction along these fibers is therefore slower (Figure 1.1; discussed further below).

Motor nerve fibers, originating from motor neurons in the spinal cord, project to the muscles in order to initiate and control movements. Again, these nerves are well myelinated and carry information rapidly. The synapse on muscle cells is specialized as the neuromuscular junction, where the neurotransmitter acetylcholine is stored and released when the action potential invades the synapse.

Finally, intact muscle is required to produce the intended movements, either intentional (voluntary) or procedural, as well as for postural adjustments in response to changes in position or to the force of gravity (further discussed in the motor section in Chapter 2).

1.2.2 The Autonomic Nervous System

The ANS has two functional divisions, the sympathetic and parasympathetic (see Figure 1.1). The sympathetic portion functions in circumstances of stress, for example, those requiring "fight or flight" reactions (increased adrenaline, sweating, mobilization of glucose). Its ganglia are for the most part located alongside the vertebral column, and are often referred to as the paraspinal ganglia. The sympathetic outflow from the CNS is from the spinal cord, from T1 to L2 (see below).

The parasympathetic division is concerned with restoring energy, and functions in quiet periods (such as after a big meal). Its ganglia are situated mostly nearer the target organ (e.g., the gut, glands and the bladder wall). The parasympathetic outflow from the CNS is from the brainstem (see below; with cranial nerves III, VII, IX, and X), and from the sacral division of the spinal cord (S2, 3, 4).

1.2.3 The Spinal Cord

The spinal cord is intimately connected with the peripheral nervous system but is part of the central nervous system. Functionally, the spinal cord is responsible for receiving sensory (including muscular) input from the limbs and the body wall and for sending out the motor instructions to the muscles. It adds the functional capability for reflex activity, in response to information from both the muscles and the skin, and can organize some basic movements (e.g., walking). In addition, it carries ascending and descending axonal pathways.

The spinal cord is an elongated mass of nervous tissue with attached spinal roots that is located in the vertebral column; it ends normally at L2 (second lumbar vertebral level) in the adult (Figure 1.3A). Although the spinal cord is uninterrupted structurally (Figure 1.3B) it is organized segmentally, with each segment responsible for a portion of the body peripherally: a sensory area supplied by a segment is called a dermatome, and muscle supplied by a segment is called a myotome. The spinal cord segments are named according to the level at which their spinal nerves exit the vertebral column. There are eight pairs of cervical spinal nerves (and spinal cord segments; C1–C8), twelve thoracic (T1–T12), five lumbar (L1–L5), five sacral (S1–S5) and one coccygeal.

Because the spinal cord is shorter than the vertebral column, a spinal cord segment responsible for a patch of skin and certain muscles in the limbs and the periphery does not correspond exactly with the vertebral level, with the exception of the upper cervical spinal cord. A lesion of the spinal cord is described as the level of the cord that has been damaged, not the vertebral level; therefore, knowing which part of the body is supplied by a spinal cord segment has clinical importance (see Table 2.1).

Each spinal cord segment has a collection of sensory and motor nerve rootlets that coalesce to form a single sensory (dorsal) and motor (ventral) nerve root on each side of the cord; the dorsal and sensory roots combine to form a spinal nerve (Figure 1.3C). Transverse sections (cross-sections) of the spinal cord reveal a core of gray matter (neurons and synapses) in a butterfly-like configuration, surrounded by white matter, consisting of tracts, also called pathways. The dorsal aspect of the spinal cord gray matter has sensory-associated functions; the sensory input is carried via the dorsal root. The ventral portion of the spinal cord has the motor neurons, known from a functional perspective as the lower motor neurons (also called the alpha motor neurons). The axons of these motor cells leave the spinal cord (and change the nature of their myelin sheath as they do so) to be distributed via the ventral root to the muscles.

1.2.3.1 Sensory Aspects Peripheral nerves carry the sensory information from the two modalities to the spinal cord where the two systems follow quite different pathways (tracts). The discriminative touch modality ascends and stays on the same side throughout the spinal cord (Figure 1.4A), whereas the pain and temperature fibers synapse and cross to the other side (Figure 1.4B) and then ascend. After the thalamic relay both reach the cortex, where further elaboration and identification of the sensory information occurs, most of this being in the realm of consciousness (discussed in further detail in Chapter 2 and also in Chapter 5). Other information (known as proprioception), derived from special sensory units responsive to stretch in the muscles (muscle spindles, see below) and movement detectors in the joints, is also conveyed to the CNS via the discriminative touch pathway; much of this information does not reach the level of consciousness.

1.2.3.2 Motor Aspects Neurons in the cerebral cortex (and brainstem) control the activity of the motor neurons in the spinal cord. One major pathway descends from the "upper" levels of the nervous system (the cerebral cortex; see below) to the spinal cord, and is termed the corticospinal tract (Figure 1.5A; also discussed in Chapter 7). These neurons are known as the *upper motor neurons* (UMN); the spinal cord neurons are therefore known as the *lower motor neurons* (LMN). The control of muscle activity is funneled through the lower motor neuron which is functionally the final common pathway for motor activity. The lower motor neuron, its motor axon and the muscle fibers it supplies are collectively known as the *motor unit* (see Figure 4.2).

1.2.3.3 Reflexes One of the most interesting aspects of muscle activity is a feedback mechanism whereby the muscle informs the nervous system about its degree of stretch. Receptors that gauge the degree of stretch are located among the regular muscle fibers; they are known appropriately as stretch receptors. These receptors are spindle-shaped and are thus called muscle spindles. The afferent information from the muscle spindles is carried by a peripheral myelinated nerve fiber, which gives off a collateral branch in the spinal cord (see Figure 1.5B). This fiber synapses directly, by way of only a single synapse (i.e., monosynaptic), with a lower motor neuron supplying the muscle from which it originated. Activation of these receptors will normally lead to a reflex contraction of that very same muscle. Other reflexes that have a protective function (such as the response to touching a hot surface or stepping on a sharp object) involve more than one synapse.

This reflex circuit, known as the (muscle) *stretch reflex*, the deep tendon reflex (DTR) or myotatic reflex, is tested clinically by tapping on a tendon (e.g., the patellar tendon at the knee), which stretches the muscle and thereby activates the muscle spindles (Figure 1.5B). The muscle contracts (in this instance producing extension of the knee). This monosynaptic reflex requires the following elements:

- The muscle spindle with an intact functioning peripheral (myelinated) nerve carrying the afferent information
- The spinal cord, where afferent interfaces with efferent
- A motor neuron (a generic lower motor neuron) at the appropriate (lumbar) level of the spinal cord, with only a single synapse
- A (myelinated) motor axon travelling as a peripheral myelinated nerve, returning to the same muscle (with a neuromuscular synapse) to effect a reflex contraction

Reflex contraction of the muscle is graded in a standard way (see Table 2.3).

The sensitivity of this reflex circuit is influenced by neurons located in the brainstem reticular formation (discussed below). Therefore, the assessment of reflex activity in the clinical setting not only is one of the most significant tests for motor functionality as well as spinal cord integrity, but is extremely important for overall assessment of the nervous system. Note that intact myelinated peripheral nerve fibers and functional neuromuscular junctions are also required for the reflex arc. Finally, healthy intact muscle, appropriate for the size and age of the individual, is needed for a response.

One of the remarkable features of the muscle spindles is their capability for resetting their sensitivity to the stretch stimulus; each spindle has within it a few muscle fibers that will reset the length of the spindle and thus alter its responsiveness. Specific neurons (called gamma motor

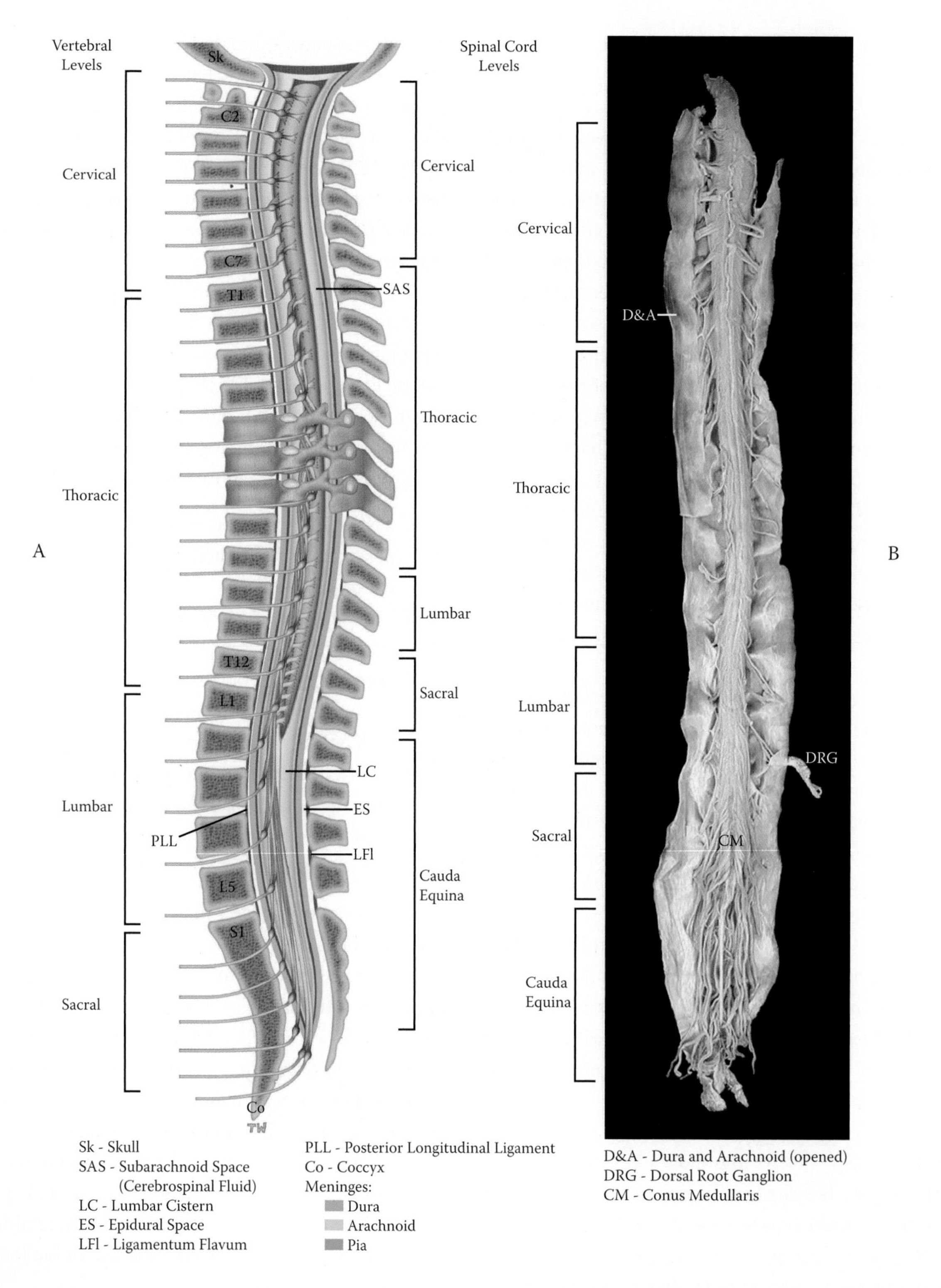

FIGURE 1.3 (**A**: left panel) The spinal cord, with its meninges, is situated within the vertebral canal. The cord terminates at the level of L2, the second lumbar vertebra (in the adult). The subarachnoid space (SAS) and lumbar cistern (LC) with cerebrospinal fluid is shown (pale yellow). Note the ligamentum flavum (LFl) and the epidural space (ES). (**B**: right panel) A photographic view of a (human) spinal cord with attached roots (dorsal and ventral) demonstrating its unsegmented appearance. The meninges (dura and arachnoid) have been opened and are displayed; note the dorsal root ganglion (DRG).

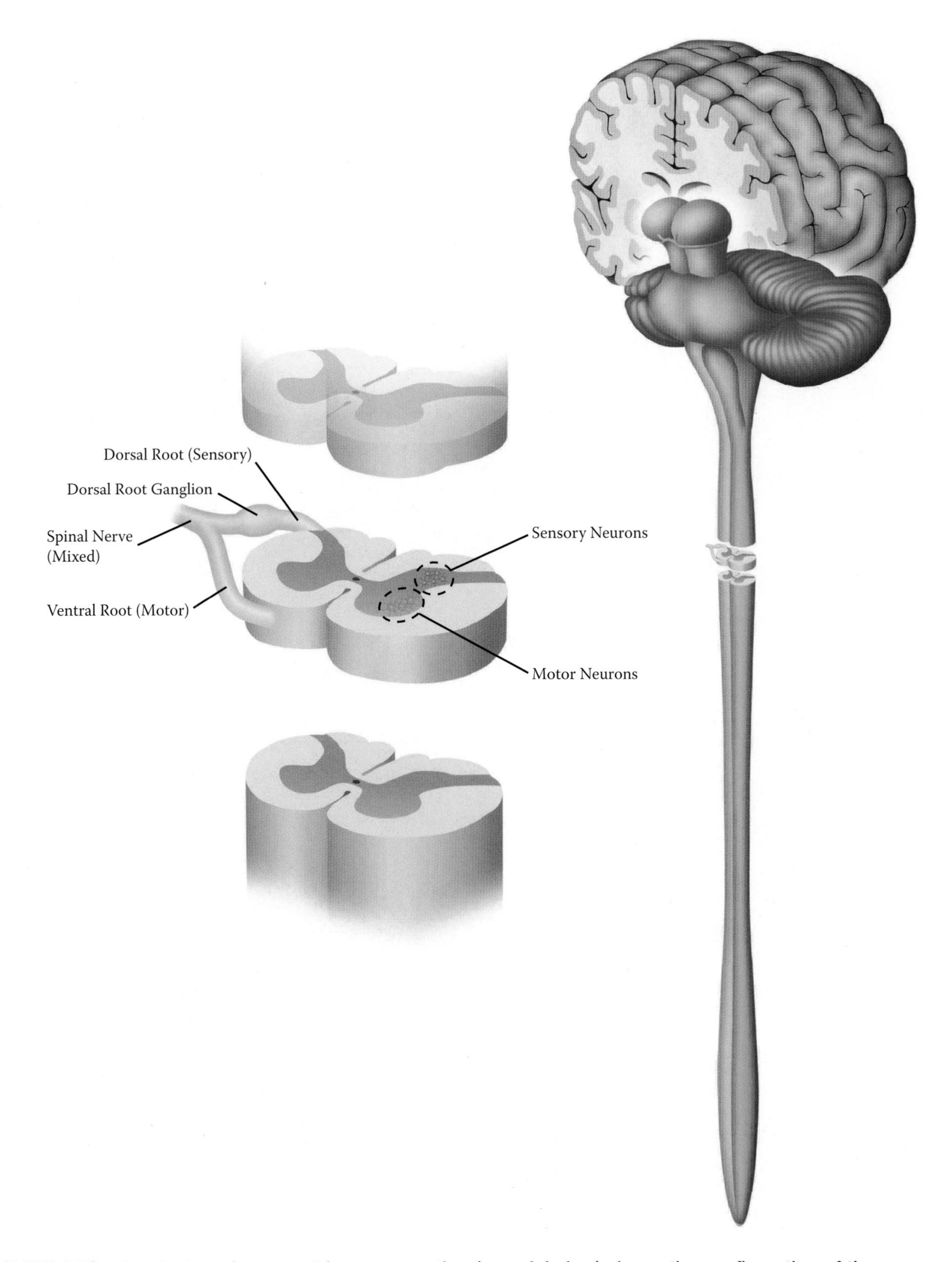

FIGURE 1.3C A spinal cord segment in cross-section (an axial view) shows the configuration of the gray matter (neurons and synapses); sensory neurons are located in the dorsal horn and motor neurons in the ventral (anterior) horn. Ascending and descending tracts (pathways) are found in the surrounding white matter. The dorsal root with the dorsal root (sensory) ganglion and ventral (motor) root are also seen, forming the (mixed) spinal nerve.

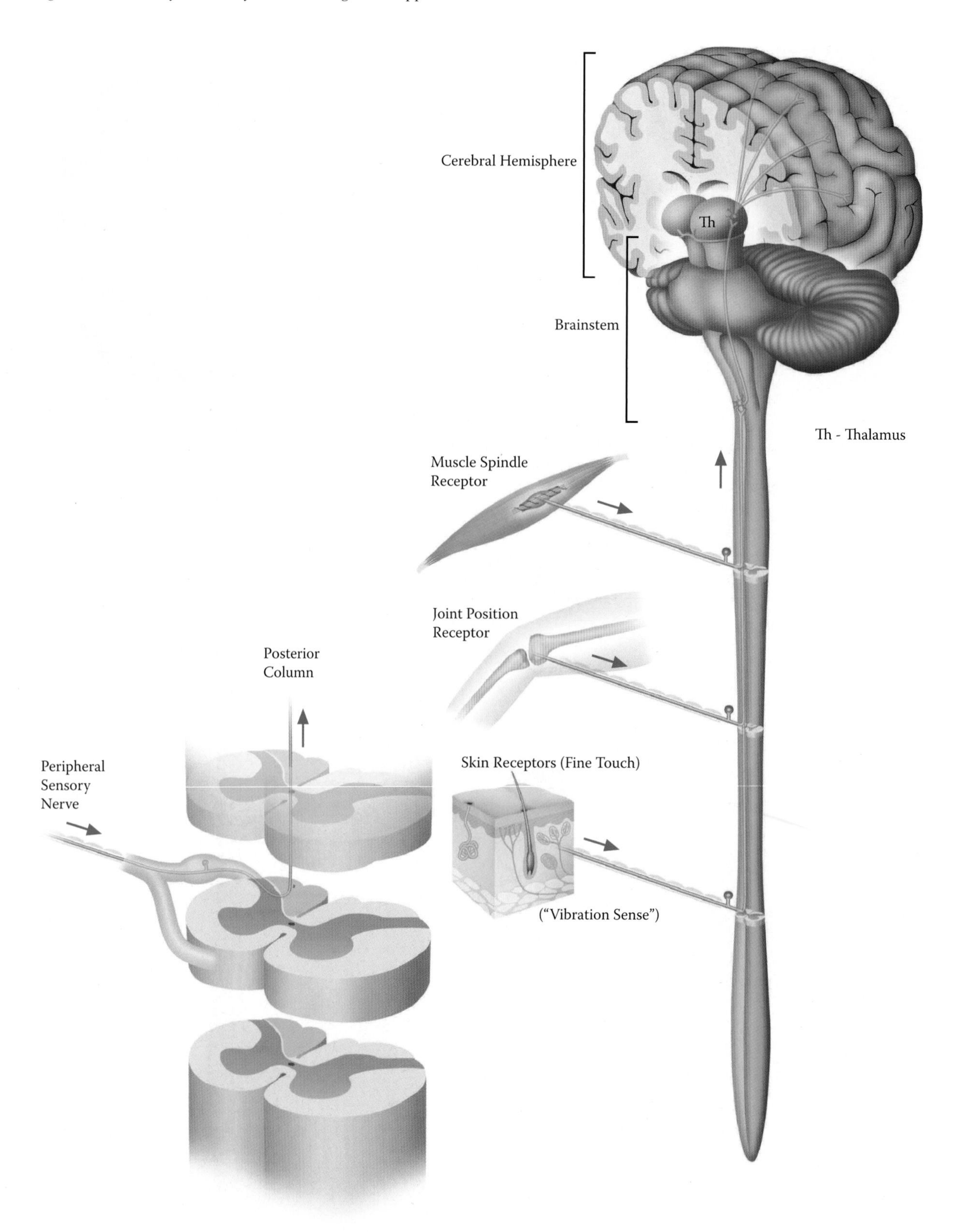

FIGURE 1.4A Sensory information from muscle spindles and joint receptors and for discriminative touch and vibration sensation enters the spinal cord and ascends on the same side, crosses and is distributed via the thalamus to the postcentral gyrus.

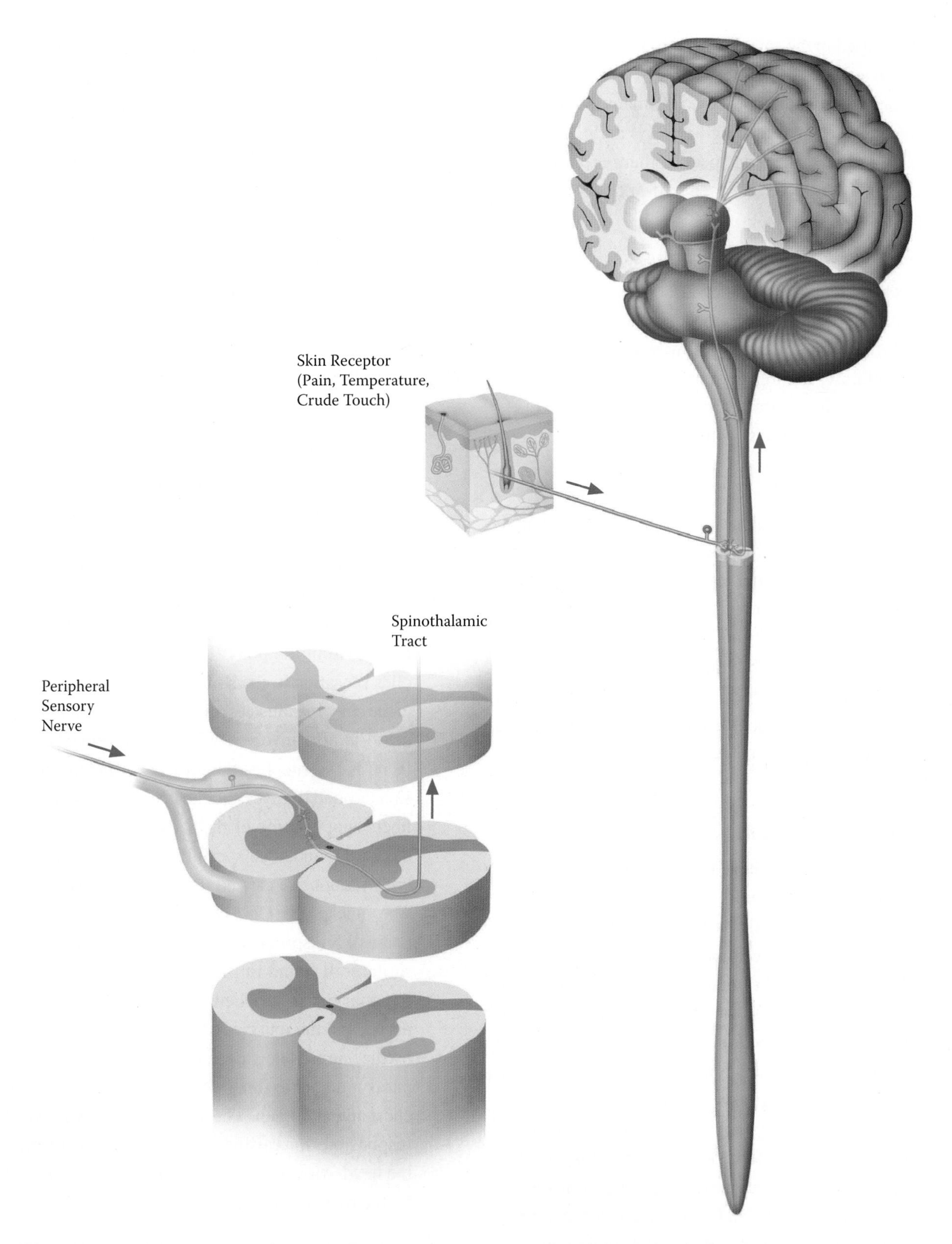

FIGURE 1.4B Sensory information for pain and temperature enters the spinal cord, crosses and ascends, and is distributed via the thalamus to the postcentral gyrus, as well as to other areas of the cortex.

neurons) located in the spinal cord supply the muscle fibers within each spindle.

To recapitulate, although there are many pathways in the spinal cord, both ascending (sensory) and descending (motor), from the clinical perspective there are three that are highly relevant (note that each half of the cord has all three pathways):

- Two sensory pathways that ascend to higher levels—the posterior (dorsal) column (for discriminative touch, proprioception and vibration) and the spinothalamic pathway (for pain and temperature; Figures 1.4A and 1.4B)
- One major motor tract, the corticospinal (for voluntary motor actions), which descends from the cortex (Figure 1.5A)

1.2.3.4 Autonomic Aspects Preganglionic sympathetic neurons are located in the "lateral horn" of the spinal cord, a small gray matter excrescence located between the dorsal (sensory) and ventral (motor) horns, from T1 to L2. Their axons exit with the ventral root and the fibers synapse in the paraspinal sympathetic chain. The parasympathetic preganglionic neurons are found in the sacral cord, S2, 3, 4, in the region of the conus medullaris, the lowest portion of the spinal cord (see Figure 1.3B and Figure 5.1).

1.2.4 The Brainstem

The brainstem, situated above the spinal cord and within the lower region of the skull, adds control mechanisms for basic movements, particularly in response to changes in position and to the effects of gravity. Most important, the nuclei controlling the vital functions of respiration and heart rate are found in the lower brainstem. In addition, there are groups of neurons located in the core of the brainstem, collectively known as the reticular formation; some of these neuronal groups have both a general effect on the level of activation of motor neurons of the spinal cord while others modulate the level of consciousness. The three major pathways continue through the brainstem—two sensory (ascending) and one motor (descending). Finally, the brainstem is connected to the cerebellum.

The brainstem is divided into three parts, from above downward; each part is morphologically quite distinct (Figure 1.6A):

- The midbrain, a smaller portion of the brainstem, situated below the diencephalon
- The pons, marked by its prominent (anterior) bulge
- The medulla, which is a continuation of the spinal cord

Attached to the brainstem are the sensory and motor nerves supplying the skin and muscles of the head and neck, known as the *cranial nerves* (usually abbreviated CN, Figures 1.6A and 1.6B). Within the brainstem are the nuclei, sensory and motor, of these cranial nerves (Figures 1.6B and 1.6C):

- The nuclei of CN III (oculomotor) and IV (trochlear) are found in the midbrain; both of these nerves are involved with eye movements. In addition, the parasympathetic supply for constriction of the pupil runs with the CN III completing the connections for the pupillary light reflex.
- The nuclei of CN V (trigeminal), VI (abducens), VII (facial) and VIII (vestibulo-cochlear) are all found within the pons. The trigeminal nerve supplies sensation to the skin of the face and is motor to the chewing muscles (mastication). The abducens nerve is also involved with eye movements. The facial nerve supplies the muscles of facial expression (around the eyes and the lips); parasympathetic fibers supply some of the salivary glands and the lacrimal gland. The special senses of hearing and body motion are carried in the VIIIth nerve.
- The nuclei of CN IX (glossopharyngeal), X (vagus), XI (spinal accessory) and XII (hypoglossal) are all found in the medulla. The mucosa of the pharynx is supplied by the glossopharyngeal nerve, and the muscles of the pharynx and larynx by the vagus. The vagus nerve provides the major parasympathetic innervation to the organs of the chest and abdomen. The spinal accessory nerve is responsible for raising the shoulder and turning the head. Movements of the tongue are controlled by the hypoglossal nerve.

Each cranial nerve emerges from the brainstem at the level where its nucleus is located, except for CN V. Examination of these cranial nerves and the reflexes associated with each part of the brainstem are detailed

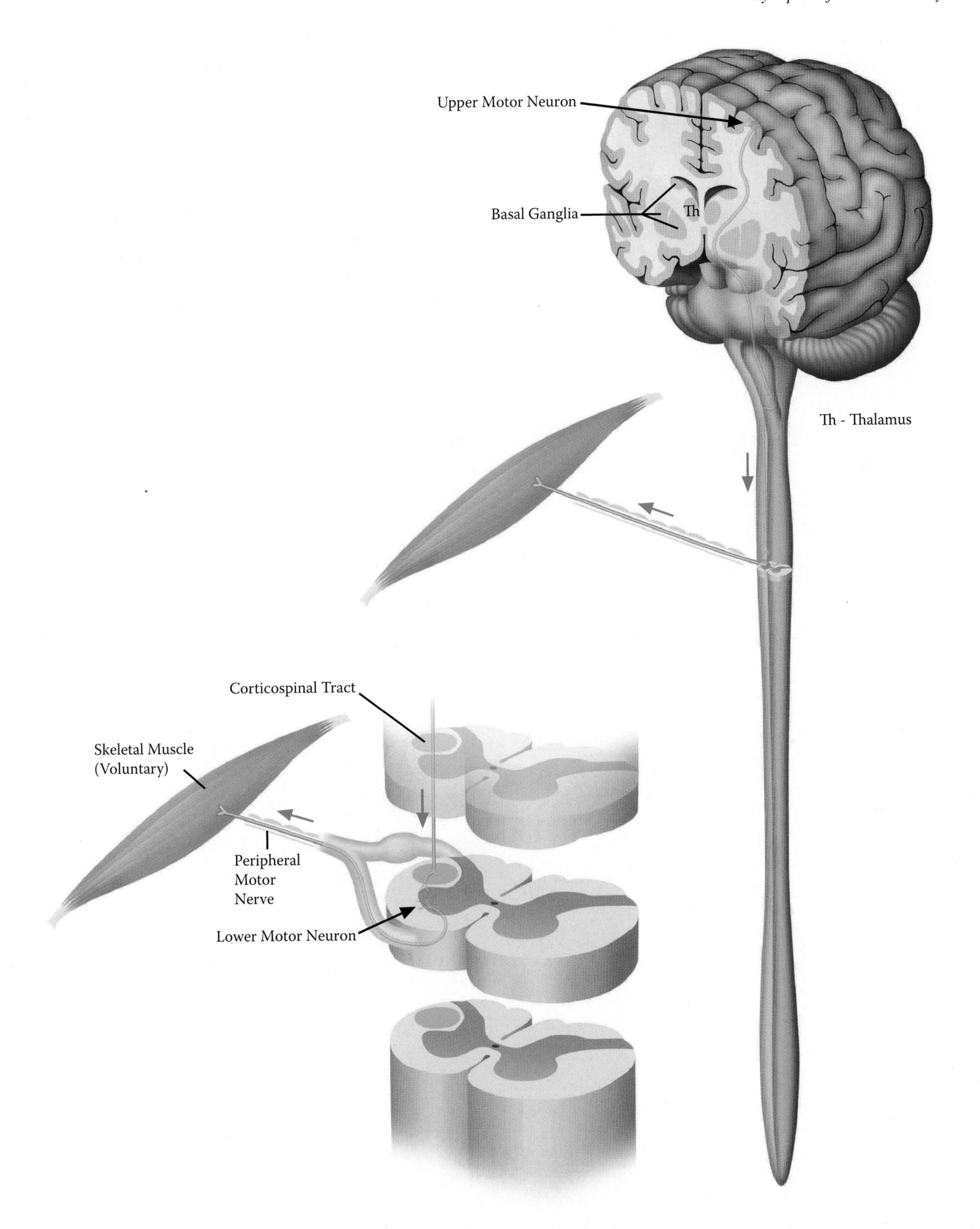

FIGURE 1.5A The motor system is organized at several levels, including the spinal cord, brainstem, and cerebral cortex. The motor neurons of the cerebral cortex are the upper motor neurons (UMN), while spinal cord neurons are the lower motor neurons (LMN). The pathway descending from the cortical neurons to the spinal cord is the corticospinal tract. The axon of the lower motor neuron innervates skeletal (voluntary) muscle.

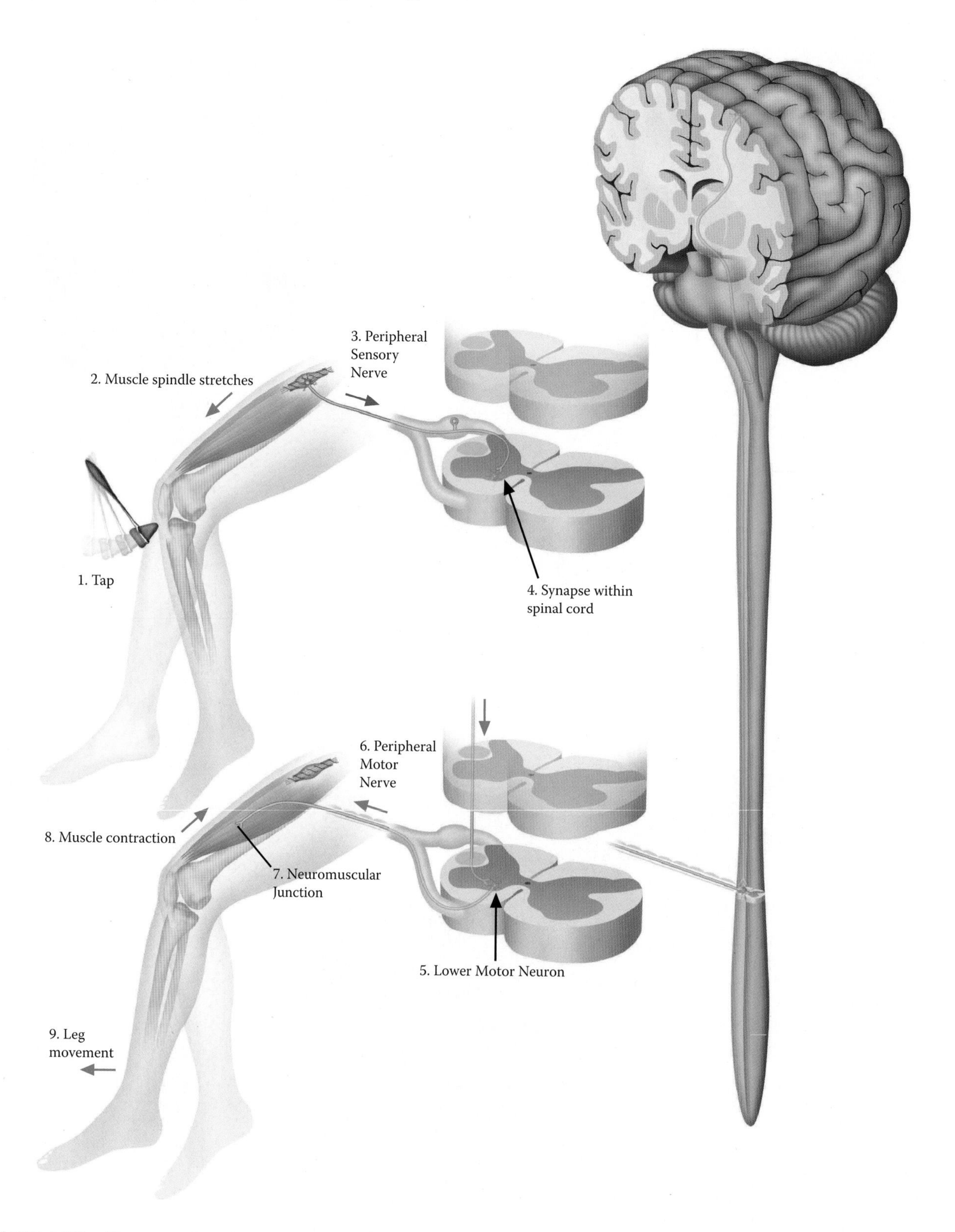

FIGURE 1.5B The stretch reflex includes the muscle spindle, an afferent fiber, a monosynaptic connection in the spinal cord with a (lower) motor neuron, an efferent fiber to the same muscle, with a neuromuscular junction, and the muscle itself. Tapping on the tendon causes a stretching of the muscle and a firing of the spindles; the afferents synapse with the motor neuron (upper illustration). The reflex response is a contraction of the same muscle (lower illustration). This reflex circuit is animated on the text DVD and Web site.

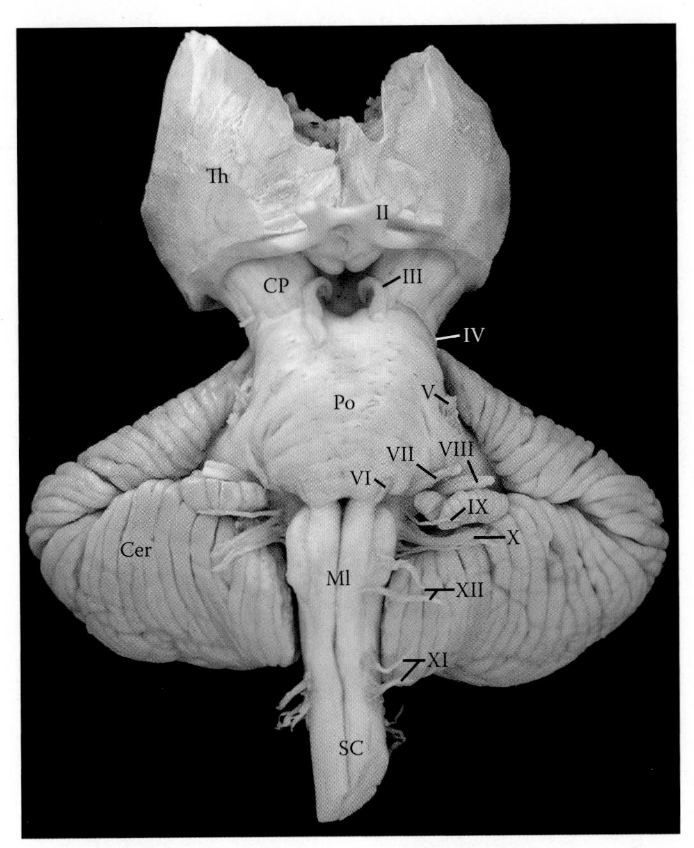

Th - Thalamus
CP - Cerebral Peduncle (Midbrain)
Po - Pons
Cer - Cerebellum
Ml - Medulla
SC - Spinal Cord

Cranial Nerves:
II - Optic
III - Oculomotor
IV - Trochlear
V - Trigeminal
VI - Abducens
VII - Facial
VIII - Vestibulocochlear
IX - Glossopharyngeal
X - Vagus
XI - Spinal Accessory
XII - Hypoglossal

FIGURE 1.6A Photographic view of the (human) brainstem from the anterior (ventral) perspective, showing the medulla (Ml), the pons (Po) and the midbrain with its cerebral peduncles (CP). Cranial nerves III to XII are attached to the brainstem. The optic nerve, CN II, also present (and chiasm and tract). The cerebellum is situated behind; the diencephalon (thalamus) is included, above.

in Chapter 2. Note that the neurons in the brainstem that innervate muscles of the head and neck are also considered lower motor neurons.

1.2.5 The Reticular Formation

The reticular formation is found within the core area of the brainstem (Figure 1.6C). It exerts a diffuse effect on neuronal activity, both downward on the spinal cord and upward toward the diencephalon and cortex. It is best to think of this group of neurons as two systems: lower and upper.

Within the lower (medullary) component of the reticular formation are the previously mentioned neurons that control heart rate and respiration; lesions here may cause respiratory and cardiac arrest. Another role of the lower reticular system (via descending motor pathways) is to modify the excitability of the lower motor neurons of the spinal cord; lesions or dysfunction of this neuronal pool will lead to changes in the responsiveness of skeletal muscles to passive stretch and altered stretch reflexes (further discussed in Chapters 2 and 7).

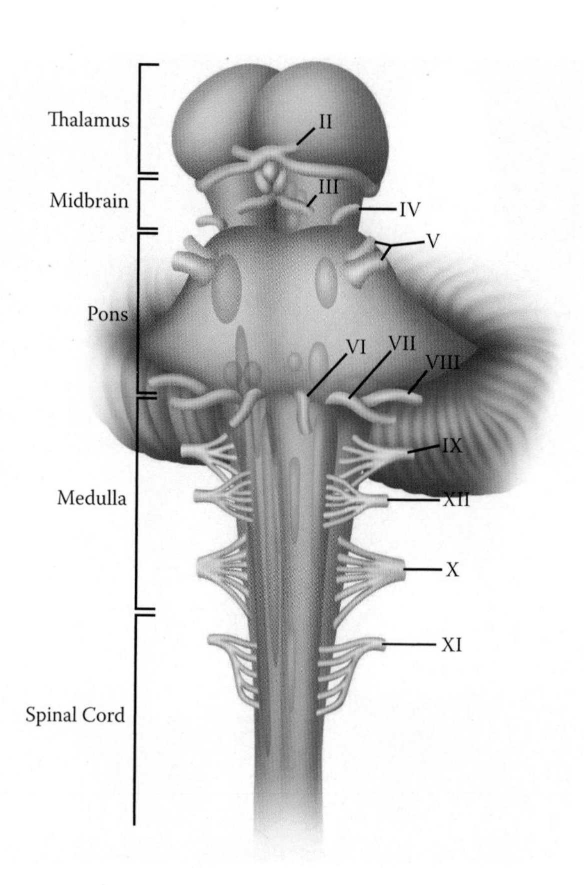

FIGURE 1.6B **The brainstem is shown with the attached cranial nerves CN III to CN XII, and their cranial nerve nuclei. The nuclei of CN III and IV are found in the midbrain; CN V, VI, VII, and VIII in the pons; and CN IX, X, XI, and XII in the medulla.**

The projections of the neurons of the upper part of the reticular formation, the upper pontine and midbrain portions, are distributed widely to the next level, the thalamus, and beyond to the cortex. Lesions here will affect the level of arousal and consciousness (further discussed in Chapter 10).

1.2.6 The Cerebellum

The cerebellum is situated at the back of the lower portion of the skull, in the posterior cranial fossa. It is positioned behind the brainstem (Figures 1.7A and 1.7B), with which it has many connections. The major contribution of the cerebellum is the facilitation of the smooth performance of motor activity and particularly voluntary motor movements (further discussed in Chapter 2).

The human cerebellum is fairly prominent, with easily observed narrow ridges of tissue, called folia. Its cortex has a unique three-layered organization, with the prominent Purkinje neurons occupying the middle layer. A set of output nuclei is located deep within its structure (the deep cerebellar nuclei). The cerebellum is connected to the brainstem by three pairs of bundles of fibers called peduncles—the inferior cerebellar peduncles to the medulla (carrying mainly afferent information from the spinal cord and medulla), the middle ones to the pons (transmitting pontine input), and the superior ones to the midbrain (carrying cerebellar efferents to the thalamus and from there to the cerebral cortex).

The midline (older, in evolutionary terms) portion of the cerebellum, the vermis, is connected to the vestibular system and is involved in the regulation of gait and balance. The lateral (newer) portions, the cerebellar hemispheres, are involved in coordination of motor activity of the limbs, receiving input from muscles and joints of the body and from motor regions of the cerebral cortex. (The connections of the cerebellum and its functional contribution to motor control are discussed in Chapter 7.) Recent evidence indicates that the cerebellum also contributes to a number of complex cerebral functions, such as learning, language, behavior, and mood stability.

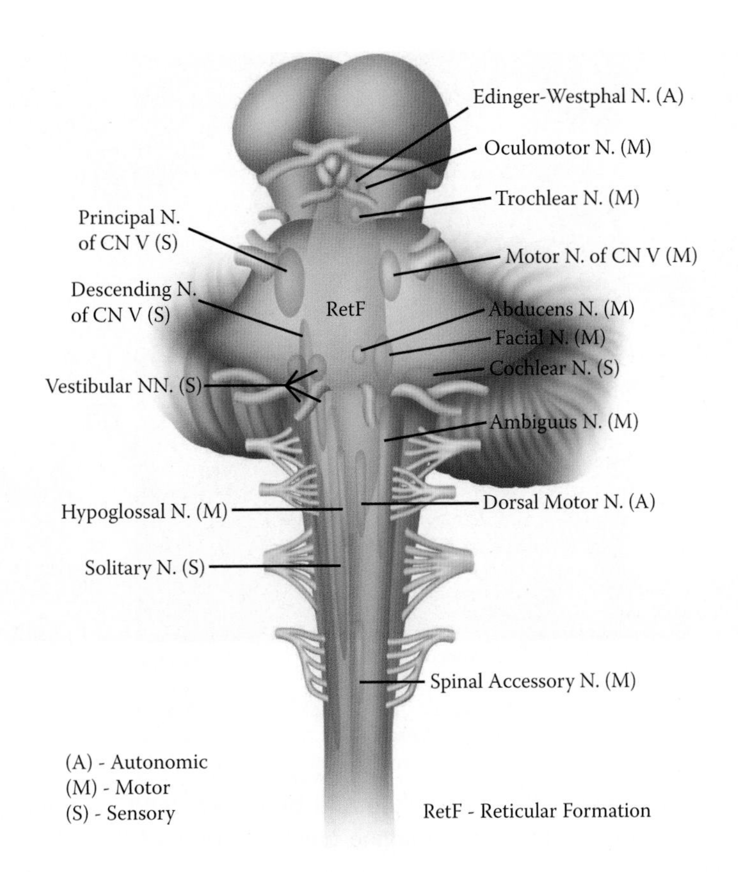

FIGURE 1.6C **The cranial nerve nuclei include sensory (S: CN V principal and descending, vestibular, cochlear, solitary), motor (M: oculomotor, trochlear, CN V, abducens, facial, ambiguus, spinal accessory, hypoglossal) and autonomic (A: Edinger-Westphal, dorsal motor). The reticular formation (RetF) is found within the core of the brainstem at all levels.**

1.2.7 The Diencephalon

The diencephalon is a small part of the brain located between the brainstem and the cerebral hemispheres (see Figure 1.1). The diencephalon is difficult to visualize as it is situated atop the brainstem (see Figures 1.6A and 1.6B; also see Figure 1.5A), deep within the hemispheres of the brain. One way of allocating its position within the brain is to locate it above the pituitary gland, which also happens to be the site of the optic chiasm (see below).

The diencephalon has two major divisions, the hypothalamus and the thalamus, both consisting of a set of nuclei. The hypothalamus controls the activity of the pituitary gland, regulates autonomic functions, including temperature, and maintains water balance, as well as organizing certain "basic" drives such as food intake; it is also involved in the regulation of sleep.

The thalamus consists of a number of relay and integrative nuclei with connections to and from (reciprocal) the cerebral cortex (Figure 1.8; see also Figure 12.3). With the exception of smell, all sensory pathways, including somatosensory, audition and vision, have their individual specific relay nucleus in the thalamus on the way to the cerebral cortex; this is the third order neuron in the sensory chain. Other circuits of the thalamus involve the integration of motor regulatory data from the cerebellum (discussed above) and the basal ganglia (discussed next). Some nuclei of the thalamus are reciprocally connected with other (association) areas of the cortex and still others have diffuse connections to wide regions of the cerebral cortex. The latter projections are crucial for the maintenance of consciousness and the regulation of the different types of sleep. The details of the connections involving the thalamus will be explained with each of the systems in subsequent chapters.

1.2.8 The Cerebral Hemispheres

The cerebral hemispheres, right and left, are the largest component of the brain; these are enormously developed

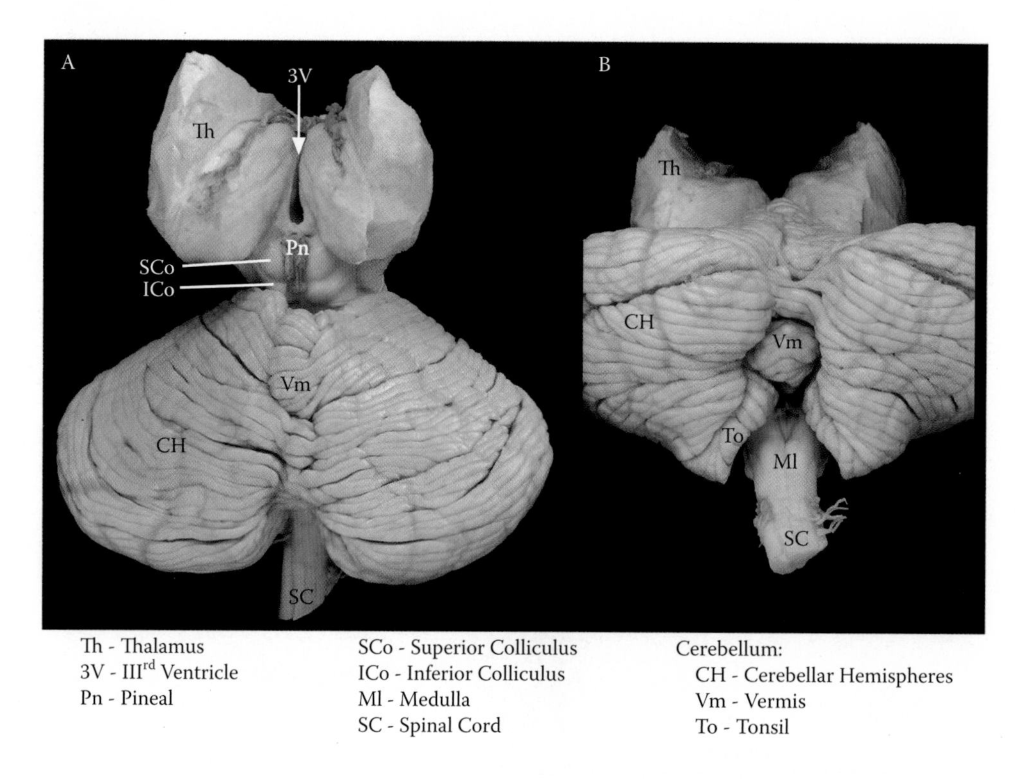

Th - Thalamus
3V - IIIrd Ventricle
Pn - Pineal
SCo - Superior Colliculus
ICo - Inferior Colliculus
Ml - Medulla
SC - Spinal Cord
Cerebellum:
CH - Cerebellar Hemispheres
Vm - Vermis
To - Tonsil

FIGURE 1.7 (A) A photographic view of the cerebellum and brainstem (and diencephalon), from the posterior (dorsal) perspective. The cerebellum includes the midline (vermis, Vm) and the lateral portions, the cerebellar hemispheres (CH). The posterior aspect of the midbrain includes the superior colliculus (SCo), and inferior colliculus (ICo); the IIIrd ventricle (3V) is also visible. (B) A photographic view of the cerebellum and brainstem, from the posterior and inferior perspective, showing the cerebellar tonsils (To) situated on either side of the lower medulla (Ml).

in primates and more so in humans. This part of the CNS, often called simply the brain, fills most of the interior of the skull, including both the anterior and middle cranial fossae. Within the central portion of the hemispheres are large collections of neurons collectively called the basal ganglia. In addition, within the hemispheres are the cerebral ventricles, spaces with cerebrospinal fluid, the CSF.

1.2.8.1 The Basal Ganglia The basal ganglia are concerned with motor activity and other brain operations, functioning in collaboration with the cerebral cortex. The motor aspects involve particularly the organization of movement sequences, the control of agonist and antagonist muscle function and the storage of procedural memory (further discussed in Chapters 2 and 7). It is hard to describe what the basal ganglia actually do in motor control until their influence is altered by diseases such as Parkinson's disease and Huntington's chorea. Then one sees movements with too little or too much amplitude, and abnormal movements such as tremor, chorea, and athetosis. It is also clear that parts of the basal ganglia participate in circuits concerned with attention and executive function, as well as with emotion.

The basal ganglia include (Figure 1.9):

- The caudate nucleus, lying adjacent to the ventricle; it has the configuration of a large head, and then a narrow body, with a trailing tail that follows the lateral ventricle into the temporal lobe of the brain (somewhat in the form of a comma; in the shape of an inverted letter C).
- The putamen, a globular nucleus located within the white matter of the hemispheres of the brain at the level of the lateral fissure (see below).
- The globus pallidus, almost almond-shaped, located medial to the putamen with two portions or divisions that are quite distinct and have different connections, external (lateral—usually referred to as the globus pallidus externus) and internal (medial—the globus pallidus internus).

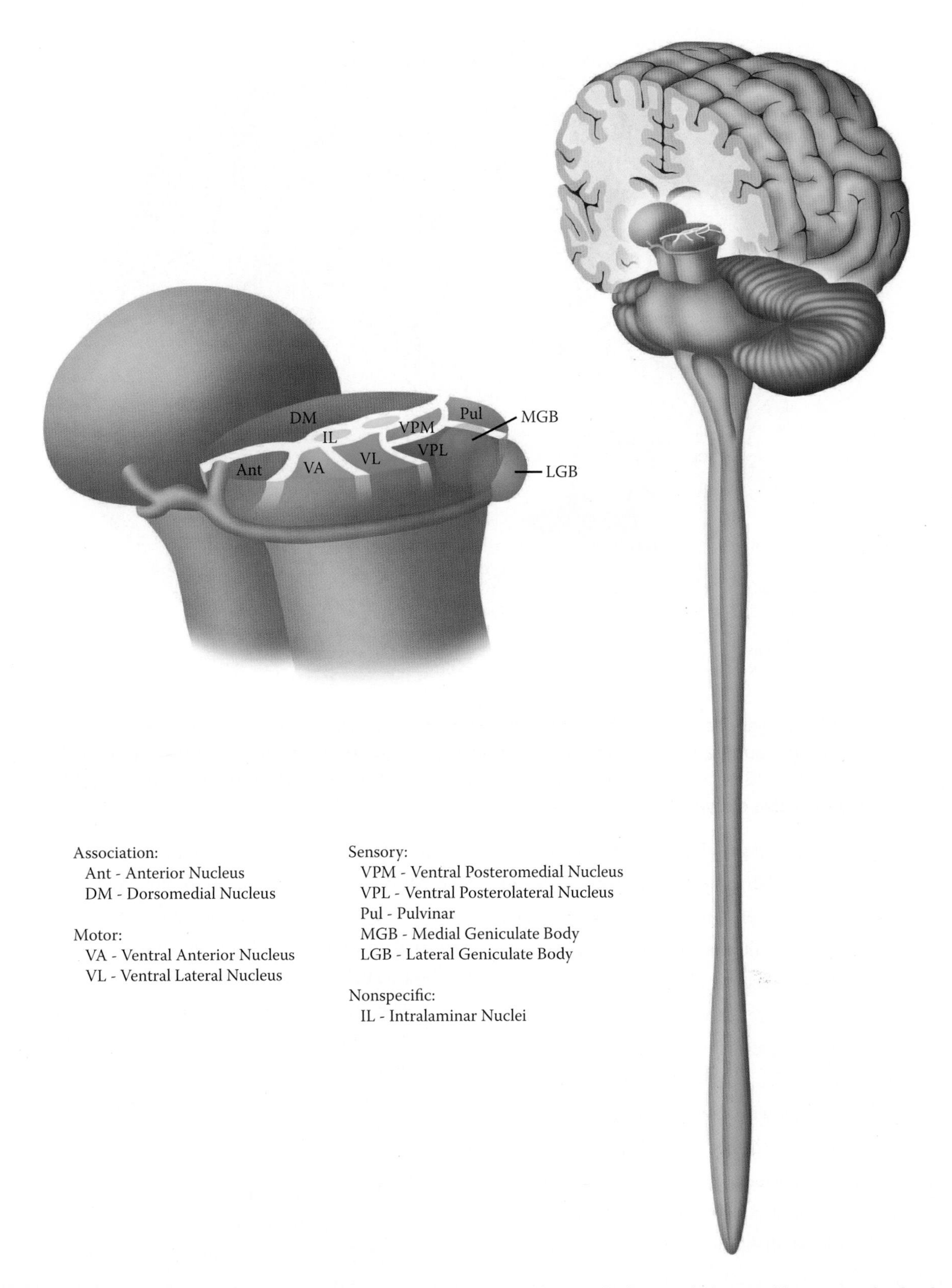

FIGURE 1.8 The diencephalon, consisting of the thalamus and hypothalamus, is located between the brainstem and cerebrum (see Figure 1.4A). The thalamic nuclei are shown on one side: there are sensory relay nuclei (VPL, VPM, MGB, LGB, Pul), motor-related nuclei (VA, VL), reticular-related (nonspecific) nuclei (IL), as well as association nuclei (Ant, DM).

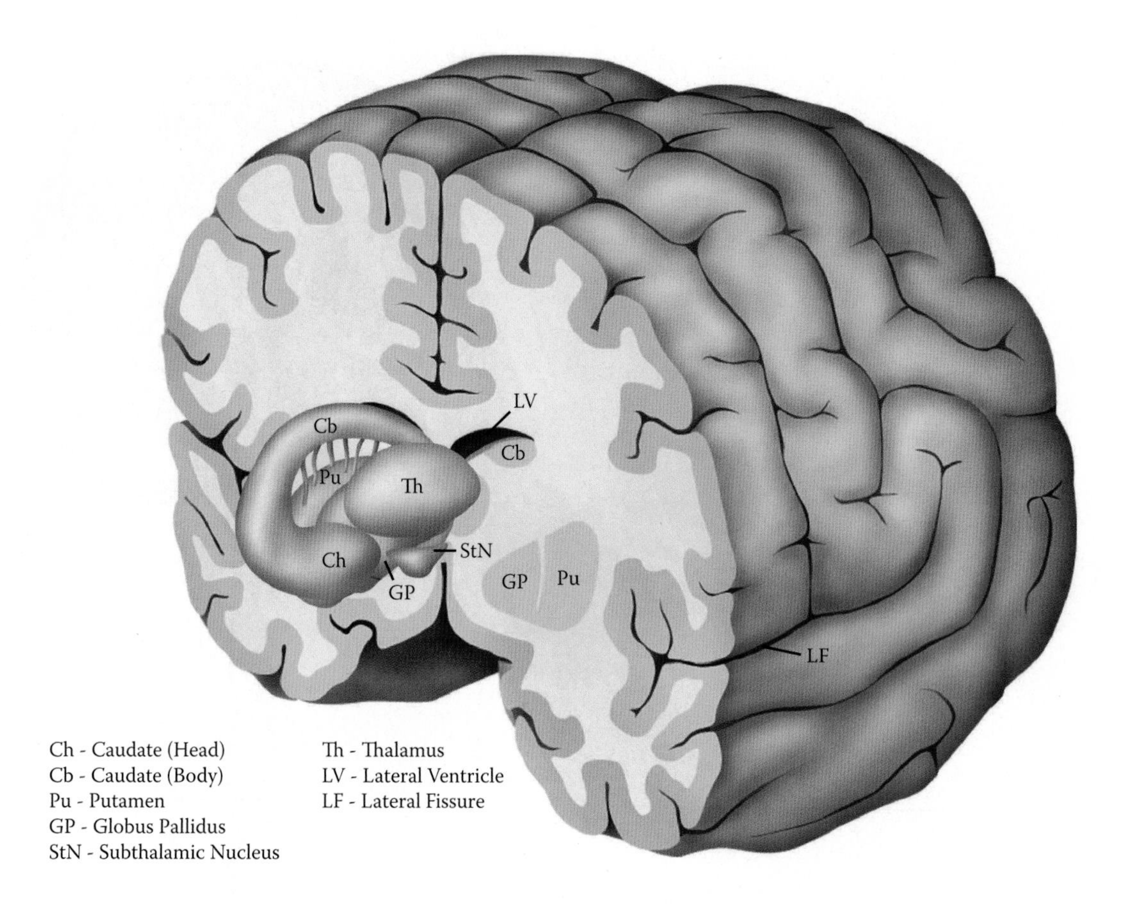

FIGURE 1.9 The basal ganglia are located within the white matter of the cerebral hemispheres. These include a number of nuclei: the caudate with its head (Ch) and body (Cb), the putamen (Pu) and the globus pallidus (GP). The subthalamic nucleus (StN), located below the thalamus, is functionally part of the basal ganglia. The cerebral lateral ventricle (LV) is also shown.

Terminology of the basal ganglia: The error in nomenclature referring to this CNS collection as ganglia should be noted; normally *ganglia* refers to collections of neuronal cell bodies in the PNS. The alternative term for these nuclei is the *corpus striatum*, a term rarely used nowadays. The term *striatum* is used and refers to the caudate and putamen together because these are structurally linked; this portion of the basal ganglia is also called the neostriatum, based on its connections with the (neo)cortex. The putamen and globus pallidus together are called the lentiform (also called lenticular) nucleus.

Visualizing the configuration of the various parts of the basal ganglia is quite challenging, even after learning the names of the component parts. Different parts of the basal ganglia can be seen when the brain is visualized with CT scans and with MRI (see radiological examples on the DVD and Web site).

In order to complete the circuitry and connections of the basal ganglia, two nuclei are now added to the functional definition of the basal ganglia: the substantia nigra, a group of cells located in the midbrain, and the subthalamic nucleus, located below the thalamus. Input to the basal ganglia comes from all parts of the cerebral cortex; its output is funneled through the thalamus and returns, in the case of motor activity, to the motor areas of the cortex. (The detailed connections of the basal ganglia will be elucidated in Chapter 7.)

1.2.8.2 The Cerebrum The cerebrum, which refers to the cerebral hemispheres without the basal ganglia, is the part of the CNS that contains the billions of microprocessors and is what most people are thinking about when talking about "the brain"; it occupies most of the interior of the skull. In fact, the two halves of the cerebrum in humans are by far the largest component of the CNS, dwarfing in size the brainstem, the cerebellum, and the diencephalon. It is this part of the brain that is needed for "thinking" (reasoning and planning), as well as language and emotion. In addition, in higher mammals, activities of the cerebrum seem to dominate and control all aspects of CNS function.

Structurally, the cerebrum consists of an outer rim of billions of neurons and their dendrites and synapses, the *cerebral cortex* (gray matter), arranged as layers of different types of neurons (see Figure 10.4). It is the cerebral cortex with its intricate synaptic connections that is the biological substrate for the highest integration of nervous activity, what we call higher mental functions, including language.

The surface of the human cerebrum is characterized by sulci (deep indentations) and fissures (even deeper), with the cortex following the surface contour; this arrangement enormously increases the actual surface area of the cortical gray matter. Although there are variations in the configuration of sulci, there are certain major fissures that divide the hemispheres into cerebral lobes: the central fissure (of Rolando), the lateral fissure (of Sylvius), and the parieto-occipital fissure. Based on these fissures, there is a standard way of dividing the cortex into lobes (Figures 1.10A and 1.10B): frontal, parietal, occipital and temporal.

The frontal lobe, the area in front of the central fissure, is generally described as having chief executive functions, in other words the part of the brain involved with major planning and decision making. The parietal lobe, between the central fissure and the parieto-occipital fissure, functions mainly in the reception and integration of the various sensory inputs and contributes significantly to spatial orientation. The occipital lobe, behind the parieto-occipital fissure and mostly located on the medial and inferior aspects of the brain, has predominantly visual functions. The temporal lobe, located below the lateral fissure, has various sound-related association tasks on its lateral portion and structures important for memory formation and emotion in its medial portion (see below).

The ridges of tissue between the sulci, called gyri, also vary somewhat but some of these have a reasonably constant location and are known to have an assigned function, motor or sensory. The most characteristic are (Figures 1.10C and 1.10D):

- The precentral gyrus, in front of the central fissure, with motor functions related to the limbs and the face, the primary motor area
- The postcentral gyrus, located behind the central fissure, with sensory functions related to the limbs and the face, the primary somatosensory area
- The primary auditory area, in the upper temporal lobe (also known as Heschl's gyrus), located within the lateral fissure
- The primary visual area of the occipital lobe, located on the medial surface along both banks of the calcarine fissure (often called the calcarine cortex)

The portions of the cerebral cortex that are not related directly to sensory and motor functions are known as association areas. Areas adjacent to the primary sensory areas are involved in the elaboration and interpretation of the sensory input to that area.

Language: The left hemisphere in humans is most frequently the repository for language function and is called the dominant hemisphere. There are two language centers, one for expressive functions (*Broca's area*) located in the inferior aspect of the frontal lobe (in the frontal opercular area), and the other for receptive functions (*Wernicke's area*) located in the superior and posterior area of the temporal lobe and also within the lateral fissure.

Memory: Memory for names, places and events is called *declarative memory*. One structure, the hippocampus, located in the medial portion of the temporal lobe, is necessary for the recording of memories of this type. *Working memory* is a term used to describe memory "in use," as a problem or task is being completed; portions of the frontal lobe are required for this type of memory function. Long-term memory is stored in widely distributed neuronal networks involving all parts of the brain (further discussed in Chapter 2).

There is a different type of memory system, called *procedural memory*, involved with performing motor acts, such as riding a bicycle, driving a car, or various sporting activities. These motor sequences are usually learned after countless repetitions. Memories of this type are "'stored" in other areas of the brain such as the basal ganglia (also discussed in Chapter 2).

Lobes of the Cerebral Cortex

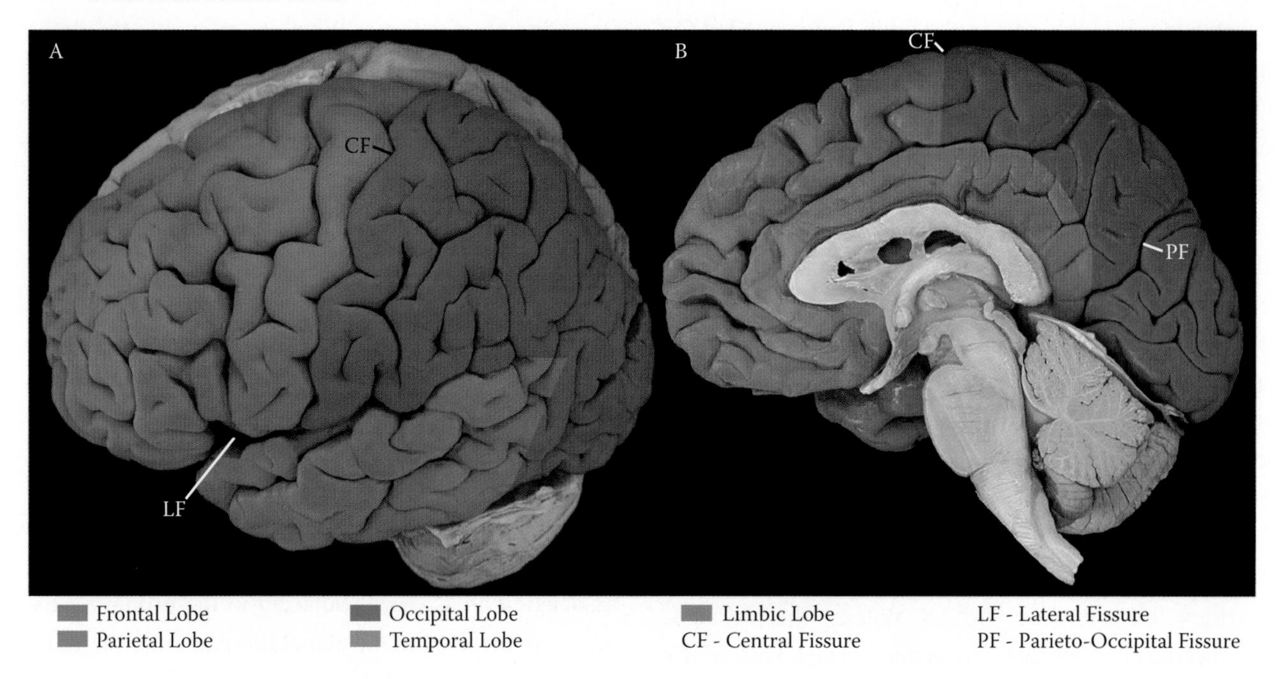

Functional Areas of the Cerebral Cortex

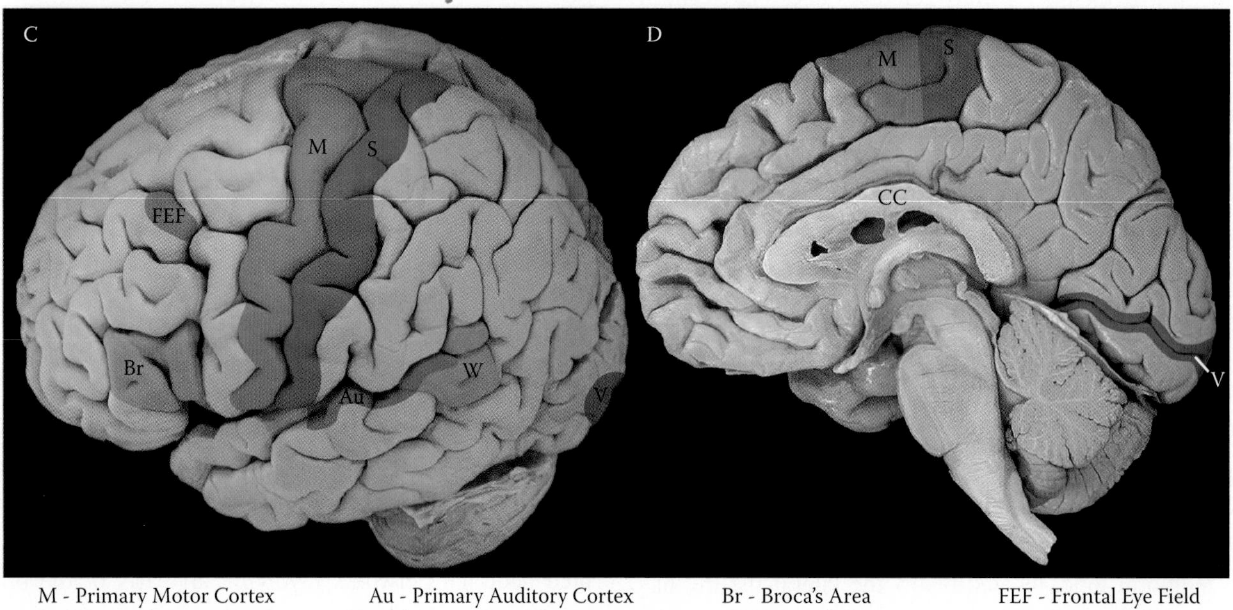

FIGURE 1.10 Photographic views of the cerebral hemispheres are shown from the lateral (A and C) and the medial (mid-sagittal) (B and D) views. (A, B) The cerebral cortex of the cerebral hemispheres is divided into lobes: frontal, parietal, temporal and occipital. The major fissures are central (CF), lateral (LF) and parieto-occipital (PF). (C, D) Gyri within each lobe have specific functions: motor (M, precentral) and somatosensory (S, postcentral), auditory (Au), and visual (V). The two language areas in the "dominant" (left) hemisphere are Broca's (Br) for expressive language and Wernicke's (W) for the comprehension of language. Voluntary eye movements are organized in the frontal eye field (FEF). Note the corpus callosum (CC).

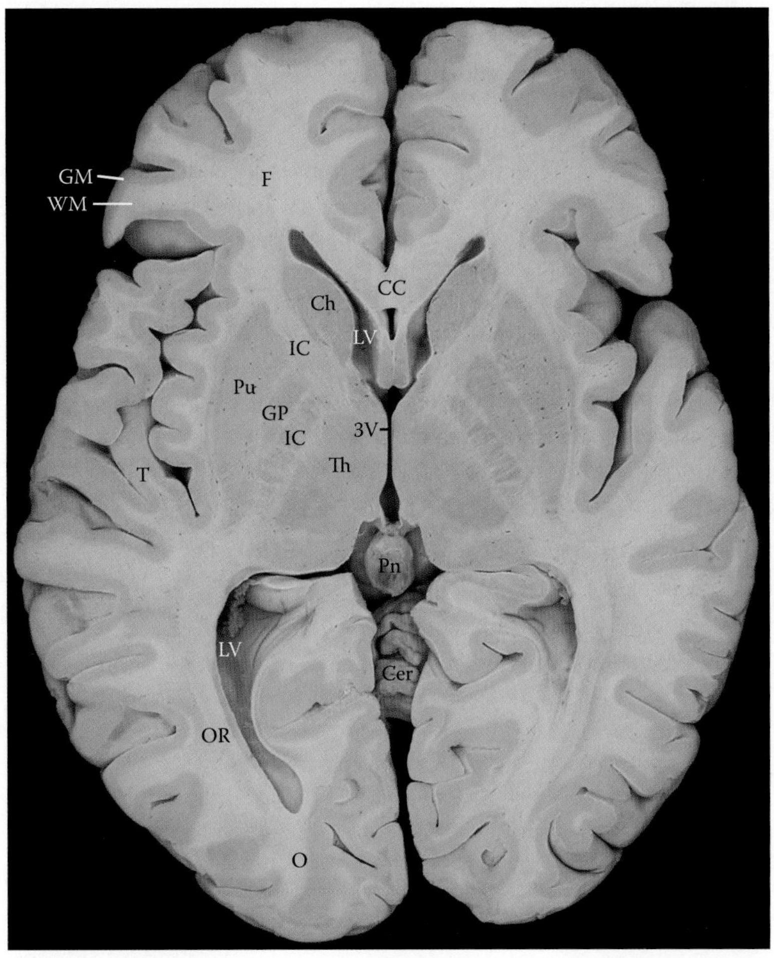

GM - Gray Matter	Cer - Cerebellum	IC - Internal Capsule
WM - White Matter	CC - Corpus Callosum	Th - Thalamus
F - Frontal Lobe	Ch - Caudate (Head)	LV - Lateral Ventricle
T - Temporal Lobe	Pu - Putamen	3V - IIIrd Ventricle
O - Occipital Lobe	GP - Globus Pallidus	Pn - Pineal
		OR - Optic Radiation

FIGURE 1.11 A horizontal section (photographic view) through the cerebral hemispheres shows areas of neurons (gray matter), including the cerebral cortex, the thalamus and the basal ganglia, and areas of axons (white matter), including the internal capsule (IC) and the optic radiation (OR). Ventricles are also seen, (LV, lateral, and 3V, IIIrd).

1.2.8.3 Cerebral White Matter Beneath the cortex and within the cerebrum are large bundles of axons, mostly myelinated, interconnecting the various parts of the brain: the white matter (Figure 1.11). This elaborate interconnectivity underlines the fact that the brain acts as a totality, with all parts contributing to the whole even while some parts are more responsible for selective functions. The fiber tracts (both sensory and motor) that project to and from the cerebral cortex funnel through a narrow channel between the basal ganglia and the thalamus, known as the *internal capsule* (see Figures 1.11 and 7.10B). This region is prone to infarction (discussed in Chapter 8). The large white matter bundle that interconnects the two hemispheres is the *corpus callosum* (see Figure 1.10 and Figure 1.11).

1.2.9 The Cerebral Ventricles

The CNS develops from the walls of the embryological neural tube. The original tube, filled with fluid, remains in each part of the nervous system. The largest spaces are found in the cerebral hemispheres; they are called the cerebral ventricles or lateral ventricles (also ventricles I and II). The fluid within the ventricular system is the cerebrospinal fluid, the CSF, which also envelops the exterior of the nervous system as part of the meningeal covering of the brain and spinal cord. (The ventricular system, the meninges, and the CSF circulation are described in Chapter 11.)

1.2.10 The Limbic System

The areas of the brain involved in the reactions to emotionally laden stimuli and to events, situations and people are collectively called the limbic system. This system gives rise to our physiological responses, (e.g., sweating, rapid pulse), our psychological reactions (e.g., fear, anger) and motor responses (e.g., fight, flight).

The limbic system includes cortical areas and nuclei in the anterior and medial aspects of the temporal lobe, including the hippocampus along with the amygdala, and subcortical nuclei within the diencephalon (both thalamus and hypothalamus) and the brainstem. (The limbic system is discussed in Chapter 12.)

1.2.11 The Vascular Supply

Brain tissue is dependent on a continuous supply of both oxygen and glucose for its function and viability. The blood supply of the nervous system will be discussed in Chapter 5 (spinal cord), Chapter 8 (cerebral), and Chapter 11 (the meninges).

Suggested Readings

Haines, D.E., Ed. *Fundamental Neuroscience*, 2nd ed. London: Churchill Livingstone, 2002.

Hendelman, W. *Atlas of Functional Neuroanatomy*, 2nd ed. Boca Raton, FL: CRC Press, Taylor & Francis, 2006.

Nolte, J. *The Human Brain*, 6th ed. St. Louis, MO: Mosby, 2008.

Steward, O. *Functional Neuroscience.* New York: Springer, 2000.

See also Annotated Bibliography.

Chapter 2

The Neurological Examination

Objectives

- To integrate the anatomical information of the nervous system with the neurological examination
- To understand the significance of abnormal signs during the neurological examination for the localization of the lesion

2.1 Introduction

There is usually a routine sequence for carrying out a neurological examination in an adult. While taking the history, the neurologist is interacting with the patient and is already assessing language and cognition, as well as emotional facets. As the examination is proceeding, the neurologist attempts to localize the problem and is considering possible etiologies. Attention may be focused on a particular area of the nervous system, based on the history and as the neurologic signs are found during the assessment. Many physicians will choose to do the motor and sensory examination first, so as to get to know the person and allow him or her to feel more relaxed, considering the unfamiliar (and confined) space of most examining rooms and the not unexpected tension associated with a physical examination of any type.

Examination of infants and children requires a somewhat different and flexible approach (beyond the scope of this book).

Formal testing begins with the cranial nerves, each of which is tested either individually or in groups. In the process one is assessing the integrity of each portion of the brainstem. This is followed by an examination of the upper limbs, both motor and sensory aspects, including reflexes—comparing one side with the other; this is followed by examination of the lower extremities. Finally, coordination and gait are tested. Fundoscopic (ophthalmoscopic) examination can be postponed until the end when the room is darkened; it should be performed after testing of visual acuity and visual fields in order to avoid dazzling the patient with bright light. Formal testing of higher mental functions can also be carried out at the end.

2.2 Systems

The nervous system is best examined from the perspective of functional systems according to the following template.

Cranial nerves:
- CN I—olfaction
- CN II—the visual system
  - Including fundoscopy
- CN III–XII
  - Including hearing (auditory) and balance (vestibular)

Motor systems:
- Voluntary (intentional, purposeful)
  - Precisional (skilled)
  - Programmed (procedural)
- Nonvoluntary (postural)

Sensory systems:
- Somatosensory—from the skin and muscles
  - Vibration, proprioception and discriminative touch
  - Pain and temperature

Mental status:
- Executive function
- Language
- Memory

Emotional status

Note: The integrity of the autonomic nervous system is typically screened by way of the assessment of bladder and bowel functions. (These are described in Chapter 13.)

2.3 The Cranial Nerves

2.3.1 *CN I and CN II*

Although always included in the list of cranial nerves, both "nerves" are in fact white matter extensions or tracts of the CNS (with the myelin of CNS). Both carry a special sense.

2.3.1.1 CN I: Olfaction (Smell) Olfaction is not highly developed in humans, yet smell is an important component of our lives and plays a significant part in the partaking of food and beverages. Any potent and distinctive aroma can be used (e.g., cinnamon or cloves). The patient is asked to "sniff" the vial with one nostril closed and to identify the nature of the aroma; then the other nostril is tested.

Pathway The sensory cells for smell are located in the roof of the nose and active sniffing is needed to get the odor to activate them. After processing in the olfactory bulb, fibers travel along the base of the frontal lobe, as the olfactory tract, to end in the anterior temporal lobe and influence the limbic system (discussed in Chapter 12). This is the only sensory system whose pathway does not pass through the thalamus.

2.3.1.2 CN II: Visual System We humans are predominantly visual creatures and visual deficits are important to detect. There are many aspects to vision. Visual acuity (discrimination of shapes and borders) and color vision are both found in the central foveal portion of the retina, where the cones are located. A Snellen chart is used for testing of visual acuity, with each eye examined separately (while the other eye is covered) and the best corrected visual acuity in each eye is recorded. Peripheral visual images including movement as well as images seen under low levels of illumination are detected by the rods, found throughout the non-foveal (peripheral) portions of the retina.

Recognition of movement is used at the bedside to detect a lesion in the visual system: retina, optic nerve, optic chiasm, optic tract, visual radiation and visual cortex. As the visual pathway spans the brain from the front to the back, its examination is most important in detecting any lesion within the cerebral hemispheres. This is done through the examination of the visual fields.

Testing is done by the examiner facing the patient, with the patient covering one eye and both of them looking at each other "straight in the uncovered eye." The examiner, with arms extended, moves one or more digits in either or both hands and the patient is asked to report when he or she sees the fingers moving on one or the other side. (This requires a fair degree of cooperation and may have to be modified for the examination of young children or adults with dementia.)

Pathway The visual fields for each eye can be charted using a perimetry apparatus (explained in Chapter 11), including the foveal region, the blind spot, and the periphery (Figure 2.1). The visual field is then divided in half; deficits in vision are described for each eye in terms of the temporal or nasal half of the visual field: a temporal (lateral) or nasal (medial) hemianopia. If the visual loss involves the visual field on the same side in both eyes, the clinical term is *homonymous* hemianopia; if not, the term is *heteronymous* hemianopia. (The details of the visual pathway are explained in Chapter 11, which will help to clarify the usage of this terminology.) A defect in the central region of the visual field that does not fit the boundary of a quadrant is called a *scotoma*.

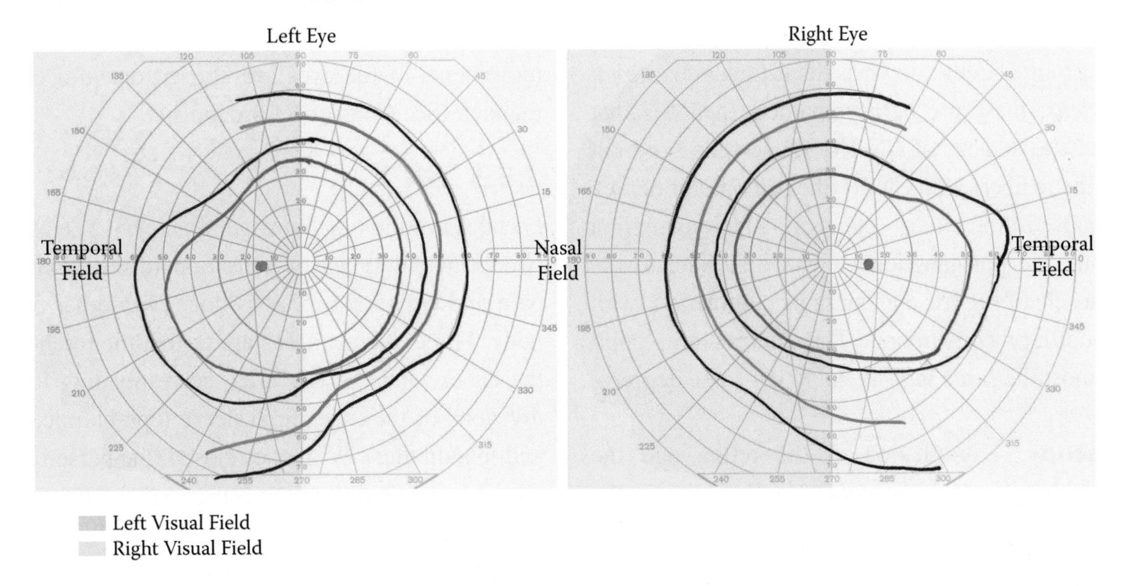

FIGURE 2.1 A perimetry chart depicts the (normal) visual fields of both eyes. The visual field of each eye is divided in half: lateral (temporal) and medial (nasal). The central area for vision is the fovea. The axons leave the eye at the optic disc to form the optic nerve; this is the site of the "blind spot" (shown in red). (Courtesy of Dr. M. O'Connor.)

The visual cortex identifies objects on the basis of shapes and borders, movement and color. Further elaboration of the visual information is performed in adjacent areas of the parieto-occipital lobe cortex. Higher mammals have a distinct region for facial recognition. Reading is accomplished by transferring the information to the posterior temporal lobe and adjacent visual association cortex in the left (dominant) hemisphere (discussed in Chapter 1).

2.3.1.3 Pupillary Light Reflex The pupillary light reflex is an essential basic reflex. This reflex is tested by shining a pinpoint light on the eye and noting the reaction of the pupil in the eye stimulated as well as that of the other pupil—the consensual light reflex. Both eyes must be tested separately. The pupillary light reflex is used in anesthetized and comatose patients to test for the integrity of the midbrain portion of the brainstem.

Pathway (Figure 2.2) The visual image, light in this case, is carried from the retina into the optic nerve; after the partial crossing of fibers in the optic chiasm, the image representation is carried in the optic tract. Some of these fibers leave the optic tract and go to the dorsal area of the midbrain, in the pretectal area, where they terminate in a group of cells called the *light reflex center*; the information is relayed immediately to the same group of neurons on the other side.

Next, the neurons in the pretectal light reflex center project to a group of cells known as the *Edinger–Westphal (E–W) nucleus*; these are parasympathetic neurons that form part of the nucleus of CN III. These cells send out parasympathetic fibers with CN III (the oculomotor nerve). After a synapse in the ciliary (parasympathetic) ganglion (in the orbit), the postganglionic fibers distribute to the (smooth) muscle of the iris and cause its contraction, with resultant pupillary constriction. The contralateral pupil reacts at almost the same instant and to the same degree.

Fundoscopy Examination of the retina and the optic disc with an ophthalmoscope usually requires the room to be darkened. The examiner employs the same eye as the one being examined (i.e. right eye for the patient's right eye) to visualize the light reflex of the patient through the ophthalmoscope, then moves closer to examine the optic disc, blood vessels and surrounding retina. A normal optic disc with the blood vessels in that area is shown in Figure 11.1.

2.3.2 CN III, IV, and VI

These three nuclei, which innervate the muscles that move the eye, are collectively called the visuomotor nuclei. Testing is done by asking the patient to follow the tip of a pencil in eight specific positions: to both sides, up and down, inward down and up, and laterally down and up, as well as in a rotatory manner. The examiner notes whether both eyes are moving symmetrically and whether there is any oscillation of the eyeball (called nystagmus) when the eye is held in position looking to either side. Movement of the eye inward (adduction) as well as upward is done exclusively by CN III; CN IV acts to move the eye down and inward; and CN VI is the nerve involved exclusively in moving the eye laterally (abduction).

Pathway The oculomotor (III), trochlear (IV) and abducens (VI) nuclei all control eye movements. CN III and IV are located in the midbrain and CN VI in the lower pons. The eyes are yoked for carrying out movements in the horizontal plane, requiring the medial rectus muscle of the eye moving inward (adduction, CN III in the midbrain on one side) and the lateral rectus muscle of the other eye (abduction, CN VI in the pons on the other side). There is a distinct pathway that interconnects these nuclei, known as the MLF (the medial longitudinal fasciculus), so that movements of the eyes are carried out in a coordinated manner. (See Chapter 6 for details.)

2.3.3 CN V

Facial sensation (via the trigeminal nerve) is tested for the two principal types of sensation (discriminative touch; pain and temperature), as is done with other parts of the body. Three distinct areas are tested, the forehead region, the cheek, and the jaw area, corresponding to the three divisions of the trigeminal nerve (ophthalmic, maxillary and mandibular). With eyes closed, the person is asked to compare the quality of the sensation on the two sides.

Asking the patient to open and close the mouth and to grind the teeth (from side to side) tests the motor portion of CN V (the muscles of mastication). The jaw reflex is tested (gently) by tapping on the point of the chin with the jaw relaxed, using a reflex hammer.

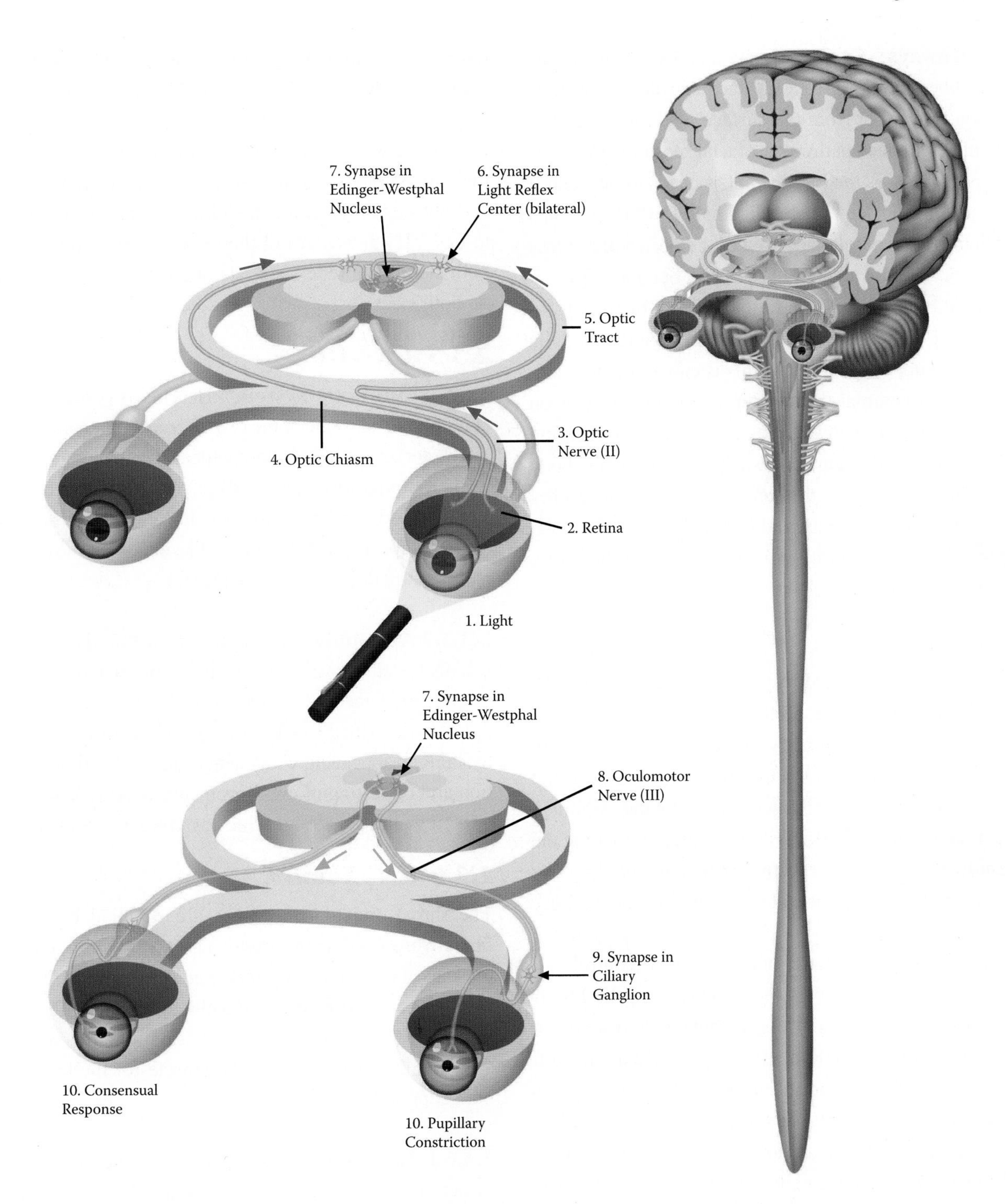

FIGURE 2.2 The pupillary light reflex. Afferents from the retina travel in the optic nerve, optic chiasm and optic tract to the light reflex center of the midbrain and synapse, with a fiber crossing to the other side. The reflex center relays next to the Edinger-Westphal (EW) nucleus, part of the IIIrd nerve nucleus. Parasympathetic fibers exit with the oculomotor nerve (CN III), synapse in the ciliary ganglion in the orbit, and supply the constrictor muscle of the pupil; the same pathway on the opposite side leads to the consensual response of the other pupil. This reflex circuit is animated on the DVD and Web site.

Pathways CN V, the trigeminal nerve, carries the sensory fibers coming from the face and lips, as well as sensation (not taste) from the surface of the tongue. The discriminative touch pathway for facial sensation synapses in the pons, crosses the midline and ascends to the ventral posteromedial nucleus of the thalamus (VPM) and then distributes to a separate face area of the postcentral gyrus (the sensory homunculus is discussed in Section 2.5.1).

Pain and temperature sensations form a distinct pathway starting in the pons, descending through the medulla and into the upper spinal cord; this is called the descending (spinal) trigeminal tract. After a relay in a nucleus on the same side, the pathway crosses and ascends, joining with the pain and temperature pathway from the body regions below the neck; the sensory information reaches the cortex via the thalamus and is distributed to the facial portion of the postcentral gyrus as well as to other areas (like somatic pain; see below).

2.3.4 CN VII

The facial nerve is tested by asking the person to look upward and wrinkle the forehead; one notes the presence or absence and symmetry of creases in the forehead. Then the patient is asked to close the eyes tightly, at first alone and then against resistance applied by the examiner. Other muscles of facial expression are tested by asking the patient to show his or her teeth and to smile, looking for any asymmetry, particularly at the corners of the mouth.

Pathway The facial nucleus, which controls all the muscles of facial expression, is located in the lowermost pons. After an unusual loop within the brainstem, the fibers leave at the cerebello-pontine angle and exit the skull to be distributed to the upper and lower face. Part of this nucleus innervates the muscles of the forehead (tested by asking the person to look upward) and the eyelid (tested by the blink/corneal reflex and by asking the person to close the eyes against finger resistance); the other part of this nucleus innervates the muscles of the lower face (around the mouth, tested by asking the person to smile).

2.3.4.1 Corneal Reflex The corneal reflex is another basic reflex that is most often tested in unconscious patients to check for the integrity of the pons (Figure 2.3). The cornea is gently stimulated with a wisp of (clean) cotton or tissue; this is an irritant to the cornea and is picked up by nociceptive fibers of CN V (the ophthalmic division).

Pathway The sensory information is carried by the pain and temperature pathway, in the descending tract and nucleus of V (in the medulla). The reflex circuit goes to CN VII (the nucleus of the facial nerve, in the pons) of both sides. The reflex consists of a closure of the eyelids, on both sides.

2.3.5 CN VIII

2.3.5.1 Auditory System Hearing is one of the special senses and is one of the two afferents that comprise the VIIIth nerve, the vestibulo-cochlear. Hearing is tested by using a tuning fork (activated) or by whispering near one ear, and asking whether the person can hear the sound(s), again comparing the two sides. (The details of the auditory pathway are described in Chapter 6.)

2.3.5.2 Vestibular System The vestibular system is usually tested only if there is a history of vertigo (dizziness) or gait imbalance. (The details of the vestibular pathway are also described in Chapter 6.) The examiner looks for nystagmus appearing with changes in head and body position and on testing of eye movements, as well as unsteadiness in walking a straight line or in making sudden turns.

2.3.6 CN IX and X

The throat and tonsillar areas are supplied by sensory fibers from CN IX; the vagus nerve (CN X) supplies the muscles of the soft palate and the constrictor muscles of the pharynx, as well as parasympathetic output to the upper gastrointestinal tract. The sensory supply to the laryngeal area and the innervation of the muscles of the larynx are both supplied by CN X.

2.3.6.1 Gag Reflex Stimulation of the area of the throat and tonsillar area with a tongue depressor leads to a gag reflex (Figure 2.4), which is a reflex to protect the upper airway. The response consists of several components, which include a "lifting up" of the soft palate (which can be seen) and in some individuals a strong gag response.

Pathway The sensory fibers from CN IX terminate in a nucleus in the medulla and connections are made with the motor nucleus of CN X, also located in the medulla, on

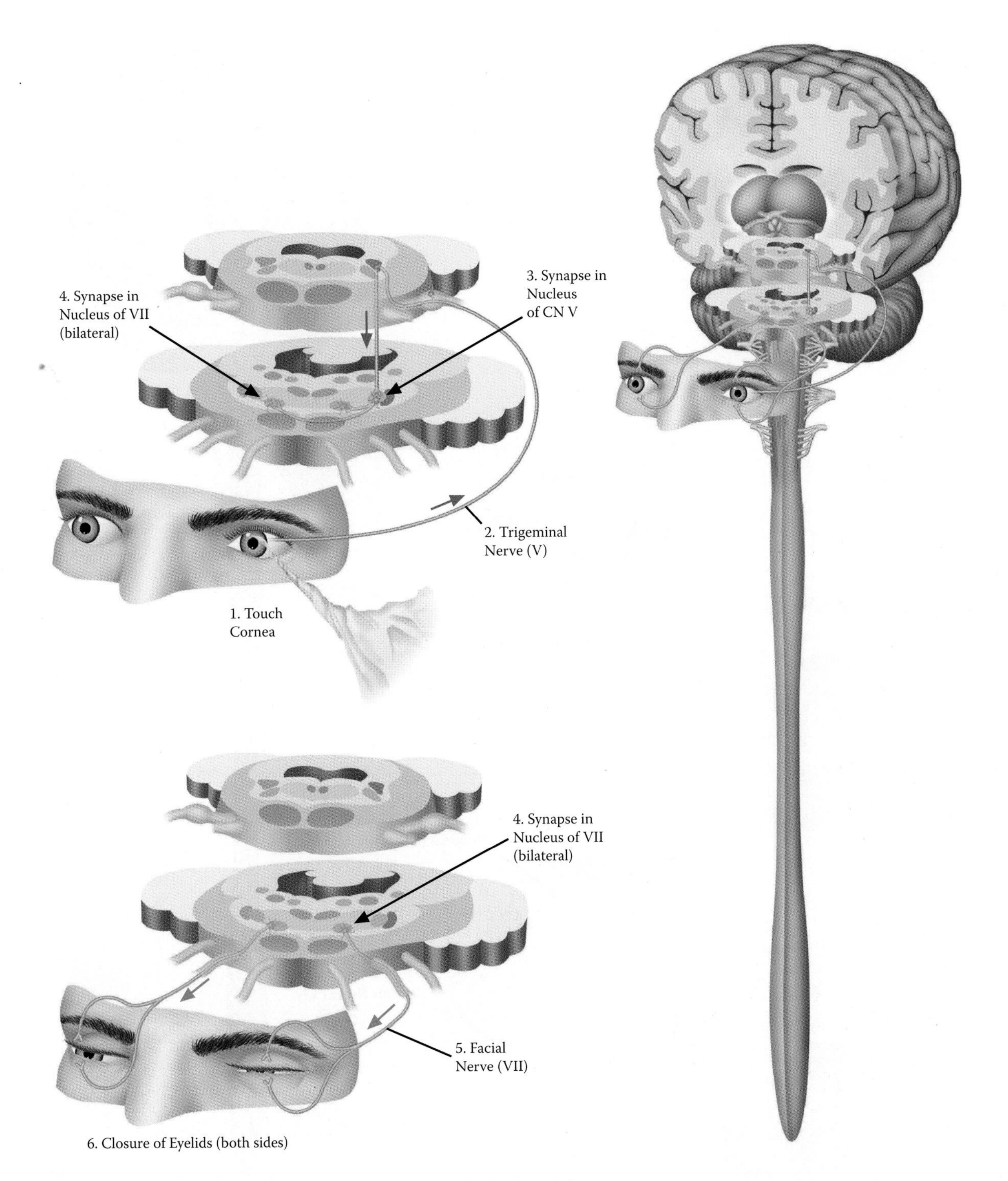

FIGURE 2.3 The corneal reflex. The nociceptive stimulus to the cornea is carried via CN V, the trigeminal nerve. These enter the pons and descend, synapsing in the descending (trigeminal) nucleus of CN V; next there is a relay to the nucleus of the facial nerve, on both sides. Fibers exit in CN VII to supply the muscles of the eyelid, which results in eyelid blinking, on both sides. This reflex circuit is animated on the text DVD and Web site.

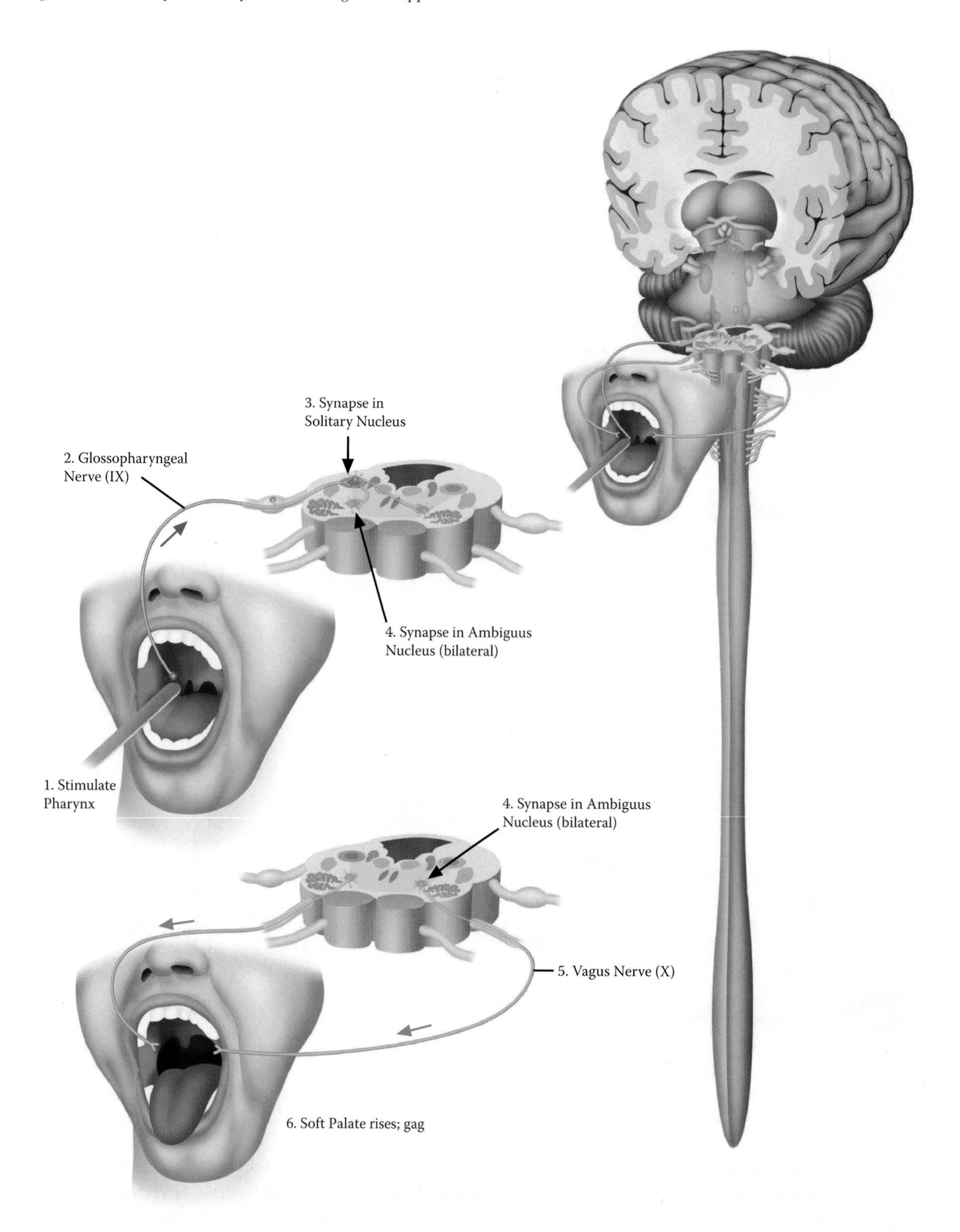

FIGURE 2.4 The gag reflex. Afferents are carried via CN IX, the glossopharyngeal nerve, to the solitary nucleus located in the medulla. The fibers relay to the nearby ambiguus nucleus of the vagus in the medulla, on both sides. Fibers exit as part of the vagus nerve, CN X, to the muscles of the soft palate. The reflex action consists of a lifting of the palate (which is shown) and a constriction of the muscles of the pharynx, that is, the gag. This reflex circuit is animated on the text DVD and Web site.

both sides. The motor response is carried by CN X, the vagus nerve. Testing of the gag reflex verifies that the medullary level of the brainstem is functioning.

2.3.7 CN XI

The person is asked to raise the shoulder, on each side, and this can be tested against resistance. This maneuver tests the strength of the trapezius muscle on each side. The person is also asked to rotate the head to one side and then the other, against resistance. This maneuver tests the strength of the sternocleidomastoid muscle on each side.

Pathway The spinal accessory nerve is in fact located in the upper part of the cervical spinal cord but, because of its circuitous route upward into and then downward out of the skull, it has traditionally been included as one of the cranial nerves. These fibers supply the upper portion of the trapezius muscle, which elevates the shoulder, and the sternomastoid muscle, which turns the head.

2.3.8 CN XII

The patient is asked to stick out his or her tongue and move it about. One looks for abnormal small worm-like movements (fasciculations). One also notes the surface of the tongue itself and whether there is any difference in appearance between the two sides, such as unilateral atrophy. Weakness of the tongue muscles can be tested by asking the patient to press the tongue inside the cheek with the examiner's finger on the outside resisting the movement. If one side of the tongue is completely paralyzed, the intact muscles on the opposite site will push (deviate) the tongue toward the weak side.

Pathway The hypoglossal nerve supplies all the muscles of the tongue as a purely motor nerve. Its nucleus is also located in the medulla.

2.4 Motor Systems

2.4.1 Motor Regulation

Before elaborating on the examination of the motor system, it is necessary to outline the overall conception of motor regulation. Control of movements is complex and requires several systems, working in parallel and in harmony. We know that many components of the motor system are involved because we see a breakdown of smooth coordinated movements and/or abnormal movements as a consequence of disease and injury to different parts of the CNS. There may be a partial or one-sided weakness (paresis) or paralysis (assuming the absence of muscle disease); movements may be too slow, too rapid, imprecise, not done in the proper sequence (fragmented, uncoordinated), or there may be movements that are not controllable (involuntary) such as tremors, twitches or flinging movements (ballism).

Three sets of muscle groups need to be considered under the term *motor regulation*:

- Distal musculature (fingers and toes)
- Proximal joint musculature (elbow and shoulder of the upper limb and equivalent muscles of the lower limb)
- Axial musculature (the spine)

(The motor system is further discussed with the problems presented in Chapters 4 and 7.)

The control of the small muscles of the hand and fingers is the most recently acquired in evolution (in primates and humans). Intentional voluntary movements need precise, rapidly implemented control of specific muscle groups that necessitates a dedicated and speedy pathway from the motor cortex to the spinal cord, the *corticospinal tract* (see below). The usual example given for skilled movements involves the distal musculature of the fingers and hand (for example, the work of a watchmaker, playing a musical instrument, writing a letter, performing robotic surgery).

In order to carry out these movements, associated adjustments of the larger joints by proximal muscle groups of the elbow and shoulder are necessary. Usually postural adjustments (e.g., of the back while playing the piano) are also essential in order for the volitional movement to be carried out successfully.

Other types of skilled motor performances require the direct involvement of these proximal muscle groups (for example, using a pottery wheel, professional tennis playing, figure skating). It is important to note that feedback mechanisms (via the cerebellum—see below and Chapter 7) are required for adjustment and refinement so that the final movement is exactly as required.

Some purposeful voluntary action sequences have become routine, such as tying one's shoelaces, driving a car, and riding a bicycle. These complex motor patterns are called into action and can be accomplished without any real conscious effort, seemingly done on "automatic pilot," although learning them has usually required countless repetitions with progressive error correction. (Note that putting a necktie on someone else requires much more concentrated effort than putting it on oneself.) The basal ganglia are thought to be involved in this type of motor action. What to name these activities is problematic and often the term used is *procedural movements*; we will use the term *programmed movements*. These can involve fine hand movements as well as muscle activity of larger joints of the elbow and shoulder, in addition to postural adjustments.

The command instructions for the voluntary control of movements of all types are also sent to the cerebellum, which then feeds back via cerebello-thalamo-cortical connections for possible adjustment of the movement. In neurologically healthy persons, this is done without conscious involvement.

All these systems need to be connected (via tracts) in order for each contributor to inform the others of its participation in planning, organizing, coordinating and adjusting the movement(s). The eventual outcome is "instructions" from the upper motor neuron, descending via various pathways to the spinal cord to connect with and control the activity of the lower motor neurons and their subsequent output to the muscles. Any motor act requires functioning motor units; as was defined in Chapter 1, these consist of a lower motor neuron, its (myelinated) axon and all its collaterals, the intact neuromuscular junctions and normally acting muscle fibers.

The control of motor activity thus requires the involvement of some or all of the following areas of the CNS:

- The cerebral cortex, for planning the action
- The motor strip, to command the desired movement, site of the upper motor neuron
- The basal ganglia, to organize the muscle groups involved
- The cerebellum, to adjust the desired movements
- The brainstem, for associated movements and adjustments to gravity
- The spinal cord and its "descending" pathways
- The lower motor neuron, as the final common pathway
- Intact peripheral nerves, neuromuscular junctions and muscle

2.4.2 Motor System Examination

Assessing the motor system necessitates a variety of different tests; this approach reflects the complexity of the control of motor functions. These include:

- Muscle inspection and palpation for signs of atrophy, hypertrophy, tremor, involuntary twitching, fasciculations, and abnormal consistency (e.g., induration)
- Passive movements of joints, both flexion and extension, performed both quickly and slowly by the examiner, for the assessment of muscle tone
- Voluntary rapid fine movements, carried out by the patient, of fingers and hand, as well as toes and foot
- Muscle strength, at various joints, usually tested against resistance
- Deep tendon reflexes (DTR)

The examination of reflexes tests for the integrity of the PNS myelinated fibers (sensory and motor), the spinal cord with its upper motor neuron modulatory activity and the sensitivity of the motor neurons.

- Coordinated movements requiring the combined actions of various muscle groups (e.g., finger-nose test)
- Balance and gait (e.g., tandem walking)

The examination usually focuses on the activity of the limbs rather than of the trunk, unless the history indicates otherwise. Table 2.1 lists the levels of the spinal cord involved in the motor control of the upper and lower

Table 2.1 Testing of Motor and Sensory Segments

Motor	Sensory
Shoulder abductors & elbow flexors = C5,6	Hand middle finger = C7
Elbow extensors = C7,8	Hand little finger = C8
Anterior thigh muscles (quads) = L2,3,4	Nipple = T4
Pretibial muscles = L4,5	Umbilical region = T10
Calf muscles = L5, S1	Knee = L3
Posterior thigh muscles = L5, S1,2	Top of foot = L5
	Bottom of foot = S1

C = cervical; T = thoracic; L = lumbar; S = sacral.

limbs. The routine includes comparing one side with the other side for motor power and reflexes. All the while the examiner is on the lookout for abnormal and/or involuntary movements.

Starting with the upper limb, the muscles are palpated and the elbow joint is flexed and extended passively, the examiner noting the tone of the muscles being stretched and whether there is any resistance to movement. Then the patient is asked to perform a precise (voluntary) movement, such as apposing the finger and thumb rapidly.

Next, movements are done against resistance, testing shoulder abduction and adduction, elbow and wrist flexion and extension, all the time comparing what one might expect from the patient (depending on gender, body size, and age). Finally the hand muscles are tested by asking the patient to squeeze the examiner's hand, and to spread the fingers against (the examiner's) resistance. Muscle strength is graded in a standardized way (Table 2.2).

The reflexes are tested again using side to side comparison: biceps to biceps, triceps to triceps and so on. A brisk tap is applied to the tendon attached to the muscle serving the reflex. Often in the upper limb, the examiner places a thumb over the tendon being tested. The examiner not only looks for the reflex response but feels for the output response. In the lower limbs, the examiner holds the lower leg or ankle when testing these reflexes not only to feel the output of the reflex but to protect against being kicked in the event of extreme hyperreflexia. There is a standardized way of recording the reflex responses (Table 2.3).

The examination follows the same pattern in the lower limb: palpating the muscles, flexing and extending the

Table 2.2 Muscle Response Grading Scale

Grade 5	Normal strength in muscle being tested
Grade 4	Relative to normal, reduced muscle strength against resistance but no difficulty moving body part against gravity
Grade 3	Marked difficulty contracting muscle against resistance but just able to move body part against gravity
Grade 2	Visible movement of body part if effect of gravity removed (e.g., movement in horizontal plane with limb supported by examiner)
Grade 1	Visible muscle contraction but no movement
Grade 0	No visible muscle contraction

Classification according to the Medical Research Council (MRC).

Table 2.3 Reflex Response Grading System

0+ = No reflex
1+ = Decreased reflex
2+ = Normal reflex
3+ = Increased reflex
4+ = Abnormally brisk reflex with clonus

knee, testing for voluntary wiggling of the toes, strength of hip flexion (and extension), knee flexion and extension, and ankle dorsiflexion and plantar flexion. Reflex testing includes the patellar and the Achilles tendons at the knee and the ankle, respectively. Rapidly jerking the foot into dorsiflexion tests for a sustained rebounding rhythmic flexion-extension, referred to as clonus; this is an abnormal response usually accompanying increased tendon reflexes (hyperreflexia, see below).

Finally, the plantar reflex is tested by stroking the outside of the bottom of the foot firmly with a blunt object. The normal response in adults is a plantar flexion of the toes, often accompanied by the person "withdrawing" the whole lower limb (so the examiner usually grasps the lower leg while doing this reflex examination). An abnormal plantar response consists of extension of the big toe and fanning of the other toes, called the *Babinski sign* or *response*; oftentimes clinicians will speak of a positive or negative Babinski. A Babinski response invariably indicates damage to the corticospinal pathway (discussed below).

Spasticity is defined as a velocity-related increase in muscle tone. The combination of velocity-related increased resistance to passive stretch of the antigravity muscles (in particular the biceps brachii, hamstrings and plantar flexors) with increased reflex responsiveness (hyperreflexia) is a sign of upper motor neuron damage. This condition is usually accompanied by clonus and a Babinski sign. Increased constant resistance to passive movement of both flexors and extensors that is not velocity dependent is known as *rigidity*. Reflexes in this state are often normal and the plantar response is usually downgoing (normal). Rigidity usually occurs in diseases involving the basal ganglia.

Motor coordination is tested in the upper limb by asking the patent to perform the finger-nose test, in which the patient is asked to touch the tip of his or her own nose with the tip of his or her finger and then to touch the examiner's finger tip (which is moved each time). An additional test for coordination involves rapid alternating movements of the hand with palm up and palm down. The lower limb is tested by asking the patient to move the heel along the front of the shin bone (the tibia) down to the dorsum of the foot. Deficits in coordination, with intact motor and sensory systems, are associated with disease processes involving the ipsilateral cerebellar hemisphere.

Balance and gait are tested by asking the person to walk along a straight line with heel touching toe (tandem walking), and by asking him or her while walking normally to rapidly turn (change direction). In addition to assessing gait stability, one also notes whether there is a normal associated swinging of the arms. The testing should also include walking on the heels and the toes, and with the feet inverted and everted.

A loss of coordination of voluntary movements is known as *ataxia*. Ataxia can occur with upper and lower limb involvement as well as with walking.

2.4.3 Pathways for Voluntary (Intentional, Purposeful) Movements

2.4.3.1 The Corticospinal Tract for Complex, Precise Movements The cerebral cortex is where all planning of intended movement occurs, starting with the "decision" to make a movement in the supplementary and premotor areas. The final command originates in the motor strip (the precentral gyrus) and usually also involves the premotor area. Control of the distal musculature is conveyed via a single uninterrupted pathway from the upper motor neuron in the motor cortex to the lower motor neuron in the spinal cord. This is the corticospinal tract, also referred to as the *direct* pathway (further discussed in Chapter 7).

Pathway (Figure 2.5) The corticospinal pathway descends through the white matter of the cerebral hemispheres. As the fibers funnel together between the thalamus and basal ganglia, the pathway becomes part of the internal capsule. It continues through the midbrain cerebral peduncle and the middle of the pontine bulge to the lower medulla, where it crosses and continues down the spinal cord as the lateral corticospinal tract (see also Figure 1.5A). This tract is responsible for "instructing" the lower motor neurons of the spinal cord regarding voluntary motor movements. For control of fine movements of the fingers, the pathway is thought to terminate directly onto the lower motor neurons. Such "instructions" require both the activation of agonist muscles required for the movement and the inhibition of antagonist muscles.

Lesions of the cortex, white matter and brainstem will cause a deficit of voluntary movement on the opposite (contralateral) side; spinal cord lesions will affect movements on the same (ipsilateral) side. A lesion anywhere along this pathway will lead to hyperreflexia and an abnormal plantar response (discussed above), the Babinski sign.

The pathway for voluntary control of the cranial nerve motor nuclei is called the *corticobulbar pathway* (the older name for the brainstem is the bulb). This pathway is involved in the control of eye movements, chewing (mastication), facial expression (except for those accompanying emotions), swallowing, tongue movements, as well as shoulder elevation and neck rotation. A unilateral lesion of this pathway will lead to weakness of the lower facial muscles, on the opposite side (further discussed in Chapter 6).

2.4.3.2 Pathways for Programmed (Procedural) Movements The basal ganglia are definitely associated with learned motor patterns that can be called into action when needed, a phenomenon known as procedural memory. This system is likely involved in organizing the sequencing of muscle contractions and relaxations, agonist and antagonists, needed for a smooth action (starting and stopping muscle action), as well as the force necessary to perform the movement.

As was mentioned in Chapter 1, the substantia nigra (in the midbrain) and the subthalamic nucleus are, from the functional perspective, part of the basal ganglia and also involved in this type of motor control; lesions involving these structures lead to problems with voluntary movements as well as the occurrence of involuntary movements (e.g., resting tremor, ballistic—or flinging—movements). The most common example of motor problems associated with degeneration of the substantia nigra system is Parkinson's disease.

Pathway The putamen is the part of the basal ganglia most involved with organizing the motor patterns. The signals are relayed via the globus pallidus and the thalamus to the motor areas of the cerebral cortex. The

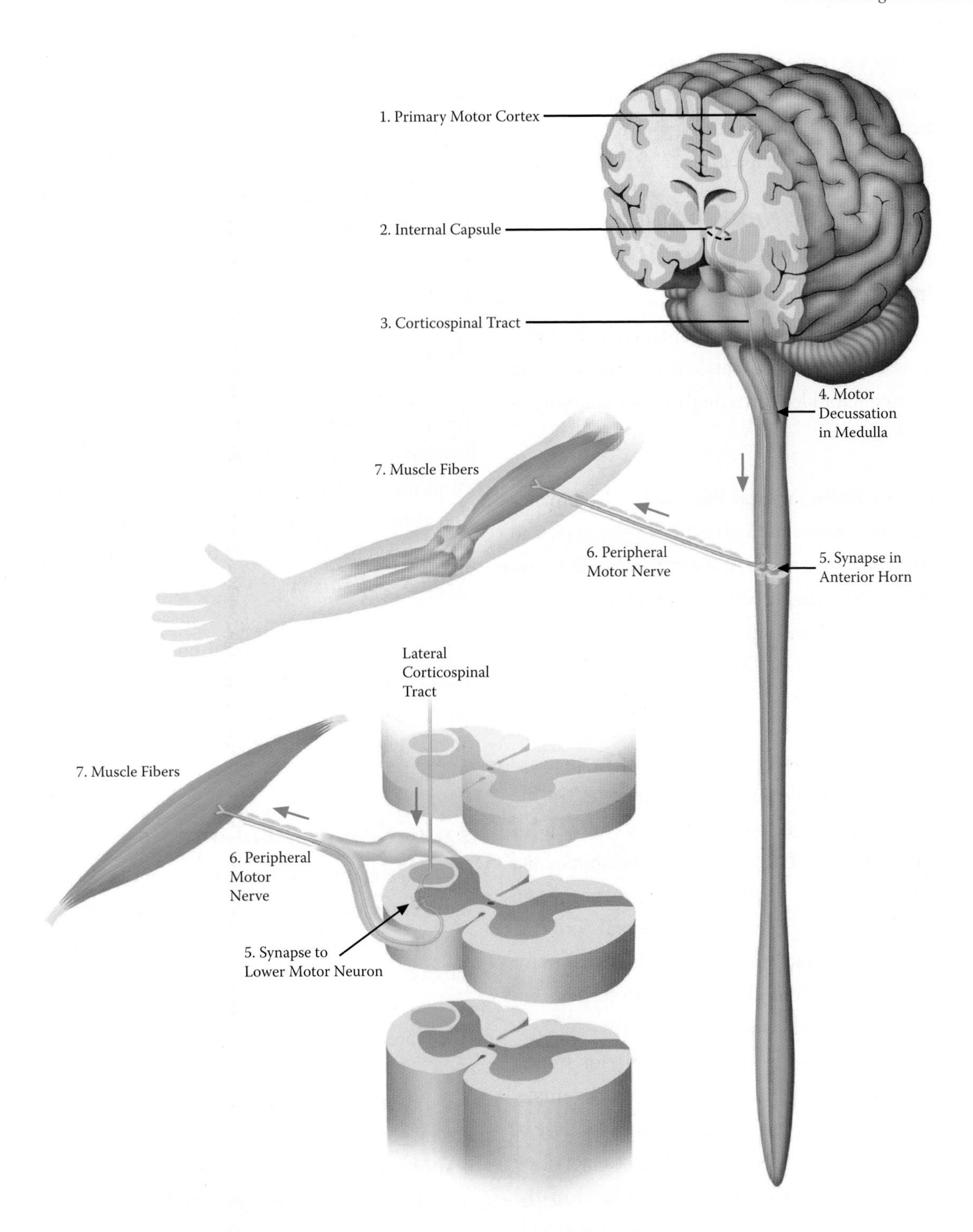

FIGURE 2.5 The voluntary direct motor pathway, the corticospinal tract, is shown, from the precentral (motor) gyrus, through the white matter of the hemispheres (via the internal capsule), crossing in the lower medulla to become the lateral corticospinal tract in the spinal cord. The synapse with a lower motor neuron is shown. This pathway is animated on the text DVD and Web site.

resulting actions are carried by the premotor area, and the fibers descend in the corticospinal tract; there are no direct connections from the basal ganglia to the spinal cord (further discussed in Chapter 7).

2.4.4 Pathways for Nonvoluntary (Associated and Postural) Movements

Most volitional movements are accompanied by complementary or associated movements, such as positioning of the elbow and shoulder (proximal) joints and adjustments in trunk posture.

2.4.4.1 Cortico-Bulbo-Spinal Pathways Commands for associated movements are carried from the cortex to the spinal cord via motor nuclei of the brainstem. Because there is a second nucleus in between the cortex and the lower motor neuron, these tracts are called the *indirect pathways* (also discussed in Chapter 7), or the cortico-bulbo-spinal tracts.

Pathways The nuclei of the brainstem that are involved in such actions include the red nucleus (in the upper midbrain) and the pontomedullary portion of the reticular formation. Tracts, named rubrospinal and reticulospinal, descend from their respective brainstem nuclei to the (lower) motor neurons of the spinal cord (see Figure 7.2). Thus, the motor cortex remains the origin for the proximal limb and trunk movement instructions, but through a two-step pathway via the brainstem.

The reticular formation is also responsible for setting the level of excitability of the motor neurons. For example, loss of this influence on the lower motor neurons because of a lesion of the pathway in the spinal cord causes increased excitability of the motor neurons below the level of the lesion, leading to muscles that are hypertonic and hyperreflexic (i.e., spasticity).

2.4.4.2 The Vestibulospinal Tract The response to changes in position in relation to gravity or to acceleration and deceleration also requires postural adjustments; these occur without conscious involvement and are totally nonvoluntary in nature.

Pathway The vestibular nuclei of the brainstem are responsible for these adjustments and give rise to a pathway that descends to the spinal cord to control the axial musculature of the spinal column; this system is under the influence of the cerebellum. Other output from these nuclei controls the adjustments of the eye muscles, forming the basis for the vestibular-ocular reflex (discussed in Chapter 6).

2.4.5 Coordination of Movement

The command for a voluntary movement is also sent from the motor cortex to the cerebellum. As the movement is carried out, the cerebellum itself receives input from the periphery (the muscles and the joints involved) and detects errors in the performance of the movement that are fed back to the cerebral cortex for an appropriate "online" adjustment.

Pathway The lateral portions of the cerebellum, the cerebellar hemispheres (also called the neocerebellum), are involved in this aspect of motor control. (The detailed pathway for accomplishing this is explained in Chapter 7.)

2.4.6 Gait and Balance

Gait is a complex activity requiring the collaboration of the indirect motor pathway, the cerebellum, the visual system and the vestibular apparatus. We have already discussed the examination of gait and tandem walking earlier in this section. (It should be noted that children under the age of 5 or 6 are unable to tandem walk.)

Pathway Medial (older) parts of the cerebellum are involved with balance and gait (i.e., the central portion or vermis). Lesions in these parts of the cerebellum or in the pathways from the cerebellum to the brainstem (medullary portion) may cause problems with balance and gait (ataxia).

The accepted test for balance is called the *Romberg test.* The patient is asked to stand with feet together and told to look at a distant point. If able to accomplish this without beginning to lose his or her balance, the patient is then asked to close his or her eyes and balance is reassessed.

The examiner should be prepared to assist should the person begin to fall.

Pathways The two components of the Romberg test evaluate both the vestibulo-cerebellar apparatus and the posterior columns of the spinal cord. If the patient is unable to stand with feet together, eyes open, a vestibulo-cerebellar lesion is strongly suspected. On the other hand, if the patient has no trouble standing feet together with eyes open, then begins to fall with eyes shut, a posterior column lesion is likely. Should there be a lesion affecting the latter pathway, unconscious proprioception (for limb position) does not reach the cortex (discussed next) and therefore visual guidance is needed for balance; when the eyes are closed this guidance is lost and the person "loses his or her balance," that is, a positive Romberg.

2.5 Somatosensory Sensory Systems

The sensory system is examined in a systematic fashion, including the two major sensory modalities—vibration, proprioception and discriminative touch on the one hand, pain and temperature on the other. Various levels of the spinal cord are examined (see Table 2.1).

2.5.1 *Posterior Column Pathway: Vibration, Proprioception, and Discriminative Touch*

The ability to sense vibration is tested by placing a vibrating tuning fork over bony prominences. Joint sensation can be evaluated by asking the patient to identify (with the eyes closed) which way the finger or toe joint is being moved (up or down). Discriminative touch is tested by asking the person whether he or she feels the examiner touch selected skin areas (usually with a wisp of cotton or tissue) and whether the sensation is the same on both sides.

Identifying objects placed in the hand with the eyes closed (stereognosis) and numbers stroked on the skin (graphesthesia) test whether the cortical areas involved with interpreting these sensations are intact.

Pathway (Figure 2.6A) These modalities of vibration, proprioception and discriminative touch are conveyed from their specialized receptors in the skin, joints and muscle and carried rapidly in thickly myelinated fibers to the spinal cord (see also Figure 1.4A). The fibers enter the cord and give off collateral branches for the deep tendon (myotatic) reflex and other reflexes, then turn upward, without synapsing, on the same (ipsilateral) side, forming the posterior (dorsal) columns in the spinal cord. A lesion of this pathway in the spinal cord interrupts the sensory information from the same side of the body.

The fibers ascend the spinal cord and synapse in the lowermost medulla in the posterior column nuclei. The fibers of the second-order neurons cross and form a new pathway, the medial lemniscus, which ascends through the brainstem, terminating in the ventral posterolateral (VPL) nucleus of the thalamus; from here fibers are distributed to the postcentral gyrus, the primary somatosensory cortex. (The details of this pathway are also discussed in Chapter 5.) This gyrus of the brain has different portions dedicated to each part of the body, called the sensory homunculus (similar to the motor homunculus, see Figure 4.1). Cortical processing in this gyrus allows for the localization of the stimulus.

Identification of the nature of the stimulus or object is accomplished by further processing of the sensory information in the adjacent areas of the superior parietal lobe, and from there the patient, when asked, can name an object held and manipulated in the palm of the hand. This is called *stereognosis*. Connections with other association areas of the cortex may lead to memories about the nature of the object. Loss of ability to recognize the significance of sensory stimuli, even though the primary sensory system is intact, is called an *agnosia*.

2.5.2 *Spinothalamic Pathway: Pain and Temperature*

This is usually tested by using the tip of a clean new pin or safety pin. The examiner can press very lightly or more heavily on the pin so that the degree of sensation of pain can be graded. (One must never press so heavily as to draw blood!) Again, the two sides are compared. Occasionally, the blunt side of the safety pin is used just to make sure that the person is responding correctly. The safety pin is to be used for that patient only and must be discarded in the designated needle box afterward.

It is often useful to calibrate the degree of sensation using a sensitive area on the face or nape of the neck and to ask the patient to close his or her eyes and image in the mind

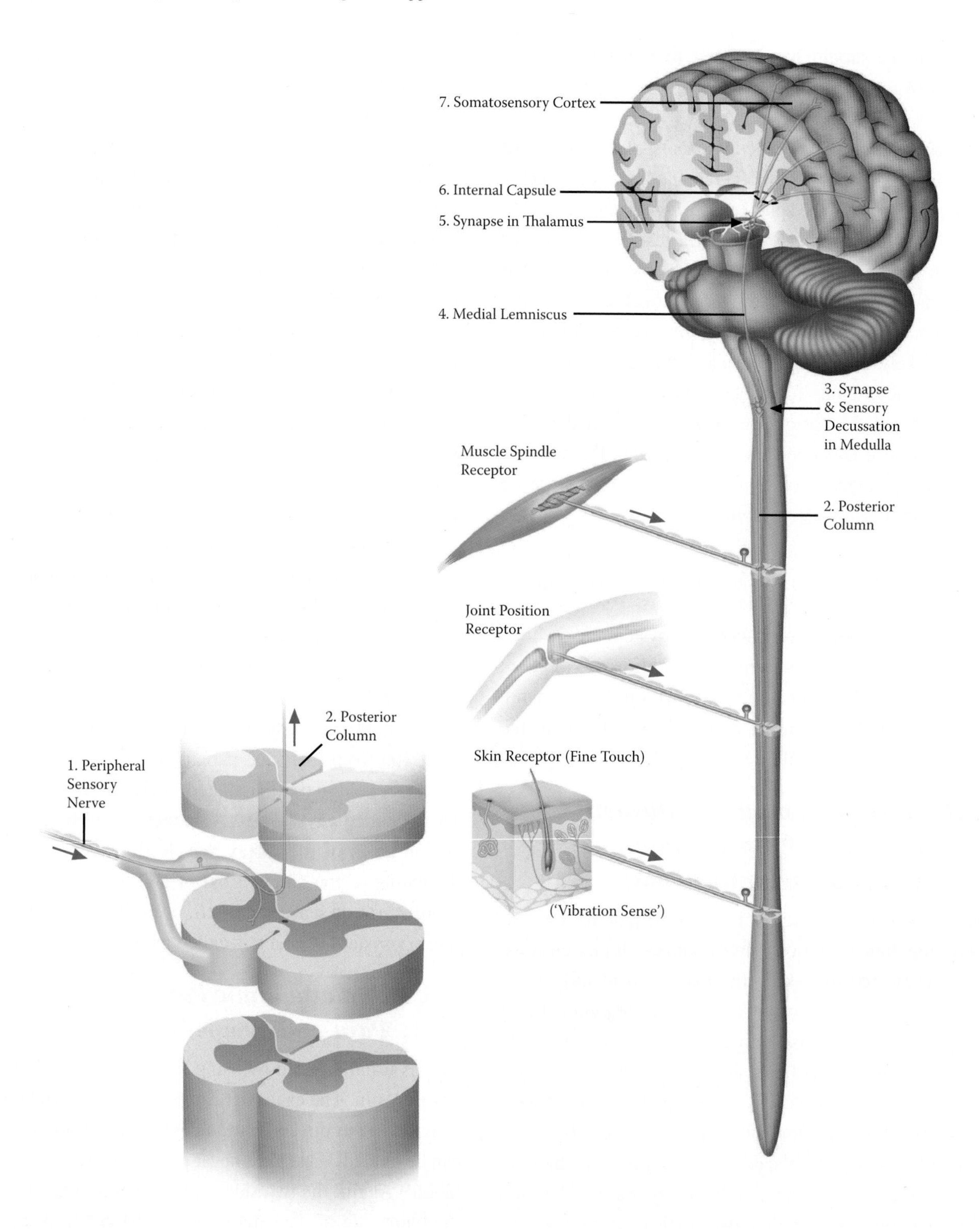

FIGURE 2.6A The pathway for discriminative touch, muscle spindle and joint afferents, and the sense of vibration is shown from the periphery to the sensory cortex. The fibers enter the dorsal horn and ascend in the posterior columns, on the same side. The first synapse occurs in the lowermost medulla, then the fibers cross and traverse the brainstem as the medial lemniscus. The second synapse occurs in the specific relay nucleus of the thalamus (VPL), following which the information is relayed to the somatosensory cortex (via the internal capsule), the postcentral gyrus of the parietal lobe. This pathway is animated on the text DVD and Web site.

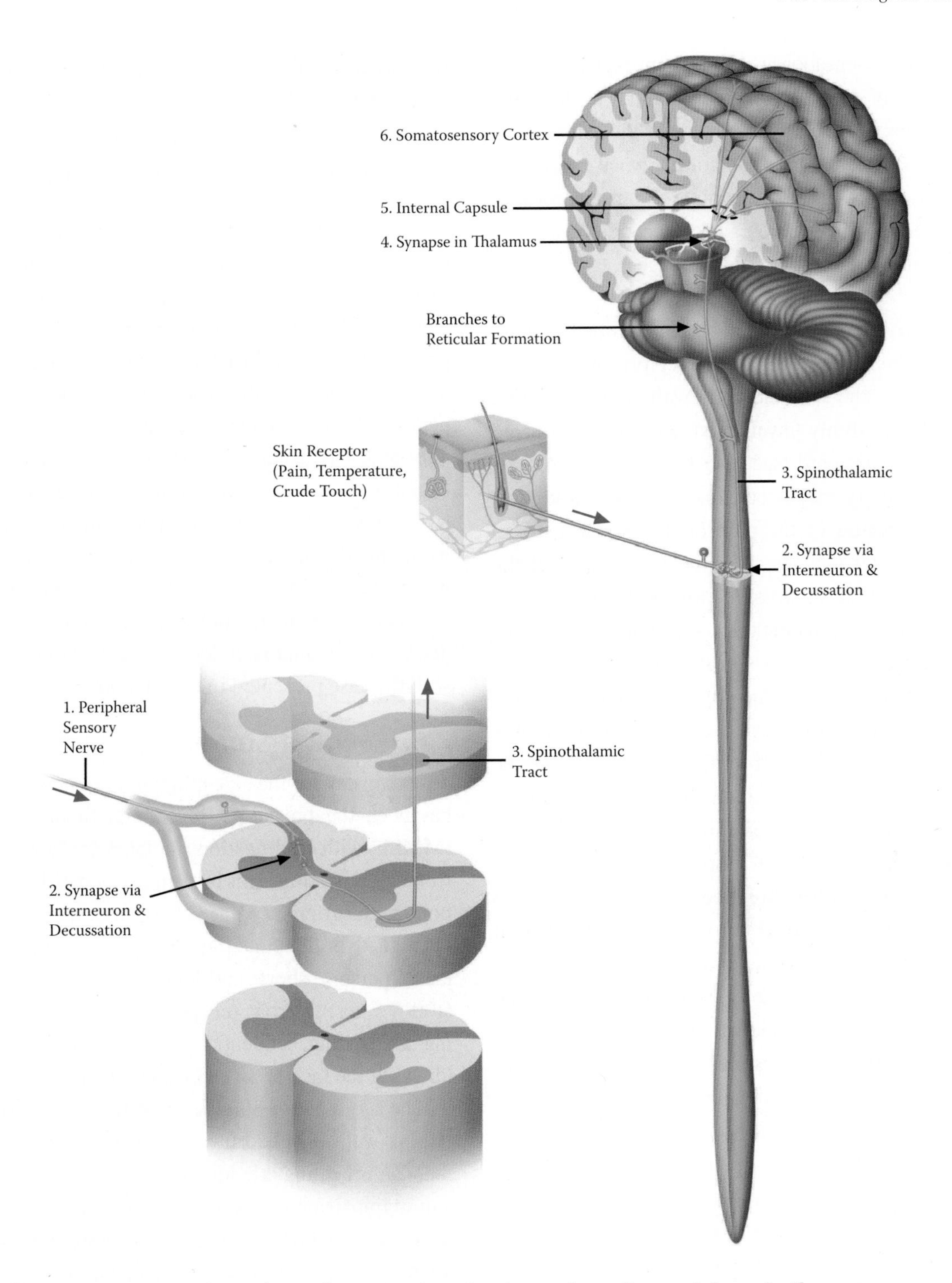

FIGURE 2.6B The pathway for pain and temperature is shown from the periphery to the sensory cortex. The fibers enter the cord, synapse and cross soon after entry, just anterior to the central canal. They ascend in the (lateral) spinothalamic tract on the opposite side, and traverse the brainstem, giving off collaterals to the reticular formation. The next synapse occurs in the specific relay nucleus of the thalamus (VPL) following which the information is relayed to the somatosensory cortex, the postcentral gyrus of the parietal lobe, and other areas. This pathway is animated on the text DVD and Web site.

that the pinprick sensation is 100 percent picky. Then as one tests areas of corresponding dermatomes in the upper and lower limbs from side to side, one asks the patient if the sensation is 100 percent picky and, if not, what percentage of "pickiness" is present. The examiner can then record the degree of sensation by location, which can then be used by future examiners to test for clinical changes.

Pathway (Figure 2.6B) These modalities are conveyed from their nonspecialized receptors, mostly free nerve endings in the skin and body wall, and are carried more slowly in thinly myelinated and nonmyelinated fibers to the spinal cord (see also Figure 1.4B). At this level, reflex activity may occur, for example, withdrawal of the limb because of the pain of the pin. The fibers synapse soon after their entry in the gray matter of the dorsal horn of the cord, cross and ascend the spinal cord as the (lateral) spinothalamic tract. A lesion involving this pathway on one side of the spinal cord therefore interrupts these sensations from the opposite side of the body. The pathway continues through the brainstem, where it is joined by the pain pathway from the opposite side of the face. Collaterals from the pain and temperature pathway enter the reticular formation (see Figure 2.6B) where they may cause a general "alerting" response.

The pain and temperature fibers again relay in the thalamus and are distributed not only to the postcentral gyrus, but also to other areas of the cortex. It is at the cortical level where the source of the pain is localized and the nature of the stimulus is identified. Other connections including to limbic structures may lead to other associations and possibly to an emotional reaction.

In summary, the various sensory modalities from the skin and muscles follow two distinct routes in the spinal cord. Lesions of one-half of the cord will lead to a dissociated sensory loss involving loss of discriminative touch sensations on the same side and loss of pain and temperature on the opposite side. Knowing which areas of the body are supplied by what level of the spinal cord will help determine the level of the lesion (level of the cord, not the vertebral level). A lesion at the brainstem level may affect both pathways leading to a loss of all sensations on the opposite side, since both pathways have already crossed. (These lesions and others will be discussed further in Chapters 5 and 6.) A lesion may also involve the corticospinal tract, thereby causing a loss of voluntary movement (see above).

2.6 Higher Mental Functions

The assessment of higher mental functions includes the evaluation of attention, executive function, language, memory, visuospatial ability, as well as personality (which embraces the unique emotional and reactive aspects of a person). It is often difficult to recognize and classify disorders associated with cortical dysfunction. Very often the person affected also loses insight into this loss of functionality and it is usually the spouse, children, relatives and friends who become aware of these changes. The challenge has been to understand what changes in mental functionality, if any, should be associated with normal healthy aging, which indicate the recently described syndrome of mild cognitive impairment, and which are indeed indicative of disease states, that is, a dementia. (Dementia is discussed with the case presented in Chapter 9.)

2.6.1 *Mental Status*

The mainstay for the screening of higher mental functions has been the Folstein Mini-Mental Status Examination (MMSE) with modifications for language, visuospatial and frontal lobe testing. This instrument, consisting of ten questions with different point values, tests the basic elements of higher brain function: attention, orientation (time, place, person), short- and long-term memory, calculation (100 − 7 test), verbal reception and verbal output, reading and writing, sequential processing, a visuospatial task (drawing the face of a clock), and integration.

The score out of 30 gives an estimate of cortical function as well as subcortical connectivity. A score of less than 27, providing attention is intact (see below), indicates a mild impairment; lower scores need more thorough testing depending on the examiner's assessment of the patient, the history, and the concerns of the family.

The mental status examination has to be tailored to age, education and the developmental level of the patient. In the pediatric population age-referenced milestones are available in chart form, such as the Denver Development Scale.

Recently, other tests of mental capacity have been devised as our population ages and with the emergence of newer concepts such as mild cognitive impairment or dysfunction. One that has been verified in the literature is the Montreal Cognitive Assessment (MoCA), designed as a rapid screening instrument for just such a condition. The various functions tested include visuospatial, executive, naming, memory, attention, language, abstraction, delayed recall, and orientation (to time and place).

2.6.2 Cortical Functions

The cerebral cortex with its microprocessor array and exquisitely complex interconnections is where our thinking occurs, as well as our language function, our memory for events and places, and our ability to plan, to react, and to interact with others (see Figure 1.10). No part of the brain can function in isolation, without the participation of many other parts. Using current technology, including functional MRI (fMRI, which shows changes in regional blood supply) and positron emission tomographic studies (PET, using radioactive tracers), certain parts of the brain can be seen to "light up" during task-related activation of specific areas. With all the "background" brain activity removed these images are now the conventional way of presenting information regarding regional brain functionality.

2.6.2.1 Executive Functions In the normal course of our lives, we must decide on what has to be done and in what sequence. Decisions have to be made as to priorities and judgments are made about the importance of the task and its ethical dimension or its social appropriateness. This ordering of our tasks is usually attributed to the function of the frontal lobes of the brain, the area in front of the motor regions, sometimes called the prefrontal cortex, particularly the dorsolateral aspects. Often, this region of the brain is called its CEO, its chief executive officer. It is this function of the brain that is impaired, progressively, in the cortical dementias (e.g., frontotemporal dementia, see Chapter 9).

2.6.2.2 Attention A particularly important feature of the brain is the ability to focus on certain happenings, whether a conversation or a task. This ability to attend is very important while doing any task (e.g., reading or a surgical procedure) or in any learning situation. Areas of the brain involved with attention include the frontal lobe as well as the anterior portions of the cingulate gyrus.

The assessment of attention has to be made first in the mental status exam as the degree of attentiveness will influence all of the subsequent testing. Often an inattentive patient will score very poorly on a mental status exam. This result overestimates the patient's deficits in other areas.

In addition, deficits in vision, hearing, proprioception and motor function can also influence the score obtained on the mental status examination. It is hard to draw a clock face if you cannot see the result properly. It is therefore often useful to perform the other parts of the neurological examination first and then come back to the mental status testing.

2.6.2.3 Language Our ability to communicate using symbols is perhaps unique to humans. The left hemisphere is usually the hemisphere specialized for language—expression and comprehension, both written and spoken. As noted in Chapter 1, two areas have been identified (see Figure 1.10C):

- Broca's area in the lower frontal lobe for expressive language
- Wernicke's area in the posterior-superior temporal region (near the auditory area) for the comprehension of language

2.6.2.4 Memory Memory for events, names and places is the memory that we use most frequently. Usually attention to the event is important for recording a memory. The essential structure needed for memory formation is the hippocampus, a neuroanatomical structure located within the temporal lobe of the brain in its medial aspect (see Chapter 12). Once the memory is "registered," if it is considered important, it is "stored" in various areas of the cerebral cortex.

While engaged in a task or event (for example, going to fetch something, remembering a name), the information is in use and needs to be retained for short periods of time, a process called *working memory*. It is thought that this aspect of memory is carried out in the frontal lobes. Our long-term memory retains information, names, places and events; in order to use this data the memory trace has to be retrieved in a timely fashion. Broad areas of the cerebral cortex are involved in the storage of this type of memory. Various memory disorders affect the registration, storage or retrieval of memory.

The second type of memory is called *procedural*, which involves wide areas of the brain, particularly the basal ganglia. We use this type of memory for all kinds of complex motor tasks, which, once learned, can be retained and used over a person's lifetime (e.g., riding a bicycle, performing a piece of music). This type of memory is not usually affected in the dementias, disorders that affect cognitive functions and memory for events, names, and places (discussed below and in Chapter 9).

2.6.2.5 Visuospatial The task most often used for testing of this function is drawing a clock face with a certain time indicated. The association area of the lower parietal lobe in the nondominant hemisphere, situated between the somatosensory, visual and auditory areas, is thought to be the region of the brain most involved with visuospatial tasks.

2.6.2.6 Impairment of Function Impairment of function can occur in any of these domains, separately or in combination. One of the impairments most challenging to detect, particularly for a physician meeting a patient for the first time, is executive dysfunction, problems with weighing information (the basis for decision making) or with attention. Changes in personality often accompany the deterioration of higher mental functions (with dementia, discussed below), and often the person is able to cover up deficits or lapses for a time by using his or her personal charm or social graces to slide through an awkward moment.

A language disorder is called an *aphasia*, defined as an acquired disruption, disorder or deficit in the expression or comprehension of spoken or written language. More specifically there can be an expressive or receptive aphasia, or a global aphasia affecting all language functions. Difficulty with articulation or enunciation of speech is called a *dysarthria*. *Alexia* is the term used for word blindness; this is the loss of ability to grasp the meaning of the written word due to a central (not visual) lesion. Inability to write owing to a lesion of higher centers, even though muscle strength and coordination are preserved, is called *agraphia*.

Difficulty with short-term memory is one of the most common concerns, particularly as people age (in fact, some older people call these lapses "senior moments"). This concern is to be appreciated because short-term memory is one of the functions most affected in a common form of dementia, Alzheimer's disease. Dementia is defined as a progressive brain disorder that gradually destroys a person's memory, starting with short-term memory, and cognitive abilities (including judgment).

A specific disorder of cortical function is the loss of ability to carry out a purposeful or skilled movement (e.g., buttoning a shirt), despite the preservation of power, sensation and coordination; this dysfunction is called an *apraxia*. A visuospatial impairment is usually difficult to detect until formally tested in the manner described.

2.6.2.7 Emotions Although most of our cerebral functions are associated with analyzing our external environment, other (older) parts of the brain have been found to "react" to these happenings. A reaction may include alterations in pulse, respiration and blood pressure, and changes in other autonomic functions (e.g., gastrointestinal). Over the short and long term, there may be hormonal changes.

The areas of the brain associated with these emotional reactions are located for the most part in the temporal lobe. The relevant structures include the amygdala and the hypothalamus, as well as other nuclei, and the cortex of the insula. Parts of the frontal lobe are also involved in emotional reactions, mostly the ventromedial aspects (in particular, the cingulate gyrus) and the orbital portion (sitting above the orbits; see Chapter 12). These brain structures are collectively called the limbic system and are considered in more depth in Chapter 12. Some limbic system components such as the septal nuclei and the nucleus accumbens (beneath the anterior corpus callosum and caudate head, respectively—see Chapter 12) are also involved in behavior associated with rewards and punishment.

2.7 Human Consciousness

One particular aspect of higher mental function is consciousness, wherein the individual is in contact with those around him and reactive to outside events (beyond himself). A unique characteristic of humans, so it is thought, is self-awareness, considered by many as an aspect of consciousness. The exact structure or structures necessary for consciousness have not been clearly defined, but involve

Table 2.4 Glasgow Coma Scale

Best ocular response	opens eyes spontaneously	4
	opens eyes to voice	3
	opens eyes to pain	2
	no ocular response	1
Best verbal response	oriented, conversing normally	5
	confused, disoriented	4
	says inappropriate words	3
	incomprehensible sounds	2
	no sounds	1
Best motor response	obeys commands	6
	localizes painful stimuli	5
	withdraws to pain	4
	decorticate response to pain	3
	decerebrate response to pain	2
	no response to pain	1
Maximum score		15
Minimum score		3

the upper part of the reticular formation, the thalamus (the nonspecific nuclei) and the cerebral cortex. Perhaps, in fact, consciousness parallels total brain activity of a specific type, involving all these structures.

Consciousness can be impaired (e.g., a confusional state after a major seizure, or with drug or alcohol toxicity) or lost after a major head injury (a concussion). Prolonged loss of consciousness occurs after severe brain injury, most often resulting from a closed head injury, and after vascular ischemic lesions affecting the upper part of the reticular formation. Of course, general anesthesia does away with consciousness temporarily. A prolonged state of unconsciousness is called a *coma*.

Consciousness is assessed by a widely accepted set of criteria, collectively called the Glasgow Coma Scale, involving ocular, verbal and motor responsiveness. Its scoring (see Table 2.4) is important for instituting therapeutic measures and sometimes for prognosis (further discussed in Chapter 13). The verbal output responses are clearly limited if the patient is intubated.

Other important items in the assessment of the comatose patient include the position of the limbs. With decorticate posturing, the arms are in the flexed position and the legs extended; this is seen with bilateral cerebral pathology above the red nuclei. With decerebrate posturing both arms and legs are in the extended position; this is seen with midbrain pathology at or below the level of the red nuclei. In addition, the reflex responses of the cornea and the pupil, and the response to testing of the oculo-cephalic reflex for pontine integrity (the caloric test, described in Chapter 6) would be important aspects of the neurological examination.

Additional testing, including brain imaging and an electroencephalogram (EEG) may be needed to determine whether the comatose person is in epileptic status, whether there is brain death or a condition called persistent vegetative state.

2.8 Sleep

Sleep, a normal brain function, is a reversible behavioral state of perceptual disengagement from and unresponsiveness to the environment. Most people enjoy an awake daytime and a night-time sleep, a diurnal rhythm. Neuronal networks located in the diencephalon (the hypothalamus) and the brainstem reticular formation control this diurnal rhythm.

The amount of time spent in sleep changes with age and differs among individuals, although there is a minimum amount of sleep time which almost all adults require. An infant spends most of its time in sleep; as it matures sleep tends to occur during definite time periods which eventually (by the age of about six months) become regularized into daytime and night-time. There are different stages of sleep and the proportion of these stages changes with age.

There are two principal states of sleep: *rapid eye movement (REM) sleep*, which has a waking form of EEG and occurs with dreaming, and *non-rapid eye movement (NREM) sleep*, which has specific EEG patterns of sleep known as K-complexes, sleep spindles, and slow waves. The restorative value of sleep depends on the ability of an individual to maintain four to five sequences per night, each lasting 90 to 120 minutes, of NREM followed by REM sleep. We usually remember our dreams only if our REM sleep phase is interrupted.

The biologic purpose for sleep has been hotly debated, although most agree that a good sleep is restorative and that some form of memory consolidation occurs during sleep. Restorative sleep is necessary for the normal function of an individual.

It is commonly agreed that up to 25 percent of the adult population suffers from some form of sleep disorder at some time or another. As people age, sleep generally becomes more disrupted and fractionated, and less restorative. This often spills over into daytime activities and can interfere with daytime function, or cause irritability and social problems. Besides poor sleep hygiene, the most common sleep disorder is obstructive sleep apnea.

Loss of sleep, even for a few days (such as with airplane travel and jet lag), does affect a person's ability to perform normal daily activities and clearly affects the quality of higher mental functions, including attention and decision making. Shift work (e.g., in the case of nurses) can severely affect a person's life. Physicians in training are equally vulnerable as are physicians on call who need to respond to complex issues and questions when they are awakened in the middle of the night.

Suggested Readings

Bickley, L.S., Ed. *Bate's Guide to Physical Examination and History Taking*, 9th ed. Philadelphia, PA: Lippincott Williams & Wilkins, 2007.

Blumenfeld, H. *Neuroanatomy through Clinical Cases*. Sunderland, MA: Sinauer Associates, 2002.

Schapira, A.H.V., Ed. *Neurology and Clinical Neuroscience*. St. Louis, MO: Mosby, 2007.

Weiner, W.J. and C.G. Goetz, eds. *Neurology for the Non-Neurologist*, 5th ed. Philadelphia, PA: Lippincott Williams & Wilkins, 2004.

See also Annotated Bibliography.

Chapter 3

Clinical Problem Solving

Objectives

- To provide a framework for clinical reasoning when presented with a patient having neurological symptoms
- To provide a framework for localization of lesions within the nervous system
- To provide a framework to correctly identify etiological diagnoses
- To review how the data gathered by neurological history and physical examination are synthesized in a logical process to allow for localization and etiological diagnosis

3.1 Introduction

When confronted with a patient with neurological symptoms, the questions that a clinician must ask while conducting a neurological history and physical examination include the following:

1. Is the problem truly neurological? Or putting it another way, is the problem due to real organic pathology? Alternatively, could the patient's symptoms be due to a disease process located outside the nervous system?
2. Where is the lesion in the nervous system? (Where is the problem?)
3. What is the pathophysiological cause or etiology of the lesion? (What is the problem?)

The neurological examination is most important in determining the localization or location of the problem in the nervous system. Each element of the neurological exam tests the various systems involved in normal neurological function and therefore is used to detect dysfunction. Chapter 2 has provided the template for a basic screening neurological examination.

The history is probably most important in determining the underlying disease process by carefully retracing the steps through which the illness evolved. Did it come on suddenly or slowly? What was the patient's health like in the preceding days and weeks?

This chapter presents a system of sequential analysis of codified troubleshooting of the problems within the nervous system; it is focused on determining first the localization and then the etiological diagnosis. Knowing where the problem is often makes the determination of the etiological diagnosis easier. At first glance it might seem that this approach is contrary to the classical clinical method of first eliciting a history for etiological diagnosis and then performing a physical examination for localization. Practically speaking the order of operations is not so important as the history and physical are usually gathered in one session. As will be seen in this chapter, it is the use of the *worksheet tools* provided and the analytic sequence that are important, and not necessarily the order in which the clinical data are gathered.

This process of analysis will now be illustrated using a typical sample case. The process includes gathering the history and performing a clinical examination, including the derivation of a minimum dataset for the neurological examination.

The data from the neurological examination will be codified in an Expanded Localization Matrix to assist in working out the location of the patient's lesion(s). Then, turning to the etiological question, the illness timeline will be plotted on a History Worksheet; this visualization of the history will then be applied to the completion of the Etiology Worksheet. In the process, a likely disease location and disease type will become apparent.

3.2 Is the Problem Truly Neurological?

The usual approach of most clinicians is to assume that the symptoms have an organic basis until further information is gathered. Certain aspects will emerge during the course of the history and neurological exam which will alert the physician to suspect that the symptoms do not have an organic basis and other causes need to be explored.

The features that would make one suspect that the symptoms are not organically based would be neurological symptoms which shift from side to side, and a lack of consistent hard neurological findings. One must approach all patients with neurological symptoms with an open mind. A known psychiatric patient can develop neurological symptoms and abnormal behavior which might turn out to be a brain tumor or herpes encephalitis and not merely another episode of psychiatric illness.

SAMPLE CASE

This 45-year-old man presented to the local emergency department having difficulty with speech and right-sided weakness.

Two months previously, he had experienced two attacks one week apart during which he suddenly lost vision in his left eye. He described a pattern in which the upper part of the visual field in the left eye would become gray, then black (as in a curtain dropping down over his vision). This phenomenon lasted ten minutes at its maximum following which his vision would slowly return in reverse sequence over five to twenty minutes. He was sure that it was his left eye in that he remembered covering each eye separately and noticed that the disturbance was only in his left eye.

The present episode occurred after the evening meal when he experienced a sudden onset of difficulty speaking, which he described as knowing what he wanted to say but being unable to output more than single disjointed words in a telegraphic fashion.

At the same time, he noticed weakness of his right arm while his wife noticed that the lower part of his face was drooping on the right. By the time he reached the emergency department and was seen, over four hours had elapsed. He was now unable to output any speech and the weakness of his right arm and right side of the face was worse; there was also mild weakness of his right leg. However, he was able to understand questions and obey commands using gestures.

His neurological examination showed a blood pressure of 170/100 in both arms; he had normal heart sounds with a left carotid bruit. His heart rate was 84 and regular. Mental status examination was limited due to the speech difficulty but he was alert and able to answer questions and to obey commands.

Cranial nerve examination showed normal extra-ocular movements and fields; there was weakness of the lower part of the face on the right. Motor exam showed a moderate right upper limb weakness, mild right lower limb weakness and hyperreflexia of the right upper and lower limbs with an extensor plantar response on the right (positive Babinski sign). Sensory examination was normal to pin prick and vibration on both sides; coordination testing was limited on the right due to weakness and normal on the left. He was able to walk with assistance and tended to circumduct the right leg.

3.3 Where Is the Lesion in the Nervous System?

This is the most important question in that the location of the lesion in the nervous system will often give context to the underlying neuropathological process.

One approach to answering this question is to use an established troubleshooting technique used in electrical engineering. This process relies on the principles of testing the sequential functioning of pieces of hardware. The system is simple: one starts where the system exerts its final output and works backward toward the source of the input to determine where the fault is.

As we have already mentioned in the Introduction to this chapter, the data gathered in the neurological examination provides the most important information for determining the location of the problem in the nervous system. Each element of the neurological exam tests the various systems involved in normal neurological function and therefore is used to detect dysfunction. From Chapter 2, you will recall that the neurological examination includes the following components:

- Mental status
- Cranial nerves
- Motor
- Reflexes
- Sensory systems
- Coordination
- Gait

The **Basic Localization Matrix** consists of a knowledge cube with the neurological examination on one axis and the major functional elements of the nervous system on the other axis, as shown in Table 3.1.

The process of determining the location of the disease process causing the nervous system to fail requires certain

Table 3.1 Basic Localization Matrix

	Muscle	Neuromuscular Junction	Peripheral Nerve	Spinal Cord	Brain-stem	White Matter	Cortex	Basal Ganglia	Cerebellum
Mental Status Exam									
Cranial Nerves									
Motor Exam									
Reflexes									
Sensory Exam									
Coordination									
Gait									

basic knowledge of the functional parts of the nervous system. This knowledge includes the different types of functional tissue types, their function and their connections. These functional elements are listed across the top of the matrix and include:

Muscle
Neuromuscular junction
Peripheral nerve
Spinal cord
Brainstem
Deep white matter and thalamus
Cortex
Basal ganglia
Cerebellum

This matrix can be expanded to provide further detail of both the neurological examination and the functional elements. A more detailed model of the matrix, incorporating the data from the sample case, the **Expanded Localization Matrix**, is shown in Table 3.2.

For instance, the mental status examination can be expanded to include language, memory and visuospatial testing, each corresponding to different areas of the cerebral cortex. The structures relevant to the mental status examination can be expanded to include frontal, temporal, parietal, and occipital cortices, as well as subcortical structures (basal ganglia, thalamus) with which the cortical areas are connected.

Many disease processes involve the destruction of a specific component of the nervous system or its connections to or from other parts of the nervous system.

For many nervous system functions, there are several parallel pathways that can be sequentially tested; this is what we do when we examine the human nervous system.

On completion of the neurological examination as described in Chapter 2, the pertinent findings can be inserted into the appropriate rows and columns of the matrix using other correlative factors to help determine the most likely localization.

For the sample case, the clinical findings included an episode of transient blindness in the left eye followed by an episode of expressive aphasia (difficulty of expression of language) and right-sided weakness, and hyperreflexia including a positive Babinski sign.

The spaces in the localization matrix that correspond to positive findings are denoted by a black X. When, as all the localization data is considered together, a potential localization becomes impossible or unlikely, the X is changed from black to red.

The full matrix is completed first; then, starting from muscle and moving up to neuromuscular junction, peripheral nerve, spinal cord, brainstem, deep white matter, cortex, basal ganglia and cerebellum, each potential location is either confirmed or eliminated.

This general approach is especially important when there may be more than one localization: a typical example is multiple sclerosis, where one might have a lesion in the spinal cord causing damage to a posterior column (and sensory loss on one side) with another lesion in the deep cerebral white matter causing sensory loss on the opposite side. Some diseases such as B12 deficiency cause concurrent localizations with a combined myelopathy and neuropathy. In these cases, it is important to keep an open mind and leave the various localization options open; these then can be confirmed by imaging, neurophysiology or laboratory testing.

The final localization hypothesis—or top two localizations—is made based on the remaining anatomic structures most likely to explain the patient's signs and symptoms. When working on any future clinical service, the maximum benefit to the student for consolidating the clinical reasoning system presented here is to complete a matrix for the patient being assessed and to decide on the probable disease localization before looking at the CT or MRI scan results.

Making a localization based on the clinical findings and this system of clinical reasoning, then having it confirmed by the neuroimaging or neurophysiology test results, is extremely satisfying for students, teachers and, of course, the patient.

We will now review the localization implications for each anatomical level in the matrix, considering the findings of the neurological examination and utilizing the information in the sample case.

3.3.1 *Muscle Disease*

As the function of muscle is specifically motor output, the correlative factors for muscle disease relate to motor function or dysfunction. Motor dysfunction can originate either in a lower motor unit (which includes the alpha motor neuron, its axon, neuromuscular junction and muscle fibers) or in an upper motor neuron unit (which includes the pyramidal cell in the cortex, its axon and the latter's variably complex connection to the lower motor neuron in the spinal cord).

The important clinical distinction is that lower motor neuron unit disease causes weakness, muscle wasting and *hyporeflexia*, whereas upper motor unit disease causes weakness, spasticity and *hyperreflexia*. The upper motor neuron tracts, the corticospinal tracts, are somatotopically organized such that the pattern of muscle weakness can give information pertaining to the precise localization of the lesion in the corticospinal tract. The term *somatotopic* means that, for example, corticospinal tract axons destined for the lumbosacral cord region travel together in the outer portion of the tract while axons destined for the cervical cord travel in the inner portion, closer to the gray matter column.

Careful observation will determine if there is wasting or hypertrophy of muscle. The pattern of wasting is important in terms of etiology. Is it located in the face or limb girdles; is it symmetric? The combination of wasting, weakness, hyporeflexia, and lack of sensory findings represents the most important clinical features of muscle disease. Many primary muscle diseases have a genetic basis and a careful family history is required (Table 3.3).

3.3.1.1 Investigations Investigations for muscle disease include serum assay for creatine kinase (CK) and aldolase. These are enzymes released from muscle when it has been damaged. A high serum level of either of these enzymes thus implies muscle tissue damage. Different isoforms of CK are contained in skeletal and cardiac muscle, thus making it possible to differentiate skeletal muscle damage (elevation of the MM isoform of CK) from heart muscle damage (raised MB isoform levels).

The electromyogram or EMG is a technique for investigating peripheral neuromuscular disease by recording electrical activity from muscle, by the use of either needle or surface recording electrodes. Using this technique, it is often possible to distinguish between myopathic and neuropathic causes of muscle weakness.

In the past the gold standard test for muscle disease was to perform a biopsy, which then showed the nature of the muscle pathology. More recently DNA probes have been developed for many of the most common genetically inherited muscle diseases.

In the sample case outlined in Table 3.2, muscle disease was considered due to weakness. It was annotated in the Expanded Localization Matrix with an "X" but ruled out and changed to red because the weakness was unilateral and associated with sensory changes and hyperreflexia. An extensor plantar response, a classic sign of upper motor neuron dysfunction, also makes the localization of muscle disease unlikely.

3.3.2 *The Neuromuscular Junction*

The neuromuscular junction (NMJ) is the penultimate step in the process of motor output. It requires intact presynaptic, synaptic and postsynaptic function. The presence of a decrease in muscle strength with sustained effort (fatigue) in the absence of reflex changes or sensory signs is highly suggestive of a postsynaptic NMJ disease such as myasthenia gravis (Table 3.4).

3.3.2.1 Investigations The investigations for abnormalities of the NMJ include the Tensilon test, repetitive nerve stimulation, single fiber EMG and assays for serum autoantibodies against acetylcholine receptors on the muscle side of the neuromuscular junction. In specialized clinics, biopsies of the neuromuscular junctions can be performed and analyzed.

Table 3.2 Expanded Localization Matrix

	Findings		Expanded Localization Matrix						Cortex					
			Muscle	NMJ	Peripheral Nerve	Spinal Cord	Brain-stem	White Matter	Frontal	Temporal	Parietal	Occipital	Basal Ganglia	Cere-bellum
Mental Status Exam														
Attention	Normal													
Short-Term Memory	Present													
Long-Term Memory	Present													
Calculation	Unable to test													
Verbal Reception	Present													
Verbal Output	Decreased						×	×	×					
Written Input	Present													
Written Output	Not tested													
Sequential Processing	Unable to test													
Visuospatial	Normal													
Integration	Unable to test													
Cranial Nerves	**Right**	**Left**												
Smell CN I	Not tested	Not tested												
Visual Acuity CN II	Normal	Normal												
Color Vision CN II	Unable to test	Unable to test												
Visual Fields CN II	Normal	Normal												
Pupillary Reflex CN II–III	Normal	Normal												
Extra-ocular movements CN III, IV, VI	Normal	Normal												
Fundoscopic CN II	Normal	Normal												
Facial Sensation CN V	Normal	Normal												
Jaw Movement CN V	Unable to test	Unable to test												
Corneal Reflex CN V–VII	Normal	Normal												
Facial Movement CN VII	Decreased lower	Normal	×		×		×	×	×					
Hearing Acuity CN VIII	Normal	Normal												
Doll's Eye/Caloric CN VIII–III, IV, VI	Not tested	Not tested												
Gag Reflex CN IX–X	Weak	Normal	×		×		×	×						
Sternocleidomastoid Trapezius CN XI	Weak	Normal	×		×		×	×	×					
Tongue Movement CN XII	Decreased	Normal	×		×		×	×	×					
Apnea Test Medulla														

	Findings		Expanded Localization Matrix						Cortex					
			Muscle	NMJ	Peripheral Nerve	Spinal Cord	Brain-stem	White Matter	Frontal	Temporal	Parietal	Occipital	Basal Ganglia	Cere-bellum
Motor Exam	**Right**	**Left**												
Involuntary	None	None												
Tone	Increased	Normal	×		×	×	×	×	×					
Bulk	Normal	Normal												
Power	Decreased arm > leg	Normal	×		×	×	×	×	×					
Reflexes	Increased arm and leg	Normal	×		×	×	×	×	×					
Fatiguability	None	None												
Sensory Exam	**Right**	**Left**												
Pin/Temp	Normal	Normal			×	×	×	×			×			
Vibr/Propio	Normal	Normal			×	×	×	×			×			
Stereognosis	Not tested	Not tested												
Graphesthesia	Not tested	Not tested												
Two Point	Not tested	Not tested												
Coordination	**Right**	**Left**												
Upper limb	Decreased	Normal	×		×	×	×	×	×				×	×
Lower limb	Decreased	Normal	×		×	×	×	×	×				×	×
Trunk														
	Right	**Left**												
Gait	Circumduction	Normal	×		×	×	×	×	×				×	×
Walk	Circumduction	Normal	×		×	×	×	×	×				×	×
Stand	Unsteady	Normal												
Tandem	Not tested	Not tested												
Romberg	Not tested	Not tested												

Table 3.3 Muscle Localization: Signs and Symptoms

Symptoms	Signs	Investigations
Weakness	Weakness	Elevated CK
Family history	Wasting/hypertrophy	Myopathic EMG
Difficulty with climbing stairs, getting out of chair	Limb girdle or other regional distribution	Abnormal muscle biopsy
	Decreased deep tendon reflexes	
	Gower's sign	
	Lack of sensory signs or symptoms	

Table 3.4 Neuromuscular Junction Localization: Signs and Symptoms

Symptoms	Signs	Investigations
Weakness	Fatigue	Normal CK
Fatigue	Lack of wasting	Positive Tensilon test
Diplopia	Involvement of CN III, IV, VI, VII, X, XII	Abnormal SFEMG (single fiber EMG)
Dysphagia	Normal deep tendon reflexes	Positive Ab against ACh receptor
	Lack of sensory signs or symptoms	Thymoma on chest CT

The Tensilon test is a technique that temporarily increases the amount of acetylcholine present in the synaptic cleft to overcome the block at the neuromuscular junction. This test uses a medication called Tensilon™ or edrophonium bromide, a mild, reversible pseudocholinesterase inhibitor. The test is positive if the Tensilon reverses the neurological deficit for five to ten minutes.

Repetitive nerve stimulation is a technique by which a repetitive electrical stimulus is applied to a peripheral nerve and the output is measured over the muscle it innervates (a compound muscle action potential or CMAP), noting whether there are changes in CMAP amplitude as the stimulation continues. Normally there are virtually no changes in CMAP amplitude after ten to fifteen seconds of rapid (2 to 3 Hz) stimulation. A decremental response (steadily decreasing CMAP amplitude with repetitive stimulation) suggests a fatiguable NMJ and therefore a postsynaptic problem. An incremental response (increasing CMAP amplitude with repetitive stimulation) suggests a presynaptic problem.

The single fiber EMG (SFEMG) is a test of the integrity of the neuromuscular junction. The ability of a neuromuscular junction to transmit impulses to all the constituent muscle fibers within a given motor unit is tested and expressed as a parameter called "jitter." The test is based on measuring the variance of jitter between two neuromuscular junctions within the same motor unit.

In this case, disease of the neuromuscular junction was considered due to weakness, but ruled out for the same reasons as muscle disease.

3.3.3 Peripheral Nerve

Peripheral nerve includes motor and sensory neuronal cell bodies, their axons including myelin, their terminal boutons and the metabolic machinery that supports them all.

As the peripheral nerve serves sensory, motor and autonomic functions, destruction of the cell bodies, axons or myelin of this tissue leads to loss of function in any or all of these systems. Sensory findings are often the presenting symptom, with motor and autonomic symptoms and signs coming later. Therefore a combination of sensory, motor and autonomic dysfunction with normal or decreased reflexes suggests peripheral nerve localization (Table 3.5).

It is important to recognize that, with diffuse peripheral nerve disease, motor and sensory symptoms and signs are typically earlier and more severe and often confined to the distal extremities. The reason for this is that the longest nerve fibers, typically from the lower spinal cord to the feet, are the most metabolically at risk and therefore more sensitive to any form of damage. Therefore, it is common that symptoms of peripheral nerve dysfunction occur in the feet and hands first. This is the "stocking and glove" pattern of peripheral neuropathy.

Table 3.5 Peripheral Nerve Localization: Signs and Symptoms

Symptoms	Signs	Investigations
Weakness	Wasting, weakness	Abnormal metabolic studies
Pins and needles, numbness	Decreased reflexes unless concurrent myelopathy	Abnormal nerve conduction studies
Dry mouth, impotence, postural dizziness	Anatomic pattern for mononeuropathy	Neuropathic EMG
Family history	Stocking and glove pattern for polyneuropathy	Abnormal nerve biopsy

One must beware of combined lesions such as cord and peripheral nerve; these can give a combination of lower and upper motor neuron findings. Typical examples include amyotropic lateral sclerosis and a syndrome caused by vitamin B12 deficiency (subacute combined degeneration).

3.3.3.1 Investigations The investigations for peripheral nerve lesions include nerve conduction studies (NCS), EMG, metabolic studies for the detection of various specific nerve disorders, and in some cases biopsies of peripheral nerves.

Nerve conduction studies are a test of the electrical response of peripheral nerves to electrical stimuli. These tests help to determine the speed of conduction and the amplitude of the electrical response from peripheral nerves. This information aids in localization of dysfunction to specific peripheral nerves, cords or roots. Slowing of conduction suggests disruption of myelin; low amplitude suggests axonal dysfunction or destruction.

EMG has been described in the section on muscle disease.

Metabolic studies useful in determining the etiology of some peripheral nerve problems include serum assays for glucose in diabetes, vitamin B12, folic acid, thyroid hormone, protein electrophoresis and specialized DNA assays for genetic forms of peripheral nerve disorders.

There are over twenty different hereditary neuropathies that can be expressed with different levels of penetrance. Family history is important if positive (see Table 3.15).

In the sample case, disease of the peripheral nerves was considered due to weakness, but ruled out due to the presence of unilateral findings, expressive aphasia, and hyperreflexia. The Expanded Localization Matrix shows the peripheral nerves as being considered but ruled out as less likely, that is, Xs in red.

3.3.4 Spinal Cord

Motor disturbances due to lesions in the spinal cord depend on which tracts are involved and at what spinal cord level. Patterns of spinal cord injury relate to the cross-sectional anatomy of the cord. From the viewpoint of clinical localization, the three most important tracts in the spinal cord are the spinothalamic tract, the posterior columns and the corticospinal tract. These have been reviewed in Chapter 1 and Chapter 2.

The clinical picture in spinal cord injury depends on which tracts are damaged (Table 3.6) and which are at risk of further damage. For instance, damage to the anterior two-thirds of the spinal cord will lead to loss of function of the gray matter containing the lower motor neurons, the spinothalamic tracts and the corticospinal tracts on both sides but spare the posterior columns as will be discussed in Chapter 5. The clinical picture would consist of flaccid distal paraplegia with loss of sensation to pin prick and temperature but preserved sensation to vibration and proprioception as the posterior columns are still intact.

3.3.4.1 Investigations The single most important investigation for spinal cord localization is neuroradiological imaging. If available, MRI of the spine at the appropriate level is optimal; however, CT scan or plain films are

Table 3.6 Spinal Cord Localization: Signs and Symptoms

Symptoms	Signs	Investigations
Weakness	Focal weakness and sensory loss at level of lesion	Plain films
Spasticity	Weakness, sensory loss, and hyperreflexia below level of lesion	CT/myelogram
Numbness	Dissociated sensory loss (Brown-Séquard syndrome)	MRI
	High degree of suspicion if known malignancy	SSEP (somatosensory evoked potentials)

useful whenever MRI is unavailable or (for medical reasons such as a pacemaker) cannot be performed.

The clinical level of the spinal cord lesion needs to be transmitted to the neuroradiologist so that the appropriate level can be scanned with several segments above and below the clinical level. Other neuroradiological investigations include plain or CT myelography in which, after a lumbar puncture has been performed, an iodinated dye is injected into the subarachnoid space. The spread of the dye in the subarachnoid space is followed up to the neck as the patient is tilted head down to make the dye flow with gravity.

Somatosensory evoked potentials (SSEP) is a test that measures the response to an electrical stimulus applied to either upper or lower limb nerves in order to determine the speed of conduction from peripheral nerves through the spinal cord to the contralateral somatosensory cortex. An electrical stimulus is usually applied to the median nerves at the wrist or to the posterior tibial nerves at the ankles. Each side and limb is tested separately. This test assists in localization by assessing at what level (and structure) there is slowing or cessation of transmission of the evoked potential.

In this case, spinal cord was considered due to limb weakness, and hyperreflexia, but isolated damage to the spinal cord would not explain the right facial weakness, expressive aphasia and other cortical deficits. The Expanded Localization Matrix shows the spinal cord as being considered but ruled out as less likely, that is, Xs in red.

3.3.5 Brainstem

The brainstem includes the midbrain, pons, medulla, cranial nerve nuclei and the ascending and descending axonal tracts, which pass through the brainstem. Brainstem lesions produce motor and sensory disturbances in the face and body, along with localized cranial nerve findings.

The location of the cranial nerves originating in the midbrain, pons and medulla was reviewed in Chapter 2. The combination of isolated cranial nerve findings with long tract sensory and motor findings with or without ataxia suggests a brainstem localization.

Due to the compact nature of the brainstem nuclei and the adjacent tracts running through the brainstem, it is important when there is dysfunction in one cranial nerve or tract to look at the neighbors at the same level. For instance a patient with unilateral facial weakness or Bell's palsy may have less obvious findings in the other cranial nerves at the same level in the pons: CN V, VI or VIII. Involvement of these other cranial nerves would suggest a central, more widespread lesion such as a pontine tumor or stroke. The signs and symptoms are summarized in Table 3.7.

It is also important to look for dissociated sensory loss (i.e., isolated discriminative touch or pain and temperature) and motor signs in face and body. Similar to the spinal cord, damage to one side of the brainstem can give facial sensory loss on the same side with loss of contralateral sensation to the body due to damage of tracts that have already crossed and are damaged in the brainstem.

The presence of coma or a progression of accumulating deficits from midbrain to pons to medulla—the rostrocaudal sequence of deterioration—suggests a lesion compressing the brainstem from above.

3.3.5.1 Investigations The investigations to confirm brainstem localization include MRI of the brain. Imaging the brainstem using CT scanning suffers from artefact caused by the petrous ridges of the temporal bones and therefore MRI is preferable.

Brainstem evoked potentials (BSEP) involve testing the auditory portion of CN VIII using an auditory clicking sound and measuring the responses from CN VIII and the various brainstem structures that relay auditory information in the brainstem and the cortex. This test might show a delay or slowing of the signal evoked by the sound on one or both sides allowing for precise localization of the lesion.

The blink reflex is an electrical equivalent of the corneal reflex: it measures the latency of the input of an electrical stimulus to CN V and the resulting blink output of CN VII on both sides. If the reflex response is slow or absent, this can help localize the lesion to CN V, the pons or CN VII on either side.

In the case under discussion, brainstem damage was considered due to unilateral right-sided weakness, speech difficulties and hyperreflexia, but isolated damage to the *left* side of the pons would produce *left* facial weakness of the upper, mid and lower face along with right-sided hemiplegia. As well, extra-ocular movements are not disturbed in this case. Finally, brainstem damage would also not explain the presence of aphasia. Brainstem lesions can produce dysarthria (a disturbance of articulation of

Table 3.7 Brainstem Localization: Signs and Symptoms

General Instructions		
1. Look for neighboring signs		
2. Dissociated sensory and motor signs		
3. Presence of coma or rostrocaudal progression		
Level	**Symptoms and Signs**	**Investigations**
Midbrain	Loss of pupillary reflex	MRI/CT scan
CN III,IV	Abnormal adduction, elevation, depression of eye movements	MRA/CTA
	Diplopia	
	Contralateral hemiplegia	
Pons	Loss of corneal reflex	MRI/CT scan
CN V,VI,VII,VIII	Loss of facial sensation	MRA/CTA
	Inability to abduct eye	BSEP*
	Pinpoint pupils	Blink reflex*
	Facial weakness	ENG*
	Loss of hearing, balance	
	Loss of caloric response	
	Contralateral hemiplegia and sensory loss	
Medulla	Loss of gag reflex	MRI/CT scan
CN IX,X,XI,XII	Loss of ability to swallow	MRA/CTA
	Loss of ability to lift shoulders or turn head	Apnea test*
	Loss of tongue movement	
	Contralateral hemiplegia and sensory loss	
	Horner's sign	
	Ataxia	

* See Glossary.

speech) by involvement of CN X or its upper motor neuron connections. Therefore, localization of the disease to the brainstem is less likely. The Expanded Localization Matrix shows the brainstem as being considered but ruled out as less likely, that is, Xs in red.

3.3.6 *Deep White Matter and Lateral Thalamus*

Ipsilateral lesions of the deep white matter and lateral thalamus tend to give hemi-motor or hemi-sensory symptoms and signs on the opposite side of the body. For larger lesions both motor and sensory findings can occur. The trunk does not tend to be spared as occurs with cortical lesions (Table 3.8).

Table 3.8 Deep White Matter and Lateral Thalamus: Signs and Symptoms

Level	Symptoms and Signs	Investigations
Internal capsule	Loss of motor function on contralateral side	CT/MRI scan
Lateral thalamus	Loss of sensory function on contralateral side	CT/MRI scan

3.3.6.1 Investigations High-resolution CT scan or MRI scanning with angiography of the brain are the preferable modes of investigation to localize the lesions in this area. Clinically it is impossible to differentiate a small hemorrhage from a small infarction. Treatment decisions with respect to thrombolysis need to ensure that there is no evidence of hemorrhage prior to the initiation of thrombolytic therapy (for further explanation of this intervention, see Chapter 8).

In this case, white matter damage in the left hemisphere was considered due to unilateral right-sided weakness and hyperreflexia. Isolated white matter damage usually does not cause Broca's aphasia. White matter injury usually causes dysarthria (difficulty in articulation) but not deficits in the formulation of language.

Large areas of damage can obviously affect both cortex and white matter in this area at which point therefore it is difficult to distinguish between the two types of damage. The Expanded Localization Matrix shows the deep white matter and thalamus as being considered as a possibility but as less likely, that is, Xs are left as black.

3.3.7 Cortex

Cortical lesions cause dysfunction of complex processing of information, the type of deficit depending on the cortical region involved. Lesions of the speech centers impair input or output or conduction of speech. Lesions of a specific area such as the angular gyrus can affect calculation and left/right discrimination. Often there is underlying white matter dysfunction causing long tract sensory or motor dysfunction from regions adjacent to the cortical lesion.

For instance, the processing of visual information for the purpose of generating visual imaging such as visual acuity, and visual fields localize above the tentorium involving both cortex and white matter tracts transmitting visual information from the eyes to occipital cortex.

Table 3.9 lists the various cortical functions by location and the effects of damage to these areas.

3.3.7.1 Investigations Some clinical tools used for testing cortical function include the Mini Mental Status Exam (MMSE) and the Montreal Cognitive Assessment (MoCA), both of which are discussed in Chapter 2. These bedside tools help with localization in areas of attention, memory, executive function, calculation, praxis and language, but are not as useful in cases of mild to moderate cognitive decline.

Detailed neuropsychological testing can be performed by a licensed neuropsychologist, who can administer a battery of cognitive tests to help in localization.

High resolution CT scan or MRI scanning and angiography of the brain are the preferable modes of investigation to localize the lesions in this area.

The electroencephalogram (EEG) is a useful tool to assess the background rhythms generated by the interplay of the cortex and thalamus. Localized slowing on the EEG can reflect focal cortical damage, whereas diffuse slowing can represent widespread injury such as in dementia or metabolic/toxic injury which affects the whole brain.

In the case under discussion, cortical damage in the left hemisphere was considered due to unilateral right-sided weakness, and hyperreflexia. Left frontal lobe damage would explain the expressive aphasia, lower right facial weakness, right upper limb hemiparesis, hyperreflexia with Babinski sign.

The expressive aphasia is due to destruction of Broca's area in the operculum of the left frontal lobe in a right-handed person.

The right hemiplegia is due to damage to the left precentral gyrus, which is the source of the upper motor neuron input to the facial nerve nucleus on the right side of the brainstem and right half of the spinal cord that controls movement on the right side of the body. Damage to these upper motor neurons leads to weakness and hyperreflexia.

Left temporal lobe damage would be suspected in the case of a receptive aphasia. Wernicke's area is centered in the superior temporal gyrus in the dominant hemisphere.

Sensation on the right side is intact, which implies that the left parietal cortex and underlying white matter connections have been spared.

The Expanded Localization Matrix shows the frontal cortex as being considered as most likely, that is, the Xs are black.

3.3.8 Centers of Integration

3.3.8.1 Cerebellum Lesions in the cerebellum cause either truncal unsteadiness for midline lesions of the vermis or ipsilateral limb incoordination for hemispheric lesions. The signs consist of ataxia, past-pointing and hyporeflexia. The diagnosis of a cerebellar lesion has to be made only by exclusion. An ipsilateral hemispheric cerebellar localization can only be made when the lower motor neuronal system, upper motor system and the sensory systems are intact.

A common mistake made by students seeing a patient with a lesion of the right internal capsule is to localize it as a left cerebellar lesion due to the presence of apparent ataxia in the left upper limb. In this case the left arm ataxia results from dysfunction of the upper motor neuron unit damaged

Table 3.9 Cortex: Signs and Symptoms

Location	Function	Effects of Damage	Dominance (Right Handed)
Frontal Cortex			
Frontal poles	Executive function, attention	Loss of judgment, planning, emotional disinhibition, inattention	
Frontal operculum Broca's area	Expression of language	Expressive aphasia	Left hemisphere
Precentral gyrus	Motor output	Loss of motor activity on contralateral side	Both
Cingulate gyrus	Motor planning, attention and execution	Apraxia, attention deficits	Both
Temporal Cortex			
Superior temporal gyrus	Reception of speech, lexicon for languages	Receptive aphasia	Left hemisphere
Hippocampus	Short-term memory	Loss of short-term memory	Slight left hemisphere
Amygdala	Generation of fear and sexual responses	Inappropriate sexual behavior	Both
Insular cortex	Reception of sound and vestibular input	Loss of auditory attention and discrimination	Both
Parietal Cortex			
Parietal lobe	Integration of and attention to sensory information	Loss of ability to recognize visuospatial relationships, sensory inattention	Right hemisphere
Postcentral gyrus	Reception of somatosensory information relayed from thalamus	Loss of contralateral sensation	Right hemisphere
Angular gyrus	Calculation, left–right discrimination, concept of finger position	Inability to calculate, recognize left/right, finger agnosia	Both
Occipital Cortex			
Occipital poles	Reception of visual information from ipsilateral LGB	Hemianopia, cortical blindness	Both
Parieto-occipital cortex	Recognition of visual objects such as faces	Hemi-attention, hemianopia, inability to recognize faces, cortical color blindness	Both

by the right hemisphere lesion. The clinical picture clarifies the situation in that the right hemisphere lesion will often be associated with an upper motor neuron pattern of weakness and hyperreflexia of the left upper limb.

Table 3.10 details the common signs and symptoms associated with midline and hemispheric cerebellar lesions.

3.3.8.2 Basal Ganglia Localization to the basal ganglia is suggested by hemibody signs of either increased (Huntington's chorea) or decreased (Parkinson's disease) spontaneous motor activity of both agonist and antagonist muscle groups. If both basal ganglia are involved then the symptoms and signs are bilateral. Dyskinesia, dystonia, tremor, rigidity, bradykinesia and difficulties maintaining a standing posture all suggest basal ganglia dysfunction. Table 3.11 details the signs and symptoms of disorders of modulating structures at several levels.

In the present case, damage to the modulating structures (in particular the basal ganglia) is difficult to determine as there is dysfunction of the corticospinal tracts originating from the left hemisphere.

The general rule is that localization of dysfunction to the modulating structures is made by exclusion in that all the primary motor and sensory modalities must be found to be normal before one can localize primarily to a modulating structure.

Table 3.10 Cerebellum: Signs and Symptoms

Level	Symptoms	Signs	Investigations
Hemispheric lesions Signs are ipsilateral to side of lesion	Vertigo	Dysarthria	CT/MRI
	Loss of coordination	Dysmetria	
	Difficulty with eye movements	Nystagmus	
		Dysdiadokinesia	
		Pendular reflexes	
		Low muscle tone	
		Lack of sensory, lower motor, and upper motor neuron findings	
Midline lesions	Difficulty with balance	Wide-based gait	CT/MRI
	Falls	Dysarthria	

Table 3.11 Basal Ganglia: Signs and Symptoms

Level	Symptoms	Signs	Investigations
Basal ganglia	Falls	Tremor	MRI/cerulosplasmin/ heavy metals/genetic
	Tremor	Bradykinesia	
	Slowness of action	Rigid muscle tone	
		Cog-wheeling and loss of postural reflexes	
Caudate	Involuntary Movements	Chorea	MRI/cerulosplasmin/ heavy metals/genetic
	Psychiatric disturbances	Dancing gait	
		Progressive dementia	
Midbrain	Falls	Loss of vertical and lateral eye movements	MRI/cerulosplasmin/ heavy metals/genetic
	Difficulty with eye movements especially down gaze	Increased extensor tone	
	Slowness of actions		

The Expanded Localization Matrix shows the cerebellum and basal ganglia as being considered but ruled out as less likely, that is, the Xs are red.

3.3.9 *Localization Shortcuts*

Patients with isolated signs and symptoms can sometimes be quickly localized using the following shortcuts. These rules are generalizations which have to put into the context of a given patient.

- Findings of sensory loss rule out muscle disease and disorders of the neuromuscular junction as the primary localization.
- Disturbances of visual acuity due to neurological disease can be localized to those structures above the brainstem that transmit information from the retina to the occipital cortex areas dedicated to the production of visual imagery.
- Seizures arise from the cerebral cortex or thalamus; this means that seizures without other deficits can be localized to the cortex or to the related subcortical modulating structures in the thalamus.
- Deficits of cognition without disturbances of consciousness or attention localize above the brainstem.

3.4 What Is the Pathophysiological Cause or Etiology of the Lesion?

Now that we have worked out the probable location of our patient's pathology, we must go back to the medical history to begin the process of determining the most likely cause of the problem.

The first step is to extract the key elements of the patient's history and enter them in the History Worksheet (Table 3.12).

Table 3.12 History Worksheet

Symptom	Event Descriptors	Time Course	**Time Zero 0**	Location	Collateral Factors	Etiological Possibilities
Pain						
Numbness						
Weakness			Sudden onset and worsening of R-sided weakness			
Loss of Vision		(3 weeks) Loss of vision left eye 5–10 minutes				
Loss of Hearing						
Difficulty with Balance						
Difficulty with Coordination						
Difficulty with Memory						
Difficulty with Speech			Sudden onset and worsening of difficulty of expression of speech			
Loss of Consciousness						
Incontinence						
Sleep Disturbance						
Loss of Milestone						

The relevant historical information has been added to the sheet based on time zero being the onset of the major presenting complaint of right-sided weakness and speech disturbance. It is then important to try and correlate the sequence and evolution of the symptoms with the etiological possibilities. To help identify the appropriate category of etiology, it is important to ask the status of the patient's health over the last weeks to months; for example, if there have been any toxic exposures, previous medical illnesses or foreign travel.

It is often very useful to ask patients exactly what they were doing at the time of onset of defining symptoms, what time of day it was, their sleep and nutritional status, as well as the names and dosage schedule of current medications and when these medications were last taken.

Determination of a range of etiologic possibilities can be accomplished by referring to the grid shown in Table 3.13 that correlates the duration of a given patient's set of symptoms with the type of disease process that is most likely to present with such a time course.

Before we proceed to use the grid to work out the etiology of the sample case, the information in the grid requires further explanation. As a rule, disorders with extremely rapid onset of loss of neurological function are paroxysmal, vascular or traumatic in nature, but can also be due to the acute effect of medications or nonmedical drugs. Disease categories evolving over a few weeks to months include toxic, infectious, metabolic, inflammatory, and neoplastic. Chronic disease processes evolving over months to years tend to be metabolic, neoplastic, degenerative, or genetic in nature.

Table 3.13 Etiology Matrix

	Neurological Disease Symptom Duration Grid						
	Symptom Length						
Disease type	**Secs**	**Mins**	**Hrs**	**Days**	**Wks**	**Mos**	**Years**
Paroxysmal (seizures, faints, migraine)	+	+	+				
Traumatic	+	+	+				
Vascular (ischemic stroke, bleed)		+	+				
Toxic		+	+	+	+	+	
Infectious				+	+	+	+
Metabolic		+	+	+	+	+	+
Inflammatory/autoimmune				+	+	+	+
Neoplastic					+	+	+
Degenerative						+	+
Genetic	+	+	+	+	+	+	+
ACUTE							
SUBACUTE							
CHRONIC							

Let us consider each disease category in more detail with reference to the sample case.

3.4.1 *Paroxysmal*

This disease category includes disorders that transiently disrupt electrical function of the brain or spinal cord.

Focal electrical discharges in a specific part of the brain can cause focal or partial epileptic seizures. These focal discharges can then spread to involve the entire brain, giving rise to generalized seizures. Primary generalized seizures can occur without focal onset.

Seizures generally are due either to an underlying genetic or metabolic/toxic abnormality, with the actual events triggered by environmental factors, or to focal brain injury. The brain injury can be from a variety of causes, such as a developmental abnormality, stroke, trauma or tumor.

All seizures have an acute time line, recurrent seizures occurring on many occasions, often over many years.

Global and focal deficits can occur after any type of seizure. Seizures cause hypermetabolism which can then exhaust cellular substrates. This metabolic exhaustion can manifest clinically as brain dysfunction either globally with generalized seizures or focally with partial seizures. In the latter instance, there may be unilateral weakness, a phenomenon known as Todd's paralysis. This metabolic exhaustion and neurological dysfunction can last up to 24 hours and can mimic stroke or transient ischemic attack (TIA).

Migraine is a form of paroxysmal transient disturbance of control of the cerebral vasculature due, at least in part, to inflammation and dysregulation of cranial blood vessels supplied by sensory branches of the trigeminal nerve. On first presentation, migraine patients having focal deficits followed by headache cannot be distinguished from those with stroke or transient ischemic attack (TIA). The availability of vascular neuroimaging is extremely helpful to rule out underlying vascular abnormalities in these patients.

In the case under discussion, the focal right-sided deficits could fit with a Todd's paralysis or migraine but the transient loss of vision in the left eye four weeks earlier would not be consistent with a cortically mediated seizure-related phenomenon or migraine. Also, there is no history of any rhythmic twitching on the right side of the body preceding the development of right-sided weakness.

Abnormal electrical discharges in the spinal cord can cause localized myoclonus, usually abnormal involuntary shock-like movements in a single limb or group of muscles.

Generalized discharges of the spinal cord involving the lower motor neurons can occur, resulting in a clinical syndrome called stiff person syndrome or Issac's syndrome. Electrophysiologically this is known as neuromyotonia;

it is believed to be an autoimmune disorder and may occasionally be triggered by occult neoplasia.

3.4.2 Trauma

This disease category includes all forms of damage to the brain and spinal cord as a result of excessive forces applied directly or indirectly to the nervous system.

Trauma can cause damage (usually in the form of hemorrhage) at every level of the nervous system. This is best remembered by tracing the structures and layers that a knife or projectile must pass through in order to enter the brain.

The outer layer is the skin and muscle of the scalp, followed by the outer skull periosteum, the skull and then the inner skull periosteum. Traumatic hemorrhage to these structures is usually externalized and causes local damage related to nearby structures such as the eyes, ears and external branches of cranial nerves.

Trauma under the inner periosteum and above the dura may cause laceration of the middle meningeal artery under the squamous portion of the temporal bone, resulting in epidural hemorrhage (further discussed in Chapter 11).

The next space into which blood can accumulate is the space between the arachnoid and the dura, the subdural space. Injury is usually to the veins bridging the space between the arachnoid and the dura causing venous bleeding. Since venous pressure is low, blood may accumulate over days to weeks causing subacute deficits due to pressure on the underlying brain.

Trauma to the major arteries that course in the subarachnoid space can cause subarachnoid hemorrhage. This can be due to either direct or indirect trauma causing shearing forces on the blood vessels and their attachments. Traumatic subarachnoid hemorrhage is clinically similar to that due to rupture of aneurysms: there is severe headache—"thunderclap headache"—and meningismus (irritation of the meninges). If a hematoma forms that extends into the brain then focal deficits may ensue. The mass effect of a large hematoma may result in decreased level of consciousness and brain herniation phenomena (see Chapter 10 e-cases).

The most serious situation is that of open head injury in which there is a skull fracture with direct damage to brain and blood vessels causing crush injury and hemorrhage to cortex and white matter. If the injury has been associated with hypoxia and hypotension as with a gunshot wound or motor vehicle accident, then there is potential for the hypoxia and hypotension to make the direct brain injury worse.

Excessive forces caused by either acceleration or more commonly deceleration can cause damage at various locations within the nervous system without actually breaching any of the protective structures. The damage caused in closed head injury includes intraparenchymal hemorrhage and shearing injury to axonal tracts: diffuse axonal injury (DAI).

The term *concussion* is used for a closed head injury with momentary loss of consciousness with no focal signs. CT and MRI scans of these individuals often show small hemorrhages in the cortex and subcortical white matter which resolve spontaneously over several weeks.

Immediate neuroimaging is the principal form of investigation for traumatic injuries. In the context of trauma, the appropriate area of the neuraxis needs to be stabilized prior to being assessed by neuroimaging to rule out any fractures or instability.

In our sample case the history and findings do not fit with a traumatic injury.

3.4.3 Vascular

This category includes disorders involving disease of large and small blood vessels supplying the brain and spinal cord.

Blood vessels usually malfunction in one of two different ways: they can burst or block. Both of these events are usually acute. Often there is a small warning blockage or leak before the complete blockage or rupture occurs. It is therefore very important to intervene at an early stage to prevent a major blockage or rupture of a blood vessel supplying the brain or spinal cord.

In general, acute vascular occlusions or ruptures occur in blood vessels on one or other side of the brain, resulting in focal symptoms on the opposite side of the body. The only exception to this rule is the cerebellum in which an ipsilateral cerebellar hemorrhage or infarction will cause clinical symptoms of ataxia and decreased coordination on the same side of the body.

The time course of blockage or rupture of blood vessels is always acute, sometimes in a recurrent fashion, with recovery between events, at least initially. Because of the potentially devastating consequences of a vascular event, an unexplained acute focal neurological deficit must be taken seriously and the precise cause identified in order to prevent eventual serious permanent damage.

The term *TIA* or transient ischemic attack refers to the situation of a temporary interruption of blood supply to an area of the brain, eye or spinal cord, causing a neurological deficit lasting less than 24 hours but usually less than 30 minutes. In neuroanatomic terms, this usually represents the effect of a small embolus which has become lodged in cerebral vessels, causing ischemic changes and neurological deficits related to the area of brain supplied by that vessel, then the embolus breaks into smaller fragments resulting in resumption of blood flow and reversal of the deficits. In the sample case, there was a TIA involving the left eye due to a small embolus to the left ophthalmic artery four weeks before the second event.

The term *ischemic stroke* refers to the situation as described above except that the embolus does not move or fragment, causing irreversible ischemic damage to the brain supplied by the vessel.

With the easy availability of CT or MRI angiography in most major centers, immediate neuroimaging of the cerebral vascular anatomy of patients with cerebrovascular disease has become the standard of care. With this modality of imaging, it is possible to visualize the exact cerebral vessel that has been occluded so that the clinicians and interventional neuroradiologists can decide on the best form of therapy.

Sustained systemic hypotension that is eventually reversed may cause differential ischemic damage to areas of the brain and spinal cord located between the territories of major supplying vessels (the watershed or arterial border-zone areas).

A complete discussion of cerebrovascular anatomy as well as ischemic and hemorrhagic stroke is detailed in Chapter 8.

The timeline for the case under discussion (with symptoms related to the left eye followed two months later by symptoms related to the left cerebral hemisphere) fits with a vascular etiology most likely involving the territory of the left carotid artery.

3.4.4 Drugs and Toxins

In general, toxic and drug-induced neurological problems cause nonfocal deficits, meaning that there is no difference in the degree of functional disturbance from one side of the body to the other. For instance the pupillary reaction to light may be abnormal but in a symmetrical fashion.

Another general rule specifically concerning "recreational" drug abuse is that drugs that stimulate the nervous system (amphetamines, cocaine) tend to produce large but equal pupils during acute exposure, whereas drugs that depress the nervous systems (narcotics) tend to produce small but equal pupils.

A common cause of acute neurological symptoms is a change in medication or dosage. It is extremely important to review every medicine that the patient is taking by inspecting the actual bottles of each medication and reviewing doses, times and recent changes. In considering the possibility of drug exposure, it is important to have relatives check the medicine cabinet and the trash can, as well as to ask specific questions about hobbies, for example, gardening or soldering.

There is a wide list of possibilities (see Table 3.14); thus the medication and recreational drug histories are essential in honing the list down to a few agents that are most likely. For nonmedical or recreational drug exposure, it is often necessary to obtain collateral history, with the patient's consent if possible, from other family members or friends.

Chronic exposures to substances such as heavy metals can lead to slow deterioration of neurological function of both brain and spinal cord. When these are suspected, blood and urine assays are available to confirm or refute the presence of heavy metal poisoning such as lead or mercury.

Screening tests of urine and blood for medications and recreational drugs involved in overdose situations are available in most emergency departments.

Table 3.14 Common Toxic Exposures
Ethanol
Marijuana
Ecstasy
Metamphetamine
Heroin
Cocaine
Methanol
Ethylene glycol
Inhaled solvents (toluene)
Heavy metal intoxication (arsenic, mercury, lead)

In terms of the matrix, drugs and medications can cause acute, subacute and chronic effects. In the sample case, the patient was not taking any medications and recreational drugs are an unlikely cause of the patient's focal deficits.

3.4.5 *Infections*

The most common bacterial infections of the nervous system include meningitis and cerebral abscess, the latter typically a metastatic infection from the heart, the paranasal sinuses, or penetrating injuries. Encephalitis, which is usually viral in origin, refers to infection of the brain parenchyma itself. Meningitis is infection of the arachnoid layer of the meninges. Infected material can also collect in the epidural and subdural spaces; these infections are referred to as empyemas.

The history for most of these disorders usually shows an onset with systemic features of fever, lethargy, and signs of meningeal irritation over several days. Meningeal irritation refers to inflammatory changes to the meninges due to a noxious agent such as blood, infection, or sometimes a chemical such as contrast material that has been introduced into the subarachnoid space. A clinical sign of meningismus is extreme neck stiffness due to irritation of the meninges when flexing either the head or the legs. Brudzinski's sign occurs with meningeal irritation causing reflex flexion of the legs when the head is flexed on the neck by the examiner. Similarly, Kernig's sign is reflex flexion of the head when the lower leg is extended at the knee. Neither of these signs is specific but if present should guide the clinician to look for causes of meningeal irritation by performing a lumbar puncture if there are no contraindications.

If the brain parenchyma is involved then focal signs and symptoms evolve. The commonest bacterial organisms are staphylococcus, streptococcus, *Haemophilus influenza* and *Neisseria* meningitis. *Listeria* and tuberculosis can affect the brainstem. Immunocompromised individuals can harbour unusual bacterial organisms.

Syphilis and Lyme disease are bacterial infections of the nervous system caused by spirochetes that usually have three phases. There usually is a rash or lesion at the point of entry, days to weeks later there is involvement of joints or skin, followed by the phase of involvement of the nervous system.

Brain abscesses usually evolve over a period of several weeks, typically with fever, mental status change, unilateral signs and epileptic seizures.

Acute viral infections of the nervous system include coxsackie, echovirus and herpes viruses. The most important for rapid diagnosis is herpes simplex type I which can cause a devastating hemorrhagic necrosis of the temporal lobes. Herpes zoster usually manifests as painful eruptions of vesicles similar to chickenpox but within the confines of one or two dermatomes; this represents a reactivation of latent virus in sensory ganglia from a primary chickenpox infection earlier in life. Herpes zoster can also cause diffuse encephalitis and myelitis, usually in immunocompromised individuals.

Protozoan infections of the nervous system such as malaria or trypanosomiasis have a time course of onset of days to weeks. There is often a history of overseas travel or previous exposure in immigrants from countries where these diseases are endemic.

Prions, a form of chronic infective agent, can cause destruction of brain by the accumulation of insoluable prion protein in the extracellular spaces. Prionic infections include Creutzfeld–Jacob disease (wild type or variant), kuru, fatal familial insomnia and Gerstmann–Straussler–Scheinker disease; all of these disorders have time lines of years but the phase of decompensation may be in the order of days to weeks. These disorders are uniformly fatal but have significantly different time courses.

Neuroimaging of patients with suspected infections of the nervous system should be performed before lumbar puncture if there are any focal signs or decreased level of consciousness, or if the infection has a duration of 48 hours or more. This is to prevent the possibility of a mass effect caused by an infection such as a brain abscess to result in downward herniation of the brain onto the brainstem.

Investigations for these disorders include appropriate cultures of any purulent discharge from ear, nose or throat, blood cultures, as well as gram stain and cultures of CSF. Specialized rapid assays for bacterial infection using immunofixation and ELISA tests are available to guide clinicians with respect to the offending organism.

The time lines for infections of the nervous system do not apply in the sample case under discussion.

3.4.6 *Metabolic*

This category of disorders includes disturbances of all of the basic metabolic substrates, electrolytes, hormones and vitamins necessary for life.

Acute systemic hypotension and hypoxia cause cellular metabolic failure and the start of irreversible neuronal cell damage within four minutes. These disorders cause acute neurological global dysfunction with loss of consciousness. Acute hypercarbia causes narcosis and coma and is often associated with hypoxia. Acute hypoglycemia also causes CNS dysfunction in a global fashion, but sometimes can present with focal neurological features that are indistinguishable from those of ischemia (e.g., unilateral focal weakness).

The timeline for these events is a matter of minutes; they are often reversible in the same time frame.

Other metabolic disturbances such as electrolyte abnormalities typically evolve over several days; hyponatremia is a classic example and can lead to seizures and coma. Hypokalemia usually causes muscle weakness and is relatively slow in onset unless caused by a potassium wasting medication. Hyperkalemia affects the heart long before it affects the nervous system. Hypocalcemia produces muscle irritability and tetany, whereas hypercalcemia causes an encephalopathy; both usually have time courses of days to weeks but can present acutely.

Hormonal disturbances affecting the nervous system include hyper- and hypothyroidism, Addison's disease and Cushing's syndrome (typically from a functioning pituitary adenoma); all of these take weeks to months to produce dysfunction of the nervous system and are usually diagnosed by primary care physicians, internists or endocrinologists.

Vitamin deficiencies may also cause neurological dysfunction, often with widely varying time courses. Vitamin B12 deficiency can cause peripheral nerve and spinal cord dysfunction. Vitamin B12 stores in the liver usually require three years to deplete to the point that the individual becomes symptomatic, after which the onset of symptoms may be very subtle and gradual. Strict vegans with marginal stores or intake of vitamin B12 are at particular risk. Folate stores are much more volatile, becoming depleted in weeks to months if dietary deficiency or malabsorption develops.

Thiamine deficiency has deleterious effects on selected hypothalamic and midbrain structures as well as on peripheral nerves. Thiamine deficiency in alcoholics, who are also usually malnourished, is still often overlooked in acute settings. Previous gastric surgery can interfere with thiamine absorption and lead to chronic CNS and PNS dysfunction.

Investigations of these disorders include blood tests for glucose, electrolytes, calcium, phosphate, Vitamin B12, folate; iron studies; liver, thyroid and kidney function; and blood gases if clinically indicated.

Metabolic encephalopathies can also occur from organ failure elsewhere in the body such as kidney, liver, and lung.

With kidney failure, when creatinine levels get above 500 to 700, decreased levels of consciousness are seen, with tremor. The serum levels of sodium, calcium or phosphate will often determine the degree of impairment.

In liver failure depending on the cause and rapidity, encephalopathy can manifest as progressively altered mental status, tremor, asterixis, hyperventilation, then if untreated, coma and death.

Respiratory failure leads to altered mental status due to hypercarbia and hypoxia. Anxiety and agitation due to hypoxia are often misinterpreted as a psychiatric disturbance and treated with medications such as benzodiazepines which then make the hypoxia worse to the point of respiratory arrest.

Treatment of these entities obviously targets the underlying cause of the primary organ failure.

In summary, the time course of metabolic disorders is typically days to years and, with the possible exception of hypoglycemia, all of which is too long for consideration in the sample case being discussed.

3.4.7 Inflammatory/Autoimmune

The noninfectious inflammatory disorders of the nervous system include such entities as Guillain-Barré syndrome, multiple sclerosis (MS), systemic lupus erythematosis (SLE), sarcoidosis, Sjögren syndrome, rheumatoid arthritis, postinfectious vasculitis and isolated CNS vasculitis. These disorders generally have time lines of weeks to months and often have concomitant systemic symptoms of vasculitis. Acute events such as seizures, stroke and encephalopathy can occur as complications to these disorders. The timeline in the history would then show a subacute-on-chronic picture. In the case of relapsing remitting MS, there will be attacks over many years.

These disorders do not fit with the timeline of the case under discussion.

Most of these disorders are investigated with appropriate serological tests for autoimmune diseases. MRI scanning is also important for MS and the various forms of vasculitis.

3.4.8 Neoplastic

Neoplasms cause neurological dysfunction by destruction of neural tissue by direct expansion of the tumor mass, sometimes with concomitant necrosis and hemorrhage. Mass effect refers to the displacement of normal brain tissue by a space-occupying lesion such as a tumor. Mass effect can also be caused by edema surrounding the tumor. The compression of normal structures can lead to altered function and ultimate destruction of these structures. The location of the tumor and its mass effect will determine the nature of the neurological deficits. Many tumors are initially slow-growing and are not discovered until there is a defining event such as a seizure.

Primary brain tumors originate from basic neural tissue elements: astrocytomas and gliomas from astroglia, oligodendrogliomas from oligodendrocytes, meningiomas from meninges and ependymomas from the ependyma.

Metastatic tumors arise from hematogenous spread of primary tumors elsewhere in the body. The common malignancies that metastasize to the brain originate in the breast, lung, and bowel.

In terms of time frames, there may be a subacute to chronic onset of a focal neurological deficit reflecting the slow growing mass effect of the tumor, with an acute event such as a seizure or stroke-like event; the latter represents a sudden expansion of the tumor mass due to hemorrhage or necrosis.

3.4.8.1 Paraneoplastic Paraneoplastic disorders of the nervous system are defined as pathological processes that affect the nervous system due to the remote effects of a neoplasm elsewhere in the body.

A common scenario is that of the formation and growth of a solid tumor (such as small cell lung carcinoma) which stimulates an immune antibody response that cross-reacts to tissue elements in the nervous system, thereby causing dysfunction and destruction in those elements. This is the case in Lambert-Eaton syndrome, in which the small cell lung tumor stimulates antibodies that are directed against presynaptic calcium channels in terminal boutons of peripheral nerves, resulting in proximal muscle weakness and autonomic dysfunction. This is an example of a presynaptic disorder of the neuromuscular junction. Serum titers of certain of these antibodies are now available.

Paraneoplastic autoimmune syndromes may also produce a polyneuropathy, other neuromuscular transmission disorders, cerebellar degeneration and occasionally cognitive decline.

Other paraneoplastic syndromes affect the nervous system by producing hormone-like compounds. A typical example is a type of lung cancer that produces an ADH-like hormone that induces hyponatremia with secondary effects on CNS function (see the section above on metabolic disorders).

Like brain tumors, paraneoplastic disorders evolve over weeks to months. In our sample case, therefore, the history and findings do not fit with a neoplastic or paraneoplastic disorder.

3.4.9 Degenerative Diseases

Neurodegenerative diseases take many forms and are often associated with other comorbid conditions that amplify their effect. Some tissues within the nervous system are relatively immune to neurodegenerative processes, for example, muscle, the neuromuscular junction and peripheral nerves. Most neurodegenerative processes affect neurons in the cerebral cortex, basal ganglia, cerebellum and spinal cord.

Many of the degenerative disorders may be reclassified as genetic disorders, which get worse over time. In other instances, the primary cause remains unknown but the course of the disease may be modified by genetic factors.

Most of the cortical neurodegenerative diseases represent a failure of central nervous system neurons or glia to recycle breakdown protein products into soluble forms that can be either reutilized in synthetic processes or removed from the cell. The presence of excessive insoluble proteins leads to the accumulation of those proteins and eventual dysfunction and death of the cell.

In the case of Alzheimer's disease, there is a failure of metabolism of amyloid and tau protein, in frontotemporal dementia (FTD) a failure of metabolism of tau protein, in Parkinson's disease synuclein protein.

Research has revealed that many of what were thought to be sporadic degenerative disorders of the nervous system have contributory genetic cofactors.

For instance, amyotrophic lateral sclerosis (ALS) was thought to be a purely sporadic degenerative disorder until enzyme studies uncovered kindreds with superoxide dismutase deficiency (SOD1). A further subgroup of ALS is SOD negative but is related to recycling disorders of cellular proteins such as ubiquitin.

Examples of cerebrocortical degenerative disorders include Alzheimer's disease (AD), FTD and dementia with Lewy bodies (DLB). Degenerative diseases of white matter include microvascular leukoencephalopathy; those of the dopaminergic systems in the basal ganglia include Parkinson's disease and corticobasal degeneration. Multisystem atrophy is a neurodegenerative disorder associated with the accumulation of synuclein protein which can affect cortex, white matter, basal ganglia, cerebellum, brainstem and peripheral nerves in various degrees at various rates of progression.

Investigations for these disorders include neuroimaging, mental status and neuropsychological testing. In some cases, such as patients with suspected prion disorders, brain biopsy can be considered in order to establish a definitive diagnosis.

In terms of timeline, all of these disorders are chronic, each with a characteristic collection of symptoms and signs that suggest the ultimate diagnosis. These are not applicable in the case under discussion.

3.4.10 Genetic Diseases

Genetic diseases affecting the nervous system often overlap with most other disease types in terms of time line. With the increasing knowledge derived from genetic studies, the distinctions between the categories of degenerative and genetic disorders are becoming increasingly blurred. One can consider many degenerative disorders as late onset genetic disorders with possible environmental triggers. Alternately, genetic disorders could be considered as the cause of most degenerative disorders but with varying times of onset. For instance, Tay-Sachs disease has always been considered a genetic disorder of sphingolipid metabolism, but it could be considered as a genetic neurogenerative disorder of early onset.

The genetic diseases may express themselves in terms of timeline from as early as in utero (diagnosed on ultrasound) to late adult life. They may present in an abrupt fashion (rapid onset coma in neonates due to urea cycle defects), over weeks to months (6 to 12 months for infants with Tay-Sachs disease), or many years (Huntington's disease).

The defining event for the severe forms of genetic disorders is often birth and the inability of the neonate to support itself metabolically in the absence of the placental circulation and the mother's normal metabolic support. Other genetic degenerative disorders declare themselves in infancy or early childhood with parental concerns about slowness of development such as a missed milestone which then prompts an investigation for the cause.

Using the localization system to classify the genetic disorders is helpful. Table 3.15 gives some examples of how some of the common genetic disorders show different timelines throughout life.

In terms of timing of onset, all of these disorders have their own characteristics. However, the basic principle is that the more severe and life threatening disorders will present earlier in life.

Blood tests exist for the various genetic markers for many of these disorders, which allows families to make informed reproductive decisions, especially for those disorders with long latencies such as Huntington's disease, which does not manifest until the usual reproductive period of a couple's life has passed.

3.5 Fail-Safe: Localization/Etiology Checklist

Now that the Extended Localization Matrix and Etiology Matrix have been consulted to analyse this case, it is important to review the Fail-Safe Localization/Etiology Checklist.

At each level of localization, there are several etiological diagnoses, which represent serious and often life threatening neurological disorders that must always be considered so as not to be missed. Table 3.16 lists these disorders according to their level of localization.

Case Summary

To summarize the process and conclusions for the case under discussion, let us refer back to the Expanded Localization Matrix to produce a concise summary of the case under discussion.

The Expanded Localization Matrix (Table 3.2) shows that the elements of the neurological exam (including right-sided weakness of the face and upper limb,

Table 3.15 Common Genetic Disorders by Localization and Time Course

Localization	Neonatal	Infancy	Childhood	Adolescent	Early Adult	Mid Adult	Late Adult
Muscle	CMTD	Metabolic myopathies	DMD	BMD	MERRF, MELAS		
Neuromuscular junction	Congenital myasthenia					Myasthenia gravis	
Peripheral nerve			DJS	CMT	HNPP		
Spinal cord		SMA	SMA	SCA			ALS
Brainstem		LD				MSA	MSA
White matter	Neonatal leukodystrophies	Leukodystr	Leukodystr	Leukodystr	LD Leukodystr	MVL	
Cortex	Lissencephaly and other cortical dysplasias	Tay-Sachs, Gaucher's, CLF	CLF	CLF	HC, CLF	FTD, MSA	AD, DLB, MSA
Basal ganglia		LD	Generalized dystonias	Generalized dystonias	HC	PD, MSA	MSA
Cerebellum			Dandy Walker		Chiari malformation	MSA	SCA, MSA

Abbreviations: AD, Alzheimer's disease; ALS, amyotrophic lateral sclerosis; BMD, Becker's muscular dystrophy; CLF, ceroid lipofuscinosis; CMT, Charcot-Marie-Tooth disease; CMTD, congenital myotonic dystrophy; DJS, Dejerine–Sottas disease; DLB, dementia with Lewy bodies; DMD, Duchenne's muscular dystrophy; FTD, frontotemporal dementia; HC, Huntington's chorea; HNPP, hereditary neuropathy with liability for pressure palsies; LD, Leigh's disease (this is mostly a gray matter disease); MELAS, mitochondrial myopathy with lactic acid and stroke-like symptoms; MERRF, mitochondrial myopathy with ragged red fibers; MSA, multisystem atrophy; MVL, microvascular leukoencephalopathy; PD, Parkinson's disease; SCA, spinocerebellar atrophy; SMA, spinomuscular atrophy.

hyperreflexia on the right side, difficulty with expression of speech) all point to localization in the left frontal area. Combining these findings with the sequence of events in the history and the correlative factors in the etiology checklist points to an acute vascular event involving the left internal carotid artery as the common element needed to explain all of the findings.

The history indicates recurrent episodes of visual disturbance in the left eye followed by a major disturbance to the area of the nervous system that controls speech as well as right-sided strength and coordination of the face and right upper limb. The History Worksheet (Table 3.12) and the Etiology Matrix (Table 3.13) suggest an acute etiology, such as a paroxysmal, traumatic or vascular disease. The unilateral nature of the symptoms and lack of history of seizures or trauma favor a vascular etiology.

The most likely explanation for the patient's story is repeated vascular ischemic events caused by emboli emanating from the left carotid artery. These initially caused disturbances in vision in the left eye through embolization of the ophthalmic artery, a branch of the left internal carotid. At the time of the major symptomatic event, the emboli caused disruption of flow to the left frontal cortex and white matter, resulting in the expressive aphasia and right-sided weakness.

The neuroradiological investigations can then be targeted to these structures. This would include either an MRI or MRA of the brain, or a CT scan with CT angiogram of the carotid arteries and intracranial vessels.

Figures 3.1 and 3.2 show the CT scan and CT angiogram images that show the findings for this case. They confirm the clinical analysis of a left frontal infarction and high-grade stenosis of the left internal carotid artery.

Case Resolution

The patient arrived in the ER four hours after the onset of new deficits and therefore was not considered a candidate for intravenous or intraarterial TPA.

The patient was admitted to hospital and treated with intravenous heparin for five days; he then underwent an uneventful left carotid endarterectomy. At six weeks post-op, he showed only minor word finding difficulties and slowness of rapid alternating movements on his right side.

Table 3.16 Fail-Safe Localization/Etiology Checklist

Localization	Diagnosis
Muscle	Congenital muscular dystrophy
	Polymyositis
	Polymyalgia rheumatica
Neuromuscular junction	Myasthenia gravis
	Lambert–Eaton myasthenic syndrome
Peripheral nerve	Guillain–Barré syndrome
	Diabetic neuropathy
Spinal cord	Acute spinal cord compression
	Transverse myelitis
Brainstem	Infarction/hemorrhage
	Meningitis (listeria)
	Tumor
	Fisher variant Guillain-Barré
Deep white matter	Infarction/hemorrhage
	Multiple sclerosis
	Progressive multifocal leukoencephalopathy
Cortex	Infarction/hemorrhage
	Herpes simplex encephalitis
	Subarachnoid hemorrhage
	Carotid dissection
Cerebellum	Expanding hematoma
	Alcoholic- or drug-induced degeneration
Basal ganglia	Drug-induced movement disorder
	Parkinson's disease
	Huntington's disease
	Wilson's disease

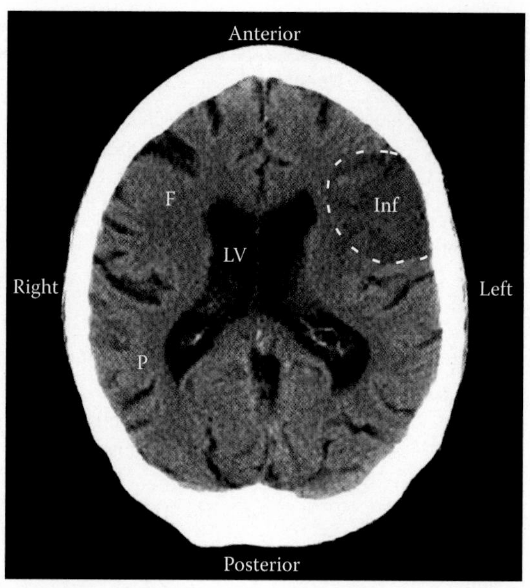

FIGURE 3.1 CT scan of the brain, showing an area of infarction in the left frontal region.

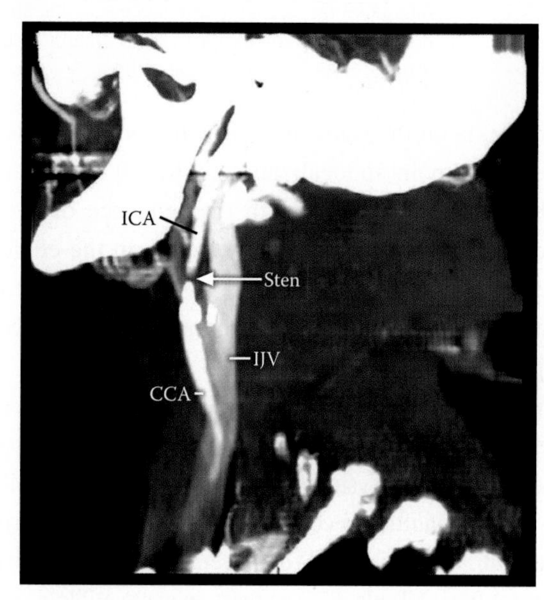

FIGURE 3.2 CT angiogram, showing a localized area of stenosis of the left internal carotid artery.

Implementation

Now that a complete example of how the system of localization and etiological diagnosis works has been completed, readers, especially medical students and residents, are encouraged to use the system by practicing it for each of the cases in the rest of the text and for the e-cases on the text DVD and Web site.

It is recommended that prior to reading each chapter and e-case that the reader print out a blank copy of the Expanded Localization Matrix, Etiology Matrix and History Worksheet which are files on the DVD and are also on the Web site. As the reader works through the case in each of the case-based chapters and online e-cases, he or she fills in the blanks for the localization and etiology using the method and criteria detailed in this chapter and also fills in the time lines of the symptoms on the History Worksheet. The conclusions reached in each case can then be compared against the discussion in the text and

the completed sheets that appear on the DVD and Web site (see http://www.integratednervoussystem.com). The Web site and DVD have been constructed with workflow logic that does not allow the user to open the completed sheets and case discussion before the history, physical exam and investigations have been reviewed.

This will assist the reader in developing the skills to use the system on patients on the ward and in the clinic.

Section II

Applying the Basics to Clinical Cases

Chapter 4

Fifi

FIFI

Meanwhile, back at home, Crash's devoted wife, Françoise, has fallen ill.

Currently thirty years old, Françoise (known as "Fifi" to her friends and relations) works as a registered nurse in the emergency department at the University Hospital. She and Crash have a son aged six, and a daughter aged five, about whom more will be heard later on.

As often happens to medical personnel working on the "front lines," notwithstanding constant hand washing technique between patients, Fifi fell briefly ill with crampy abdominal pain, low-grade fever, and recurrent diarrhea about two weeks ago. After three days at home, she recovered and went back to her regular shift in the ER.

Yesterday morning Fifi awoke with peculiar sensations in her legs: her feet felt as if they had "gone to sleep" and her calves as if there were bugs crawling under the skin. Although she felt distinctly strange, Fifi (possessing the typical Type A personality so characteristic of health professionals) got ready to go to work. On the way downstairs to the kitchen, however, she noted that her legs seemed unusually heavy, and she tripped twice on the hall carpet. While not exactly sure what was going on, Fifi thought it might be better if she called in sick.

As the day progressed, things went from bad to worse. Fifi began to have trouble walking upstairs, her legs feeling like lead, and had to proceed one step at a time, using one hand on the railing to help propel herself upward. She thought about going to a local walk-in clinic—of course it was Saturday and her family practitioner's office was closed for the weekend—but decided to stick it out for the day in the hope that tomorrow would be better. Nevertheless, Fifi had the foresight to ask her children's babysitter to stay overnight.

Today brought a marked deterioration in Fifi's condition. Her legs had become so weak that she could scarcely roll out of bed. She found that her fingers had begun to tingle and that she had difficulty lifting her arms to put them in the sleeves of her dressing gown. When she tried to head for the bathroom for a glass of water to take her

Objectives

- To learn the main anatomical components of the peripheral nervous system and their roles in the production and control of movement
- To review the microscopic anatomy and physiology of peripheral nerves
- To understand the symptoms and signs resulting from a disease process affecting the peripheral nervous system
- To understand the localization characteristics, as revealed by history and physical examination, of lesions of the peripheral nervous system affecting movement control, as distinct from lesions of the central nervous system

4.1 Clinical Data Extraction

Before proceeding with this chapter, you should first carefully review Fifi's story in order to extract key information concerning the history of the illness and the physical examination. On the DVD accompanying this text you will find a sequence of worksheets to assist you in this process.

Note: It is essential that you complete the DVD worksheets at this stage in the chapter and that you NOT skip through to the next section. The purpose of this textbook is to help you learn to navigate your way around the nervous system in clinical encounters. The only way to learn this properly is by trying to solve clinical problems and, in the process, by making mistakes. It is obviously preferable to make *virtual* mistakes now rather than *real* mistakes later on.

4.2 The Main Clinical Points

Fifi's chief complaints are that she cannot walk, is unable to swallow liquids, and is having trouble breathing.

The essential elements of Fifi's medical history and neurological examination are as follows:

- A diarrheal illness two weeks prior to the onset of symptoms birth control pill, she discovered that her legs would not support her properly; she had to shuffle slowly, holding on to the furniture. The physical effort involved made her extremely short of breath. Worse, when she tried to swallow water she choked badly, some water exiting through her nose.

Now thoroughly alarmed, Fifi had the babysitter call an ambulance and was brought to the very emergency department where she normally works.

Concerned with Fifi's appearance, the triage nurse immediately transported her to the resuscitation area and called for one of the ER physicians—this means *you*.

On your examination you find Fifi to be anxious and short of breath. She appears mentally intact and answers all questions appropriately but her voice sounds nasal, almost a whisper. Her visual acuity is grossly normal; her fundi are also normal. Pupillary size and reaction to light are equal and symmetric. Extra-ocular movements are full, with no nystagmus; she has no ptosis (droopy eyelids). She is unable to keep her eyes shut against resistance; her smile is but a weak grimace. She can protrude her tongue on request but has a poor gag reflex.

In her upper limbs there is symmetric, diffuse muscle weakness, 3/5 on the MRC scale proximally, 4/5 distally (see Table 2.2 for explanation). In the lower limbs, the weakness is more severe: 2/5 proximally and 3/5 distally. She has no difficulty initiating or stopping movement on request but cannot sustain a strong contraction of the involved muscles. Even though she is very weak, Fifi is still able to move her fingers and toes quickly, and to perform rapid alternating hand and foot movements. Passive stretch of her limbs reveals a reduction in muscle tone. Her tendon reflexes are completely absent in all four limbs. Brisk stroking of the soles of her feet with the handle of a reflex hammer produces a normal flexion response.

Despite the history of sensory disturbances in hands and feet, the only abnormal sensory finding is a decrease in vibration sense at the ankles and toes. Other sensory modalities (touch, position, pinprick, temperature) are normal.

Finally, Fifi's abdominal examination shows no bladder enlargement; she is able to provide a urine specimen on request without difficulty.

- The development of severe muscle weakness and mild sensory symptoms over a period of two days
- On examination, the presence of diffuse, symmetrical muscle weakness involving the face, swallowing musculature, and all four limbs (proximally and distally), the legs more than the arms
- Difficulty breathing and a faint voice
- Complete absence of muscle tendon reflexes
- Relatively minor abnormalities on sensory examination despite the history of sensory disturbances

With respect to localizing the disease process, it is also important to note several findings that Fifi does *not* have:

- Despite her respiratory distress she is alert, gives a coherent history and shows no sign of confusion or memory loss.
- She has no visual difficulties or gaze problems.
- She does not have difficulty in initiating movements, but rather difficulty in carrying out the movements with sufficient force.
- Her limbs are hypotonic, not stiff or rigid.
- Despite the presence of severe weakness she has no difficulty in controlling her bladder and has no signs of urinary retention or inability to void.
- Her plantar responses are flexor, not extensor (the so-called Babinski response).

In brief, therefore, Fifi has a neurological disorder primarily affecting the maintenance of muscle strength, developing over a period of just two days. In consequence, we must redirect our attention to the parts of the nervous system that are involved in carrying out voluntary movements.

4.3 Relevant Neuroanatomy

Before proceeding to attempt to localize Fifi's disease process, we will briefly review and, where necessary, expand on the anatomy of the motor system outlined in Chapter 2.

Based on the results of functional magnetic resonance imaging studies (fMRI), it appears that the process of planning a movement prior to its performance takes place in the supplementary motor area of the cerebral cortex, anterior to the cortical areas that actually send the signals downward to eventually result in the desired movement. The actual decision to move appears to reside within the complex circuitry linking the motor cortex, the basal ganglia, and the thalamus (see Chapter 7). The message to commence a specific movement is transmitted to the motor cortices, located bilaterally in the precentral gyrus, anterior to the central fissure.

The cortical area directly initiating movement of the right arm and leg is located in the left cerebral hemisphere, and vice versa for the left side of the body. The area controlling movement of the opposite leg is located near the midline, at the vertex of the brain, while arm movement originates more laterally (over the convexity of the hemisphere), and face/tongue movements originate where the inferior motor cortex dives into the lateral fissure (see Figure 4.1). This anatomical distribution of movement control constitutes the so-called "motor homunculus" originally derived by Penfield and colleagues in the context of direct motor cortex stimulation during cortical excision for the treatment of medically intractable epilepsy.

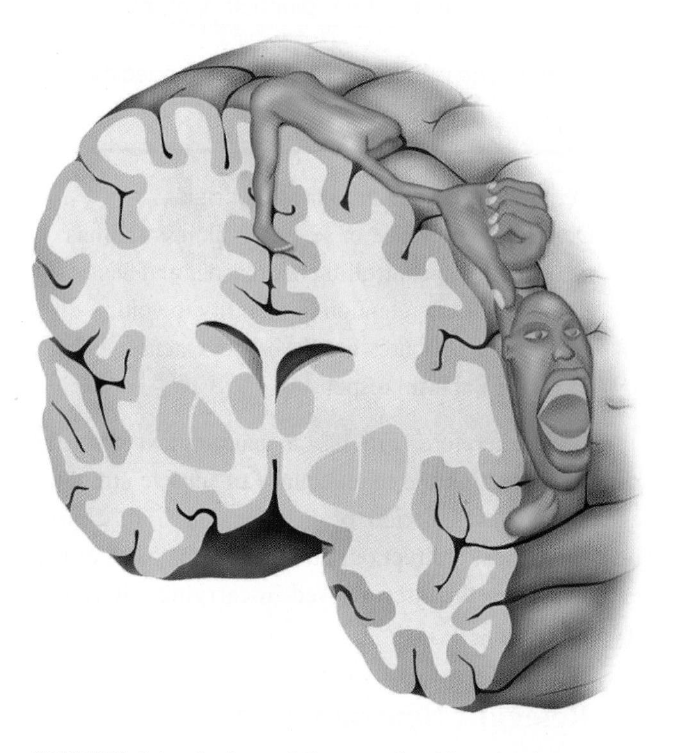

FIGURE 4.1 A view of the cerebral hemispheres sectioned coronally, just anterior to the precentral gyrus, illustrating the relative area of distribution of the motor cortex devoted to movement of different parts of the body (the motor "homunculus"). Note the relatively large areas devoted to movement of the hand, fingers, face and tongue.

Once initiated, a specific movement, in order to be carried out in a precisely controlled fashion, requires moment-by-moment sensory feedback from movement detectors in the muscles and joints of the part of the body being moved. This information, as we saw in the introductory exposition of Crash McCool's spectacular flying activities, is transmitted unconsciously to the cerebellum, where it is constantly compared and integrated with parallel information originating in the motor cortex. The ongoing integrated information is then fed back to the motor cortex to permit performance of the next epoch of the limb movement required.

As we saw in Chapter 2, the actual message, elaborated with the assistance of the basal ganglia and cerebellar motor circuitry, is transmitted to the spinal cord, thence to the muscles, by two parallel descending routes. One route is direct to the cord by single neurons without intermediate relays, the other is an indirect route involving relay centers in the brainstem. These pathways, along with the contributions of the basal ganglia and cerebellum, will be considered in more detail in Chapter 7.

The final step in the descending pathways involved in carrying out a movement command is the recruitment of motor neurons in the area of the spinal cord responsible for a specific limb—the anterior horn cells. These cells, once excited by descending motor pathways, direct action potentials down their axons, which leave the spinal cord via the anterior (ventral) roots (see Figure 2.5) and travel down peripheral nerves to specific muscles participating in the desired movement. A given motor neuron, its axonal branches and terminal synaptic boutons, and its associated cluster of muscle fibers, constitute a *motor unit* (see Figure 4.2). Any movement, whether of a finger, arm, tongue, or eye, requires the graded and coordinated activity of a number of motor units.

Finally, let us reemphasize the fact that no voluntary movement can be carried out in an organized fashion without constant feedback concerning muscle length, muscle tension, and joint position via sensory neurons originating, respectively, in muscle spindles, tendon organs, and joint capsules. In addition to sending sensory information to higher motor centers in the cerebellum (and elsewhere),

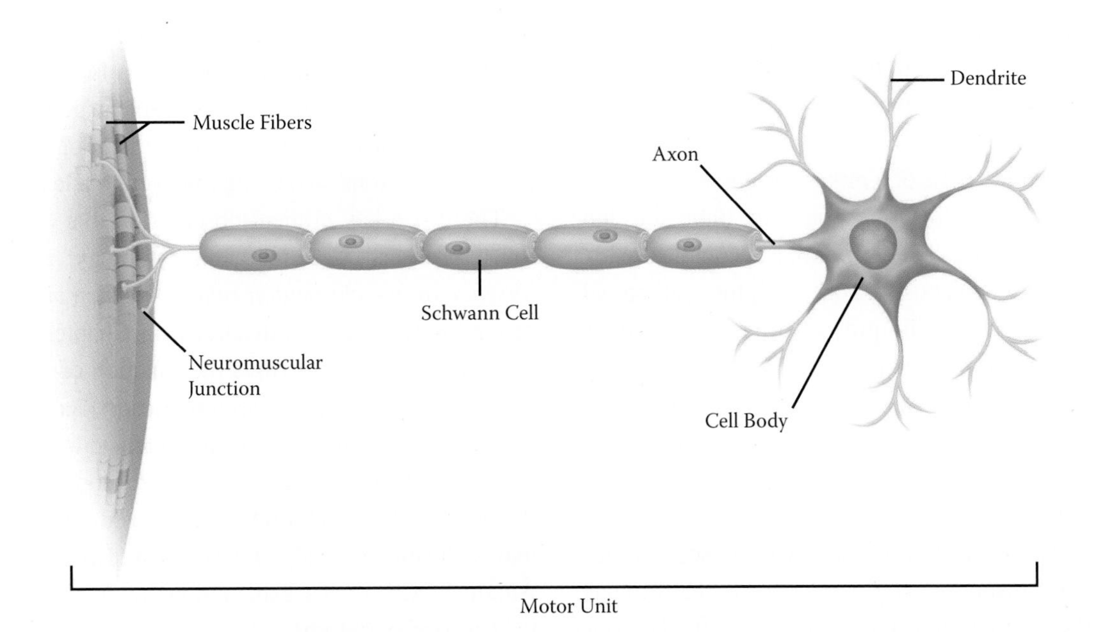

FIGURE 4.2 Drawing of the component parts of the motor unit: CNS lower motor neuron, peripheral axon (usually myelinated), neuromuscular junction, and voluntary muscle.

as previously mentioned, the same sensory fibers send branches that synapse directly or indirectly (via interneurons) with anterior horn cells busy generating the movement (see Figure 1.5B). These feedback loops form monosynaptic and disynaptic reflex arcs that are crucial in the peripheral circuitry involved with movement; they are also the anatomical basis of the reflexes tested in the basic neurological examination outlined in Chapter 2.

4.4 Localization Process

Having briefly surveyed the system presumably involved in the pathogenesis of Fifi's muscle weakness, we must now return to the clinical information provided and attempt to localize the lesion. Is the cause of her weakness located in the cerebral hemispheres, brainstem, spinal cord, peripheral nerves, neuromuscular junctions, or muscles?

If we begin at "the top," there are a number of reasons why the problem is unlikely to be located in the cerebral hemispheres, or involve the command center, as it were, of the motor system. First of all, even when profoundly weak and in respiratory distress, Fifi is alert and cooperative, with no evidence of a disturbance of consciousness or attention. Second, even if the disorder were confined to the cerebral motor system components, Fifi is not having any difficulty initiating or coordinating a movement: when she attempts to flex her elbow, for example, she does so quickly and in a smooth fashion. What she cannot do, however, is generate sufficient force to flex the elbow vigorously while you, as part of your examination, attempt to prevent her from so doing. She simply cannot generate sufficient force in the biceps brachii.

Since Fifi has weakness of facial movement and swallowing, it is conceivable that her disease could be located in the brainstem, the origin of nerves destined for facial and pharyngeal muscles. On the other hand, this possibility is made less likely by the fact that Fifi has no sign of a dysfunction clearly implicating the brainstem, such as nystagmus or a disturbance in consciousness. (The brainstem's contributions to consciousness will be reviewed in Chapter 10.)

In addition, as was mentioned in Chapter 3, clinical experience has taught us that lesions involving the descending motor control pathways above the spinal cord characteristically produce an *increase in muscle tone* below the level of the lesion. This tone increase is secondary to a disinhibition of segmental monosynaptic reflex arcs from muscle spindle to anterior horn cell back to muscle (see Figure 1.5B). Without suppression from higher centers, these reflex pathways are normally set to produce considerable reflex contraction of a muscle when it is passively stretched—the phenomenon of *hypertonia*.

The presence of hypertonia in a weak muscle is a cardinal sign, therefore, of a lesion in the descending motor pathways above the level of the anterior horn cell: the so-called *upper motor neuron lesion*.

A second characteristic of a motor system lesion at this level is the extensor plantar response or Babinski sign mentioned in Chapter 2. This response is a complex reflex involving plantar cutaneous sensory receptors and spinal cord circuitry; it is normally present in newborn infants and is modified by the developing corticospinal tract into a plantar flexion response. A lesion anywhere in the direct or indirect descending motor pathways will result in the reappearance of the primitive toe extensor response, the Babinski sign.

Thus, in the absence of hypertonia in weakened muscles and of Babinski responses, it is very unlikely that Fifi has a lesion in the motor system above the level of the anterior horn cells.

This conclusion applies just as well to the spinal cord as it does to the brainstem. There are a number of other reasons, however, why a high spinal cord lesion is unlikely. Most obviously, a spinal cord lesion could not explain the weakness of the muscles of the face and pharynx (the latter producing her swallowing problems). In addition, spinal cord lesions typically interrupt the descending control pathways for voluntary control of urination (discussed in Chapter 13). If this pathway is suddenly disrupted for any reason, the bladder becomes incapable of releasing urine and becomes extremely large, a finding Fifi does not have at the time of your assessment.

Having traveled this far down the motor control pathways without encountering a mixture of clinical findings that accounts for Fifi's symptoms and signs, we are left with a problem involving the motor units: anterior horn cells, their axons, neuromuscular junctions, and the muscle fibers they control.

In general, a dysfunction of a proportion of motor units in a given muscle will produce difficulty in generating normal amounts of tension in the muscle when an attempt is made to contract it: exactly the problem Fifi has developed. Nevertheless, it is unlikely that Fifi has a specific problem of either anterior horn cell or muscle fiber function, for the simple reason that she also has sensory symptoms. Although it is possible that her disease process might be simultaneously affecting anterior horn cells, muscles, and sensory nerve fibers, as a general rule such a scenario is unlikely. In large part, the most economical explanation for the location of a disease process is also the most likely. This being the case, the most probable explanation for the muscle weakness is a process affecting motor axons, with the sensory axons also being involved to a lesser extent.

There is a key clinical observation that also supports the hypothesis of motor axon pathology: the complete *absence of muscle tendon reflexes*. The anatomical basis of tendon reflexes was discussed in Chapter 1. Any disease process that interrupts the function of a majority of large (muscle spindle afferent) sensory axons or motor axons participating in a given tendon reflex will abolish the reflex. If a disease process were to stop the function of the anterior horn cells themselves, as distinct from their axons, the muscle tendon reflexes would also be abolished. Again, however, this explanation would not account for the sensory symptoms.

Now that we have examined the reasons why Fifi's muscle tendon reflexes are absent, we can go back and explore what would have happened to the reflexes had the disease process been located at the level of the upper motor neuron (cerebral hemispheres, brainstem, or spinal cord). Dysfunction in the upper motor neurons results in a relative lack of input from the highest levels of the motor system to the anterior horn cells. This lack of input, largely inhibitory in nature, results in the muscle tendon reflexes being exaggerated, or abnormally brisk (hyperreflexia).

At this point, because of its crucial importance in neurological localization, we repeat the mantra already chanted in Chapter 3: in a patient with muscle weakness, the three cardinal signs of an upper motor neuron lesion are hypertonia, hyper-reflexia, and an extensor plantar response (Babinski sign). In contrast, lower motor neuron lesions are characterized by hypotonia, hypo- or areflexia, and a flexor plantar response. Clearly, Fifi's clinical picture corresponds to the latter description.

We have not devoted much attention to the possibility that Fifi might have an acute disorder of neuromuscular junction function, or of muscle itself, primarily because of her sensory symptoms. As we noted in Chapter 3, neuromuscular junction dysfunction results in progressively increased muscle weakness with sustained or repeated contractions over a short period of time, the phenomenon of muscle fatigability. Although Fifi is very weak, repeated contraction of given muscles do not make them any weaker than they were at the first contraction. In addition, neuromuscular junction problems do *not* typically decrease muscle tendon reflexes. Furthermore, neuromuscular junction disorders tend to specifically target the face and extraocular muscles, producing facial weakness (which Fifi has) and ptosis plus eye movement paresis (which she does not). Finally, as we already noted for anterior horn cell and muscle diseases, neuromuscular junction disorders are not accompanied by sensory symptoms or signs.

Proceeding past the neuromuscular junction, we again note that muscle diseases produce pure muscle weakness, often associated with a modest amount of muscle wasting if the disease process has been present long enough (months). Muscle diseases (myopathies) may decrease muscle tendon reflexes but rarely abolish them.

Thus, if we review the localization possibilities given in the Expanded Localization Matrix (see Table 3.2), black Xs would have to be entered in the columns for muscle, neuromuscular junction, peripheral nerve, spinal cord, brainstem, white matter/thalamus, and frontal lobe cortex. At the end of the clinical reasoning process, however, the Xs for all the columns except peripheral nerve must be changed from black to red.

In other words, our descending voyage through the motor control system has indicated that the disease process acutely affecting Fifi probably resides in cranial and spinal motor nerve fibers and, to a lesser extent, sensory nerve fibers.

4.4.1 Peripheral Nerves

Before we address our second basic question (i.e., *what* is Fifi's disease process?), it is important to consider in more detail the structure and function of the part of the nervous system involved: peripheral nerves.

In essence, peripheral nerves are cable systems consisting of motor and sensory axons, as well as various supporting structures. The latter include *myelin sheaths*, highly organized protein-lipid layers surrounding the thicker axons, *Schwann cells* (which produce the myelin), collagen fibers, and nutrient blood vessels (see Figure 1.2C). Most peripheral nerves are mixed, containing motor, sensory and autonomic axons; a few are purely motor or sensory.

Motor axons carry electrical impulses in an outward direction from cell bodies located in the brainstem, spinal cord and autonomic ganglia. The axons terminate, usually with numerous branches, in skeletal muscle, cardiac muscle, smooth muscle of intestinal and bronchial walls, selected glands or, in the case of the sympathetic and parasympathetic systems, peripherally located autonomic neurons.

Sensory axons carry electrical impulses in an inward direction from simple and complex receptors located in skin, joint capsules, muscle spindles, tendons, oral cavity, intestines, heart, lung, meninges or the walls of large arteries (see Figures 1.4A and 1.4B). For the most part, their cell bodies are located in ganglia present in spinal dorsal roots, cranial nerves and viscera. There are central extensions from these sensory ganglia, either from the cell body or from the sensory axon itself, entering the spinal cord and brainstem to make synaptic contacts with second-order sensory neurons which convey information to higher regions of the central nervous system (see also Figure 1.2B). The organization of special senses (such as vision, hearing, and olfaction) is highly specialized and will not be considered here (see instead Chapters 6 and 11).

A magnified microscopic view of a peripheral nerve in cross-section is shown in Figure 4.3. Individual axons are of varying diameters, with a range of about 0.5 to 20 μm. All of the larger axons, whether motor or sensory, have sheaths of myelin, the sheaths being thicker for larger axons (8 to 20 μm) and thinner for smaller axons (1 to 7 μm). Many of the smallest axons (0.5 to 1 μm) have no myelin sheath whatever. As will be seen, there are significant functional differences between the large myelinated axons and the small unmyelinated fibers.

With respect to Fifi's disease process, it is important to note that there is a functional barrier between a nerve's blood vessel lumina and the extracellular space surrounding the axons. This barrier is in the form of tight junctions between endothelial cells that prevent the passage of large molecules from the blood circulation: the *blood–nerve barrier.* The barrier is relatively leaky at the level of the dorsal and ventral roots, the proximal spinal nerves (formed by the junction of the dorsal and ventral roots) and small distal nerve branches, and in the region of potential compression points such as the carpal tunnel for the median nerve (in the palmar aspect of the wrist). Please note that a similar barrier exists between central nervous system capillaries and the surrounding neuronal and glial tissue: the *blood–brain barrier.*

The development of a myelinated axon is illustrated in Figure 4.4. The myelin sheath itself is generated by adjacent Schwann cells and consists of tightly apposed

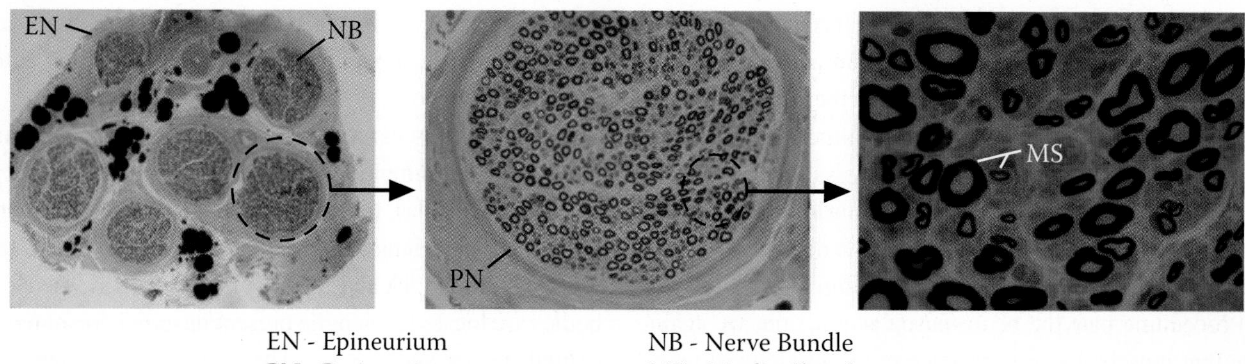

FIGURE 4.3 Photomicrographs of a cross-section of a typical mixed nerve, at different magnifications, showing myelinated axons of varying sizes, as well as unmyelinated fibers. (Courtesy of Dr. J. Michaud.)

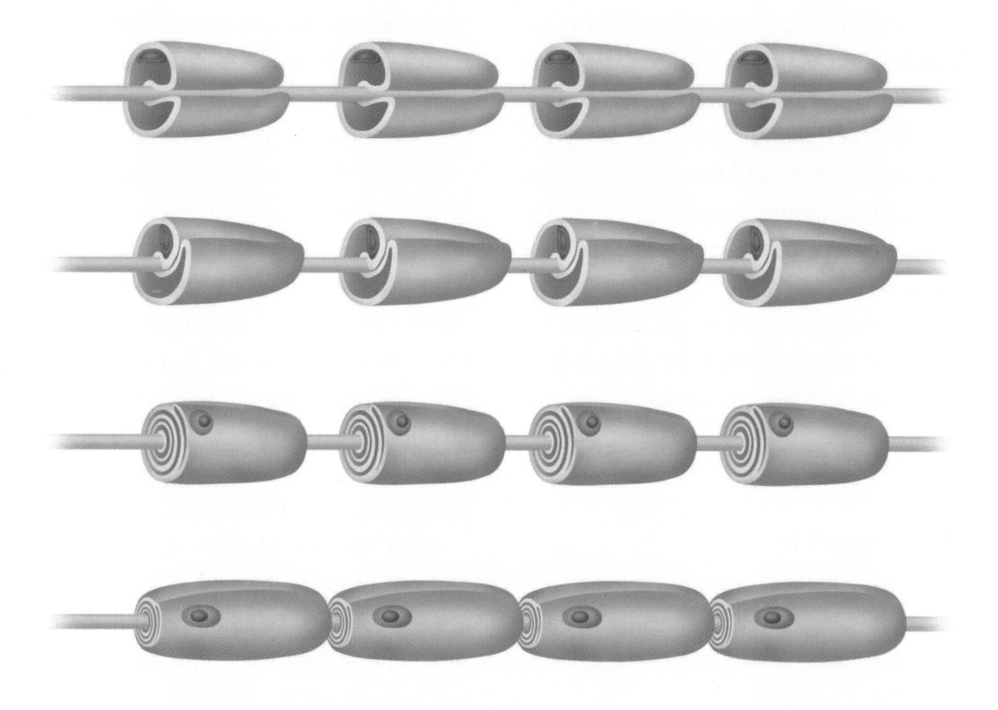

FIGURE 4.4 Drawing illustrating the formation of myelin segments around a peripheral nerve axon. Each Schwann cell has a leading edge that gradually wraps around the axon to form multiple membranous layers, or lamellae. The mature myelinated axon is at the bottom.

lamellae of Schwann cell plasma membrane. In essence, the plasma membrane has been wrapped repeatedly around the axon like a jelly roll. A given Schwann cell is responsible for the creation and maintenance of a segment of myelin for a single peripheral axon (unlike oligodendroglial cells in the brain and spinal cord, which provide myelin segments for a number of axons). A section of myelin sheath provided by a Schwann cell is called an *internode*, and is separated from the internode provided by an adjacent Schwann cell by a short section of unmyelinated axon, the *node of Ranvier.*

Myelinated axons cluster their sodium channels primarily at the nodes of Ranvier. As we saw in Chapter 1, action potentials in myelinated fibers are propagated rapidly along the axon by leaping from node to node, the process of *saltatory conduction* (see Figure 1.2C). In contrast, unmyelinated axons have sodium channels distributed evenly along their axonal membranes and conduct action potentials in a contiguous fashion by spread to adjacent membrane.

Saltatory conduction is obviously far more rapid than contiguous conduction: a large myelinated axon conducts action potentials at rates of 45 to 60 m/sec along its main trunk, while an unmyelinated axon can manage a rate of only 1 m/sec or less.

Large, rapidly conducting myelinated axons include muscle spindle afferents, sensory fibers from tendons and joint capsules, and motor axons destined for skeletal muscle fibers (α motor neurons). Intermediate-sized (8 to 12 μm) myelinated axons include many sensory fibers for touch, and motor axons destined for muscle spindles (γ motor neurons). Small myelinated axons (1 to 2 μm) include sensory fibers for fast ("sharp") pain and first-order efferent autonomic neurons (synapsing in sympathetic and parasympathetic ganglia). Finally, unmyelinated axons include those for slow ("burning") pain and second-order (post-ganglionic) autonomic fibers.

Thus, if we return to our patient, Fifi's peripheral nerve pathology appears to involve primarily motor axons originating in brainstem and spinal cord, and destined for muscles of the face, palate, pharynx, diaphragm, intercostal muscles and limbs. To a lesser extent the disease process also must be affecting some of the large myelinated sensory axons (e.g., those for touch and vibration sensation) but not to such an extent as to permit detection of a sensory deficit on clinical examination.

When considering disorders of movement in this text, we will occasionally, in an attempt to relate a complex organization to a structure encountered in daily life, use the analogy of the automobile. For example, we can liken the motor planning center in the frontal cortex to the driver of the automobile. At the "business" end we can likewise consider the skeletal muscles as the motor, and muscle energy substrates such as glycogen as one of the fossil fuels used by the engine. In Fifi's illness, we can consider the problem as residing in the cable linkage between the accelerator pedal and the fuel injection system, as well as the injection system itself.

4.5 Etiology: Fifi's Disease Process

If we examine Fifi's history, we find an evolution of symptoms from first inkling to severe disability over just two days. Referring back to Table 3.13, we can see that she is unlikely to have a degenerative or neoplastic disease because the timeline is far too short. As well, in the case of a tumor, the process seems too dispersed in space (head, all four limbs to an equal extent) to allow for such a diagnosis. On the other hand, a vascular etiology is equally unlikely as the onset is too slow.

Between these extremes we find several categories of disease that could, in principle, create a symptomatic evolution over two days: (accidental) ingestion of a neuroactive drug or environmental neurotoxin; an infectious disease with an affinity for the nervous system; an inflammatory, autoimmune disorder directed at the peripheral nerves.

For the sake of completeness, we will touch on all ten of the disease categories listed in Table 3.13, directing most of our attention to the three most likely categories.

4.5.1 *Paroxysmal*

Paroxysmal symptoms in peripheral nerve rarely occur. An inflammatory process of a proximal nerve may produce brief, spontaneous discharges of motor units, manifested as irregular focal muscle twitches—usually referred to as *fasciculations*. A multiple sclerosis plaque in the brainstem or spinal cord at the exit point of the motor nerve roots may manifest itself in this way, without measurable muscle weakness. Generalized paroxysmal disorders of peripheral nerve do not occur.

4.5.2 *Trauma*

Focal acute traumatic lesions of one or several peripheral nerves, including the brachial or lumbosacral plexi, are unfortunately all too common. Such lesions usually affect only one limb and are seen in the context of motorcycle accidents, sports injuries and war-related wounds. A generalized traumatic injury to peripheral nerves is not compatible with survival.

4.5.3 *Vascular*

An abrupt arterial occlusion in an arm or leg may produce an acute-onset regional ischemic multifocal neuropathy with predominantly distal weakness and numbness. An embolic clot from a fibrillating left atrium may lodge in a brachial artery or at the bifurcation of the abdominal aorta into the femoral arteries (a so-called saddle embolus). Isolated ischemic cranial neuropathies are relatively common, particularly in patients with diabetes; the most vulnerable cranial nerves are III and VI. Generalized peripheral nerve ischemia is a theoretical possibility in severe systemic hypotension but the patient would likely also lose consciousness.

4.5.4 Toxic

There are a large number of environmental toxins capable of selectively damaging peripheral nerves. Examples include inadvertent inhalation of volatile organic solvents, such as toluene, and ingestion of heavy metals, such as lead (contained in lead-based paints, pottery glazes and automobile batteries). In addition, particularly for a patient employed in a hospital, one must consider accidental exposure to neurotoxic drugs such as the anticancer drugs vincristine and cisplatin.

All of these agents tend to damage peripheral nerves by disrupting the transportation of nutrients from the cell body to the peripheral axonal branches by axoplasmic flow. Nutritional deprivation of the distal portions of axons results in degeneration (sometimes in just a few days) of the terminal portion of the axon, a process known as *dying-back*. The dying-back phenomenon affects longer axons earlier than shorter and, thus, will impair movement and sensation in the hands and feet, the latter more than the former.

What this means is that a toxin or drug-induced dying-back neuropathy will selectively produce weakness of distal musculature, leading to grip weakness, wrist drop and drop-foot, a phenomenon characterized by a tendency to trip constantly because of an inability to dorsiflex the feet while walking. Thus, quite apart from the fact that most toxic neuropathies require many days, if not weeks, to develop, Fifi is unlikely to have such a disease process for the simple reason that she also has severe proximal muscle weakness.

Although it is not a toxic neuropathy, we should also consider the possibility of *botulism*, a toxic disease process in which the inadvertent consumption of botulinum toxin (typically in home-canned vegetables contaminated by *Clostridium botulinum*, an anaerobic bacterium) leads to the rapid onset (one to two days) of severe generalized weakness, difficulty swallowing and speaking, and respiratory failure. This story sounds very much like the one we recounted for Fifi. Botulinum toxin—the current darling of cosmetic surgeons—produces its effects by inhibiting the release of acetylcholine at presynaptic nicotinic (skeletal muscle) and muscarinic (smooth muscle) nerve terminals. Thus, botulism is a neuromuscular junction disorder, a category of disease we have already eliminated on the grounds that Fifi has sensory symptoms and absent rather than intact tendon reflexes. Even though the degree of muscle weakness in botulism may be as severe as in Fifi's case, the toxic blockade of muscarinic nerve terminals means that there are also obvious features of parasympathetic autonomic insufficiency: constipation due to paresis of intestinal smooth muscle contraction, and dilated pupils due to paralysis of the postganglionic nerve terminals (see the discussion in Chapter 2 of the pupillary light reflex).

4.5.5 Infection

A number of infectious agents are capable of producing fairly acute peripheral nerve symptoms. Of these, the most notorious is probably poliovirus, the causative agent of acute poliomyelitis. This neurotropic virus selectively affects motor neurons in the brainstem and spinal cord, leading to rapidly progressive weakness of both proximal and distal muscles, along with complete loss of muscle tendon reflexes. Respiratory failure, such as we have encountered in Fifi, is not uncommon.

There are several reasons, however, why Fifi is unlikely to have poliomyelitis. First, the disease is rare in North America because of childhood immunization programs. Second, the virus exclusively affects lower motor neurons, so sensory symptoms would be unlikely beyond pain secondary to parallel inflammation of the meninges. Third, poliovirus (as well as other more ubiquitous viruses that occasionally mimic it—Coxsackie, West Nile) tends to produce markedly asymmetric disease. One leg may be profoundly weak while the other is virtually intact. As we have seen, Fifi's weakness is quite symmetric.

Other microorganisms capable of involving peripheral nerves include the spirochetal agents responsible for syphilis and Lyme disease (respectively, *Treponema pallidum* and *Borrelia burgdorferi*), and *Mycobacterium leprae*, the causative agent of leprosy. All three organisms tend to involve sensory axons more than motor. Furthermore, the spirochetal agents both have a constellation of non-neurological symptoms such as skin lesions and arthropathy, neither of which Fifi experienced. Leprosy follows a chronic course and has been virtually eliminated from North America.

4.5.6 *Metabolic*

Subacute-onset generalized weakness of skeletal muscle is not unusual in a number of metabolic disorders, in particular hypokalemia, hypothyroidism and Cushing disease (hypercortisolemia). A generalized neuropathy may occur in the context of chronic renal failure but typically evolves over a period of weeks to months rather than 48 hours. Diabetes mellitus may also be accompanied by an insidiously progressive distal polyneuropathy of a predominantly sensory type. Less often, patients with diabetes may develop a fairly abrupt focal motor neuropathy manifested as pain, weakness and muscle atrophy, typically in the quadriceps (diabetic amyotrophy). Finally, a generalized polyneuropathy may develop in patients with deficiencies of vitamins B1 (thiamine) and B12 (see Chapter 3), but the symptoms evolve over months to years, not days.

4.5.7 *Inflammatory/Autoimmune*

We now come to an acute-onset, fairly common disorder which, as will be seen, best explains Fifi's time course, symptoms and signs: acute postinfectious polyradiculoneuropathy, better known as *Guillain-Barré syndrome* (polyradiculoneuropathy means a generalized disease process affecting both spinal nerve roots and peripheral nerves). This disorder is believed to be generated by the body's humoral and cellular immune systems in response to infection with a variety of microorganisms. The most common trigger agents are Epstein-Barr virus (human herpesvirus 5), *Mycoplasma pneumoniae* and *Campylobacter jejuni*. These cause, respectively, infectious mononucleosis (glandular fever), pneumonia and acute diarrhea. You will recall that Fifi developed the latter symptom two weeks before her neurological illness.

The mechanism of disease in Guillain-Barré syndrome is believed to be molecular mimicry. One of the surface components of the triggering organism shares a common molecular configuration with a component of the peripheral nervous system contained either in myelin or in the axonal membrane at the node of Ranvier (or both). GM1 ganglioside, an important component of nerve cell membranes, is a strong contender as one of the molecules being mimicked. In attempting to combat the infectious agent, the immune system begins to attack peripheral nerves by at least two mechanisms (see Figure 4.5).

The first mechanism of attack is an antibody-mediated obstruction of sodium channels at the nodes of Ranvier, leading to a block in saltatory conduction. The second is a T-cell-mediated attack on myelin itself, with myelin being damaged by cytokines emanating from the T-cells, then stripped off the axon and subsequently digested by macrophages. In the latter case the "nude" axonal membrane, lacking appropriate numbers of sodium channels, is largely incapable of transmitting action potentials in the contiguous fashion of a normal unmyelinated axon. The net effect is an inability to transmit signals to skeletal muscle, thus weakness of acute onset.

This autoimmune attack on peripheral nerves has several peculiar features that help explain the specific constellation of symptoms developed by Fifi. Presumably because the distribution of the potentially antigenic compound varies from nerve to nerve, Guillain-Barré syndrome tends to involve motor axons more than sensory, yet spares motor axons in certain select nerves altogether. The best examples are the three cranial nerves that contribute to eye movement (III, IV, VI), and the nerves for voluntary control of bladder and bowel sphincters. Thus, even though profoundly weak, patients with Guillain-Barré syndrome are typically able to look in all directions and have no loss of bladder or bowel control.

The second important peculiarity of the peripheral nerve pathology in Guillain-Barré syndrome is the fact that motor axons—let us say for a proximal antigravity leg muscle—are usually not involved homogeneously throughout their courses. The antibody- and cell-mediated disease processes tend to affect those portions of the axon where the blood–nerve barrier is relatively deficient or leaky. As was already mentioned, these include the ventral spinal roots, proximal spinal nerves, distal branches and compression points. In the case of disease of the ventral roots, the "radiculopathy" component of the acute postinfectious polyradiculoneuropathy, we find the explanation for the severe proximal as well as distal weakness in Guillain-Barré syndrome. Since motor axons for both proximal and distal muscles must all pass through the ventral roots, the subsequent length of the axon is relatively immaterial if the axon has already been blockaded at the root level. Clearly this is a very different disease process from the one we encountered in dying-back neuropathy, and it accounts for the significantly different clinical picture.

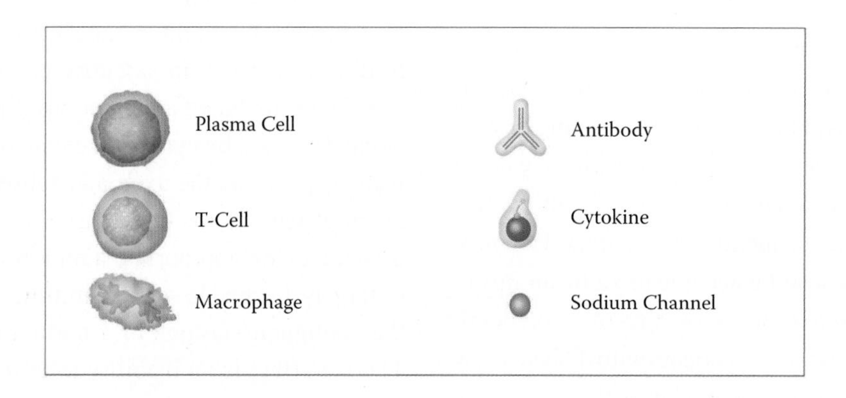

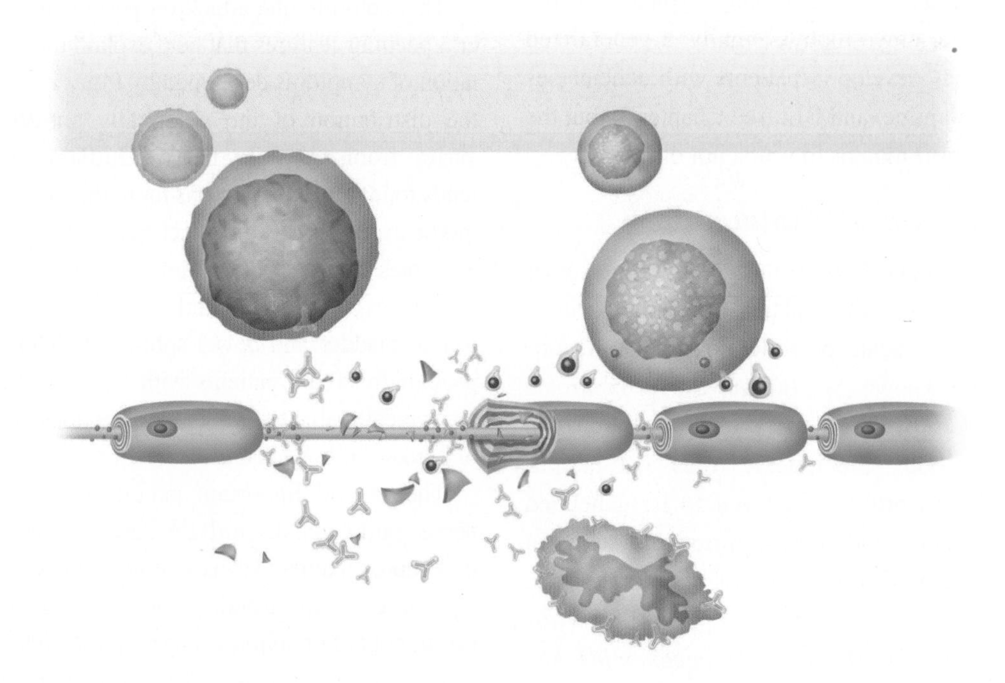

FIGURE 4.5 Illustration of the pathology in acute inflammatory demyelinating polyneuropathy. Some antibody molecules are attaching themselves to sodium channels at the nodes of Ranvier, leading to channel failure and a block in axon potential conduction. T-cells are actively damaging myelin lamellae in one of the internodes; macrophages are present to strip away the damaged myelin and to ingest the myelin fragments. The myelin from one internode has completely disappeared.

4.5.8 *Neoplastic*

Primary neoplastic lesions affecting peripheral nerves usually involve a single nerve, although multiple peripheral nerve neurofibromas are seen in type 1 neurofibromatosis, as are multiple schwannomas (Schwann cell tumors) in type 2 disease. The latter disorders are multifocal, however, not generalized. Metastatic spread of non-neurological malignancies is unfortunately a common phenomenon but, again, only affects nerves or plexi in a regional fashion.

As we saw in Chapter 3, a generalized paraneoplastic neuropathy may occur in the presence of a malignant tumor at a distant site (especially lung carcinoma); the onset of symptoms, however, is slow (weeks to months).

4.5.9 *Degenerative/Genetic*

Generalized degenerative polyneuropathies are nearly always genetic in origin, hence our fusion of the last two etiologic categories. In general, these neuropathies are of the dying-back type, thus tending to produce progressive distal muscle weakness and sensory loss, more marked in the legs than in the arms. These diseases evolve over many years and, especially if the large sensory fibers are affected, may result in sensory ataxia and eventual loss of ambulation. Inheritance patterns may be dominant, recessive or X-linked; some forms are predominantly motor in nature, others purely sensory or even autonomic. A common and very imprecise eponym employed for hereditary motor and sensory neuropathies is *Charcot–Marie–Tooth disease* (see Table 3.15).

4.6 Summary

Fifi has a relatively rapid-onset disease evolving over 48 hours; it is characterized by severe limb, facial and pharyngeal weakness accompanied by distal extremity numbness and paresthesiae. The presence of proximal and distal limb weakness, complete areflexia, and flexor plantar responses is consistent with a generalized polyradiculoneuropathy. Given the recent history of a diarrheal illness, the most likely diagnosis is an acute inflammatory demyelinating polyneuropathy, otherwise known as Guillain-Barré syndrome.

4.7 Additional Information

Fifi's illness was first described in detail, in the French literature, by Guillain, Barré, and Strohl during the First World War. In most, but not all cases there is an apparent inciting event, typically an infectious process as we have already seen, but occasionally a surgical procedure or an injury.

The onset of symptoms is usually over a period of several days, often beginning with spontaneous sensory symptoms (tingling, burning, crawling sensations) in the extremities, the legs more than the arms. There follows a rapidly progressive weakness, proximally and distally, at first in the legs, then the arms, sometimes the face and pharyngeal muscles. If nerves to intercostal and diaphragmatic muscles are also significantly involved, respiratory failure may ensue. The progression from first symptoms to greatest degree of weakness takes place over a maximum of 6 weeks, with the degree of disability extending from mild extremity weakness through inability to walk to complete paralysis. Sometimes the autonomic nervous system is also involved, the presenting feature being either arterial hypertension or postural hypotension. Recovery is almost invariable, usually over a period of weeks to an entire year, with functionally complete recovery in about 80% of patients.

The diagnosis of Guillain-Barré syndrome (Strohl's contribution has been, unfortunately, largely forgotten) is typically based on the characteristic presentation of ascending weakness, absence of muscle tendon reflexes, and preservation of flexor plantar responses. There are two useful diagnostic tests which help confirm the diagnosis. An examination of CSF characteristically shows an elevated protein level but a normal or near-normal cell count, a combination referred to as *albuminocytological dissociation*. The protein elevation results from the disease process involving ventral spinal roots that are in direct contact with CSF in the spinal subarachnoid space. It is important to recognize that, in relatively rapid-onset cases such as Fifi's, the CSF protein elevation may take several days to develop and may not be present at the time of an initial lumbar puncture.

The second diagnostic test consists of nerve conduction studies that usually reveal a marked reduction in the amplitude of compound motor action potentials, a result of extensive conduction block at nodes of Ranvier. In addition, there may be a decrease in motor conduction velocities and a marked increase in distal motor latencies, all due to patchy demyelination of motor axons.

4.8 Investigations

A lumbar puncture (LP) produced clear CSF with 1 lymphocyte/μL, 0 red blood cells, glucose 2.6 mmol/L, protein 1.26 g/L (normal <0.45 g/L).

Motor nerve conduction studies revealed a 90% reduction in the amplitude of compound motor action potentials, normal conduction velocities in the main nerve trunks and distal latencies 2–3 times normal.

A stool culture was positive for Campylobacter jejuni.

Pulmonary function studies at the bedside revealed a 75% reduction in vital capacity and forced expiratory volume.

4.9 Treatment/Outcome

Fifi was admitted to the intensive care unit for close monitoring. Increasing respiratory distress and declining pulmonary function necessitated airway intubation and mechanical ventilation within 12 hours of admission.

Within 48 hours of admission, Fifi demonstrated grade 1/5 muscle strength proximally and distally in all four limbs. She was given two infusions of intravenous immunoglobulin 1 g/kg on succeeding days beginning the day after admission. The role of the immunoglobulin infusion was to obstruct access of the circulating pathogenic antibodies to their peripheral nerve targets. Beginning about 1 week after admission, Fifi began to improve steadily and was extubated on the 15th day.

During the first few days in the ICU, Fifi developed pronounced tachycardia due to antibody-mediated blockade of the vagal (parasympathetic) preganglionic poorly myelinated nerve fibers supplying the heart; this problem was successfully treated with a β-adrenergic blocking agent.

Throughout the ICU admission, Fifi had an aggressive physiotherapy program, at first with passive range of movement and the use of ankle and wrist splints to prevent contraction deformities. At the same time, she was frequently repositioned to prevent the development of bed sores.

Once extubated and stabilized, Fifi's program of physiotherapy was intensified to work on re-establishing independent sitting, then ambulation. She continued to improve and was walking without help 6 weeks after admission. By the time of discharge at 8 weeks, she was functioning independently, her only deficit being a mild foot drop gait.

Related Cases

To improve the localization and clinical skills introduced and developed in this chapter, please work your way through the cases linked to this chapter (located on the text DVD and Web site).

Suggested Readings

Hadden, R.D.M. et al. Preceding infections, immune factors, and outcome in Guillain-Barré syndrome. *Neurology* 56 (2001): 758–765.

Hughes, R.A.C. et al. Practice parameter: Immunotherapy for Guillain-Barré syndrome. *Neurology* 61 (2003): 736–740.

Oomes, P.G. et al. Anti-GM1 IgG antibodies and campylobacteria in Guillain-Barré syndrome: Evidence of molecular mimicry. *Ann. Neurol.* 38 (1995): 170–175.

Penfield, W. and L. Roberts, *Speech and Brain Mechanisms.* Princeton, NJ: Princeton University Press, 1959.

Prineas, J.W. Pathology of the Guillain-Barré syndrome. *Ann. Neurol. Suppl.* 9 (1981): 6–19.

Yuki, N. et al. Association of *Campylobacter jejuni* serotype with antiganglioside antibody in Guillain-Barré syndrome and Fisher's syndrome. *Ann. Neurol.* 42 (1997): 28–33.

Web Sites

Anatomy of the movement control system: http://www.neuro.wustl.edu/neuromuscular/index.html (University of Washington)

Guillain-Barré syndrome: http://www.jsmarcussen.com/gbs/uk/profs.htm

Chapter 5

Cletus

CLETUS

Cletus McCool is Crash's country cousin. The son of Crash's Uncle Pierre, Cletus lives in the bush of North Fountainbleau County.

One Saturday night Cletus and his buddies were peeling up and down County Road 1, the only one in the county, in their 1969 Duster. Ajutor, the driver of the old rattletrap, was traveling too fast and missed the corner at the end of the road, hitting a large boulder of four-billion-year-old pre-Cambrian rock, causing the car to come to a sudden stop. Cletus was in the back seat with no seatbelt and never knew what hit him.

At the scene of the accident, the paramedics found Cletus to be unresponsive to commands or questions, but moving all limbs to pain; his blood pressure was very low. He was transported to the nearest university hospital.

The next thing that Cletus remembered was that he was in a hospital room with lots of beeping sounds, a tube down his throat and a big pain on the left side of his chest.

"Cletus, squeeze my fingers," said the nurse. He squeezed her fingers. "Cletus, move your legs and feet." What legs and feet? ... He could not feel them, let alone move them. He suddenly became terrified and started to grab at his tubes: then the lights went out again.

Objectives

- To learn the principal anatomic components of the spinal cord
- To learn the anatomy of the vascular supply of the spinal cord
- To learn the effects of damage to these spinal cord components and their clinical manifestations
- To learn the major spinal cord syndromes and their clinical manifestations
- To become sensitized to the medical, physical and psychosocial implications of spinal cord injury

5.1 Clinical Data Extraction

Before proceeding with this chapter, you should first carefully review Cletus's story in order to extract key information concerning the history of the illness and the physical examination. On the DVD accompanying this text and the Web site you will find a sequence of worksheets to assist you in this process.

The clinical findings as described above allow us to localize the lesion within the nervous system. The other piece of important information, of which Cletus was not aware, is the presence and characteristics of the deep tendon and superficial reflexes. For the sake of discussion, let us assume that the reflexes were normal (2+) in the upper limbs and absent in the lower limbs. The plantar responses were essentially absent: neither flexor nor extensor. Cletus was also unaware that his bladder was being drained by a urinary catheter as he had no feeling of bladder fullness or control. With this information, the neurological examination section of the Extended Localization Worksheet can be filled out.

The History Worksheet is fairly straightforward: a sudden onset of lower limb weakness in the context of an automobile accident suggests that the etiology is probably trauma. Other things to consider are collateral etiologies that occur as a result of trauma, such as systemic hypotension due to blood loss or hypoxia due to hemo/pneumothorax. However, follow the localization and etiology protocols carefully; you might be in for a surprise.

5.2 The Main Clinical Points

- Cletus was involved in a severe traumatic deceleration injury.
- Cletus was hypotensive but noted to be moving all four limbs at the scene of the accident.
- Cletus had a surgical procedure performed to his chest to repair a ruptured thoracic aorta.
- Following the procedure, Cletus was found to have lower limb paralysis and sensory loss to pinprick but not to vibration.
- Cletus also had no sensation of or voluntary control of bowel or bladder function.
- His mental status, cranial nerves and upper limb power, sensation and reflexes were all normal.

When he woke up, there was this geeky looking old guy with blond hair at his bedside, making him do a lot of weird stuff. "Cletus, follow the pen with your eyes; wrinkle your forehead; smile; stick out your tongue." He could do all these things OK.

He was able to squeeze the doctor's fingers, and bend and straighten his arms at the elbows and shoulders. When asked to move his legs and feet again, he was unable to feel or move his lower limbs. The examiner took out a safety pin and tested him on the face, arms, trunk, and legs and around the belly button. He could feel it on both sides down to the belly button but not below, on either side.

Then a strange thing happened—the guy took out this tuning fork thing that vibrated and tested him over his body. Not only could he feel the vibration over his face and arms, but also over his toes, ankles and knees!

Then he could hear them talking in the background in muffled tones. "Time is going to tell ..." he heard the surgeon talking to his mother. "The operation in his chest went very well, we had to put a patch on his aorta ... the plastic tube in the side of his chest should come out in a couple of days ... you will have to ask the neurologist whether he will walk again." Cletus could hear his mother sobbing at the end of the ICU bed.

In summary, Cletus has suffered a neurological injury related to power and sensation in the lower limbs, including loss of bowel and bladder control; these deficits developed after the surgical procedure to correct a ruptured thoracic aorta.

5.3 Relevant Neuroanatomy

The spinal cord consists of a central gray matter H-shaped column surrounded by white matter tracts that transmit information from either the peripheral nerves to the brain or information with respect to motor and autonomic function from the brain to the spinal cord. Figure 5.1 shows the anatomy of the spinal cord, its blood supply, the dura and the spinal nerves, which exit from the spinal cord at each level. Figure 5.2 shows an MRI of the spine.

There are many different tracts, ascending and descending, with differing functions; they are often not easily tested clinically. Three of the most important tracts clinically will be highlighted here. These are the the posterior columns, spinothalamic tract and the corticospinal tract as discussed previously in Chapters 1 and 2.

The gray matter column is responsible for the interface between the peripheral nerves and the peripheral nerve input or output at each segmental level, as well as connections with either ascending or descending tracts. The gray matter is functionally divided into dorsal and ventral parts. The dorsal portion is where the sensory input from the peripheral nerves interfaces with the spinal cord either to perform local activity such as reflex withdrawal or to transmit sensory information via either the spinothalamic tract for pain and temperature or the posterior columns for vibration and proprioception.

The peripheral nerve fibers serving proprioception and vibration enter the spinal cord through the dorsal gray matter: these fibers do not synapse here but ascend to the lower medulla via a medially placed fasciculus for the legs, or via a laterally placed fasciculus for the arms, to terminate in the posterior column nuclei in the lower medulla. Second-order neurons then cross to the opposite side and ascend to the thalamus. Third-order neurons then transmit this information to the somatosensory area of the postcentral gyrus of the parietal lobe (see Figure 2.6A).

The peripheral nerve fibers serving pain and temperature enter through the dorsal rami to terminate and synapse with interneurons and second-order neurons, which cross in the anterior commissure (anterior to the central canal) and ascend as the (lateral) spinothalamic tract to the lateral thalamus on the opposite side of the body from the sensation. Third-order neurons transmit the sensory information from the lateral thalamus to the somatosensory cortex in the parietal lobe (see Figure 2.6B).

The corticospinal tract is the major descending tract, which transmits information with respect to motor movement to the spinal cord. The corticospinal tract originates mainly in the frontal precentral gyrus with some contribution from the parietal cortex. These fibers descend in

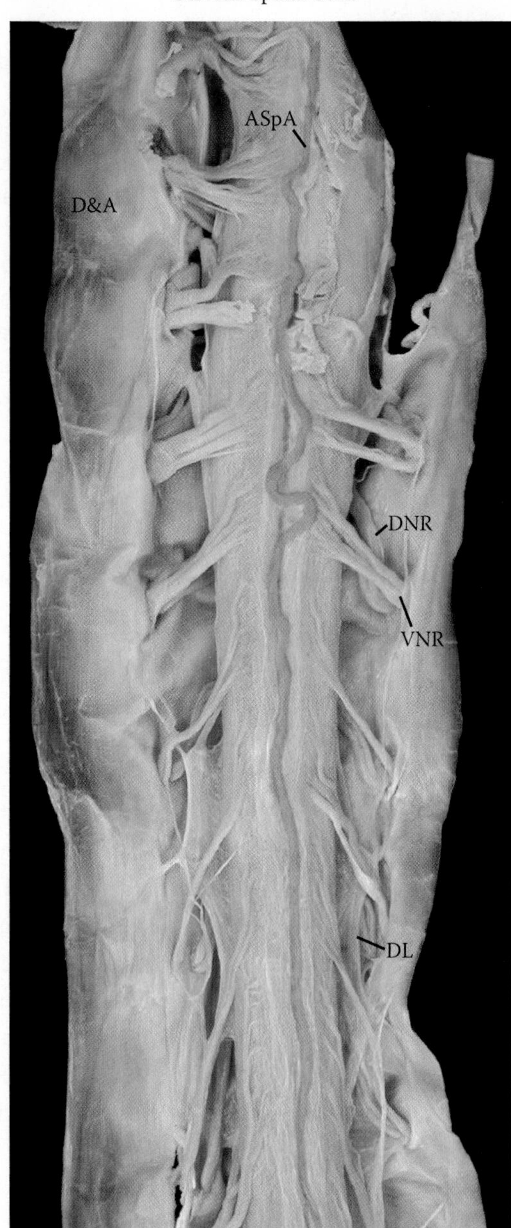

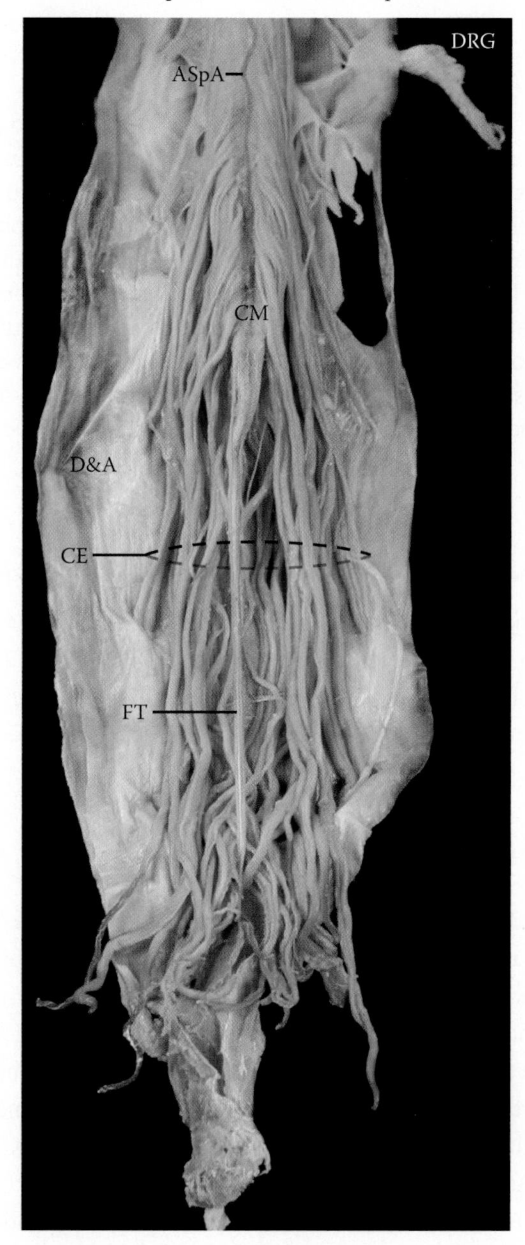

FIGURE 5.1 Photographic views of the spinal cord: The cervical level (on the left), includes the anterior spinal artery; note the dorsal and ventral roots (DNR and VNR), as well as the denticulate ligaments (DL). The sacral spinal cord (on the right) forms the tapering conus medullaris (CM); the lumbar cistern encloses the spinal roots, forming the cauda equina (CE), as well as the filum terminale (FT).

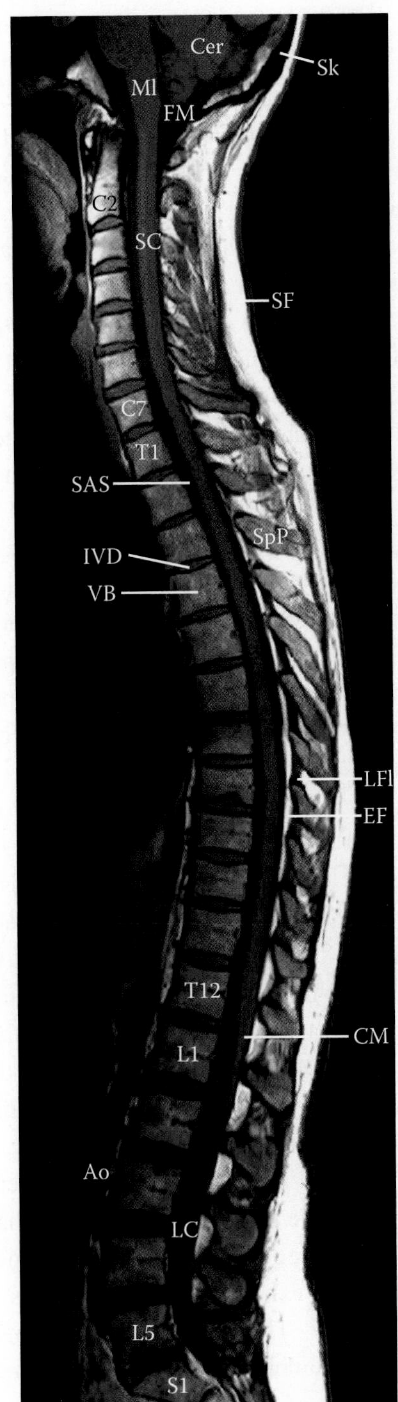

Sk - Skull
FM - Foramen Magnum
Cer - Cerebellum
Ml - Medulla

SC - Spinal Cord
SF - Subcutaneous Fat
SAS - Subarachnoid Space
IVD - Intervertebral Disc
VB - Vertebral Body
SpP - Spinous Process

LFl - Ligamentum Flavum
EF - Epidural Fat
CM - Conus Medullaris
Ao - Aorta
LC - Lumbar Cistern

FIGURE 5.2 MRI of the spinal column, showing the lower portion of the skull with the foramen magnum (FM), the spinal cord, the vertebra and soft tissue (fat). The C2 vertebra has a typical triangular configuration; other landmark vertebra (C7, T1, T12, L1, and L5, S1) are also numbered. Note the conus medullaris (CM) and lumbar cistern (LC). The ligamentum flavum (LFl) and epidural fat (EF) can also be seen.

the internal capsule, through the cerebral peduncles in the brainstem and anterior white matter of the pons. In the lower medulla, the majority of these fibers cross in a structure called the pyramidal decussation to descend in the lateral funiculus of the spinal cord as the lateral corticospinal tract, giving inputs to the lower motor anterior horn cells at each level. These lower motor neurons in turn send their axons to limb muscles to perform voluntary motor activity (see Figure 2.5).

Corticoreticular tracts, which originate in the cortex, descend to reticular formation in the lower brainstem. These nuclei then give outputs that descend to the spinal cord and modulate muscle tone and reflexes. These are called reticulospinal tracts, discussed in Chapter 7.

Disruption of any of these tracts, plus the spinal gray matter, can therefore cause symptoms and signs of sensory and motor deficits, depending on which tracts are involved and at what level.

The blood supply of the human spinal cord, unlike the spinal cords of other animals, which have branches emanating from each spinal level, is relatively tenuous. The vertebral arteries join below the origin of the basilar artery to form the anterior spinal artery from above. The great radicular artery of Adamkiewicz joins the anterior spinal artery supplying the spinal cord at a variable level between T8 and L1 as a single large branch originating from the thoracic aorta. The spinal cord is supplied by a large anterior branch whose territory is the anterior two-thirds of the cord, and by two smaller posterior spinal arteries, which have their origins at segmental levels and supply the posterior third of the spinal cord (see Figure 5.4).

5.4 Localization Process

Using the Extended Localization Matrix worksheet, the major findings to be entered are weakness, absent reflexes, sensory loss to pinprick and preserved vibration and proprioception in the lower limbs in a symmetrical fashion. Upper limb motor, reflex, sensory function, cranial nerves and mental status were all normal. The sensory deficit also involved the trunk below the level of the umbilicus. The plantar reflex was absent, but this is of little localizing value.

Muscle disease and disease of the neuromuscular junction would be unlikely in this context as these two localizations do not cause sensory findings. A crush injury to the muscles in both legs could cause weakness but there would be lots of pain.

An injury to the peripheral nerves might be another possibility. The problem with this is that Cletus's findings are bilateral below the umbilicus. Injury to peripheral nerves would have to be widespread and symmetrical below a specific delineated level, a very unlikely possibility.

The presence of a motor and sensory level below the umbilicus would be consistent with a spinal cord injury about T10 (see Table 2.1), but what about the reflexes? Shouldn't injury to the corticospinal tracts cause hyperreflexia?

In order to answer these questions we have to refer back to the diagram of the spinal cord with the spinal nerves, gray matter and the three major tracts.

Figure 5.3 shows the location of the spinothalamic tracts (which carry pain and temperature), the posterior columns (which carry vibration and proprioception) and the corticospinal tracts, which carry motor information to the lower motor neuron of the spinal cord to control movement (see also Chapter 2).

The clinical examination of Cletus indicated weakness of all muscles on both sides below the umbilicus. This finding is consistent with disruption of the corticospinal tracts at about the T10 level; T10 motor nerves supply abdominal muscles around the level of the umbilicus. Lower limb reflexes are often absent in the situation of an acute spinal cord injury due the presence of a phenomenon called *spinal shock* in which the neurons of the spinal cord are diffusely dysfunctional or stunned for an average of seven to ten days after injury.

In this case, the gray matter of the spinal cord served by the artery of Adamkiewicz has become infarcted causing the lower motor neuron output to be affected, leading to weakness and absent reflexes.

The finding of loss of sensation to pinprick below the umbilicus needs to be localized. The sensory fibers for pain and temperature originate in the periphery and enter the dorsal aspect of the spinal cord in the substantial gelatinosa. There the fibers synapse onto a secondary neuron, which projects to the opposite side through the anterior commissure of the spinal cord in front of the central canal to form the spinothalamic tract on the opposite side. The spinothalamic tract ascends to terminate in the ventroposterolateral nucleus of the thalamus. From there a third-order neuron connects the thalamus to the somatosensory cortex

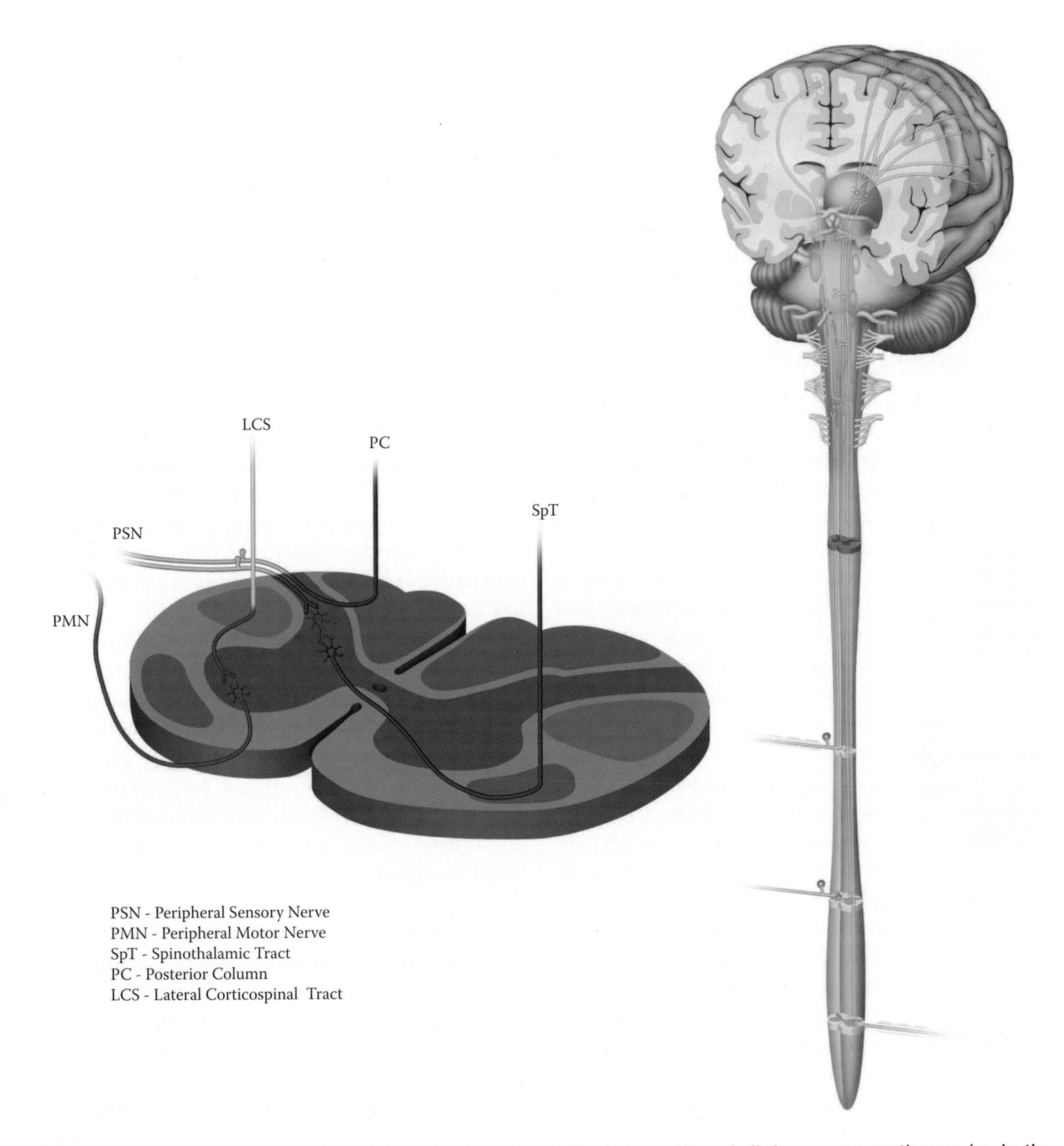

FIGURE 5.3 Complete transection of the spinal cord: note the interruption of all the sensory pathways (on both sides) and the lateral corticospinal tracts bilaterally.

in the parietal lobe (see Figure 12.3). To localize this finding there would have to be widespread symmetric damage to all peripheral nerves below the umbilicus. More likely the findings can be explained by transection of the spinothalamic tracts on both sides at the level of the lower thoracic spinal cord.

The last major finding is that the sensation for vibration and proprioception has been preserved. This suggests that the nerves and their spinal cord projections serving these modalities have been spared. This would suggest that the posterior columns of the spinal cord and their upward projections are intact.

Therefore, the localization for this combination of findings (including bilateral destruction of the corticospinal tracts and destruction of the spinothalamic tracts) suggests that the damage has been confined to the anterior aspect of the spinal cord, with sparing of the posterior columns, the level of the cord lesion being at about T10.

A complete transection of the spinal cord at that level would result in total loss of function of the corticospinal and spinothalamic tracts, and of the posterior columns on both sides. This is not the situation in the case of Cletus.

5.4.1 Spinal Cord Syndromes

The major clinical findings in spinal cord injury relate to the relative degree of damage to the three tracts and the gray matter, as well as the level at which the damage occurs. Using the Expanded Localization Matrix, we see that the probable major clinical findings are loss of sensation to pinprick and temperature; loss of sensation to vibration, proprioception and discriminative touch; weakness, reflex changes and changes in muscle tone. These findings need to be described in reference to a given dermatomal level as well as whether the findings are symmetric and replicate in both anterior and posterior dermatomes. This means that the sensory findings should correspond to equivalent levels when the patient is tested either supine or prone. Figure 13.4 provides a dermatomal map and standardized method for documenting spinal cord deficits.

Damage to the spinothalamic tract on one side will cause loss of sensation to pinprick and temperature below the lesion on the opposite side of the body. Damage to the posterior columns will cause loss of sensation for vibration, proprioception and discriminative touch on the same side below the lesion. Damage to the corticospinal tract will cause weakness, hyperreflexia and spasticity below the level of the lesion on the same side, with the exception that hyperreflexia and spasticity may not be present for seven to ten days after an acute injury due to spinal shock.

Injury to just the gray matter at a given level will cause loss of all sensation and a lower motor neuron pattern of weakness with absent reflexes at the levels of damaged gray matter.

Several terms are used to describe spinal cord injury and level. Paraplegia refers to loss of function of the legs. Therefore, the lesion can be anywhere from a cord level of L2 to the cervical levels. Quadriplegia refers to the results of a spinal cord injury above C4, with paralysis of both arms and legs. Pentaplegia is a term sometimes used for high quadriplegics in which the roots of C2, 3, 4 are affected, resulting in paralysis of all four limbs and of the diaphragm. It is more precise to state the level of neurological injury, as these terms can be ambiguous. For instance a complete C6 level cord injury results in lower limb paralysis and upper limb weakness below C6, in other words a partial quadriplegia.

Four major spinal cord syndromes are described to illustrate this anatomy in action.

5.4.1.1 Complete Transection of the Spinal Cord This syndrome is the easiest to understand in that it consists of complete destruction of all spinal cord elements at a given level. Therefore, a complete spinal cord transection at T10 will cause paralysis; hyperreflexia; extensor plantar responses; spasticity; loss of control and sensation of bowel and bladder function; and loss of sensation for pinprick, temperature, vibration; and proprioception on both sides at and below that level of the spinal cord. All tracts are involved and all are damaged (Figure 5.3).

Depending on the cause, speed and intensity of the trauma that causes the transection, spinal shock may occur—typically with more acute trauma—during which time there is paralysis of the limbs below the level of the transection but spasticity and deep tendon reflexes can be decreased or absent. After recovery from spinal shock, hyperreflexia, increased tone, and extensor plantar responses appear.

5.4.1.2 Anterior Spinal Artery Syndrome This is what Cletus has. He had an interruption of the blood supply to his spinal cord in the anterior spinal artery territory due to occlusion of its major feeding artery in the lower thoracic region, the artery of Adamkiewicz, while his aorta was being clamped. This caused an infarction to the anterior two-thirds of his cord, damaging the spinal cord gray matter, spinothalamic tracts and corticospinal tracts, with sparing of the posterior columns. This explains why he could not move his legs or feel a pinprick below the umbilicus but could feel the vibration sensation in his toes (Figure 5.4).

The extent of the spinal cord infarct can be over several segments such that the clinical level of deficit may not represent the actual entry level of the artery of Adamkiewicz.

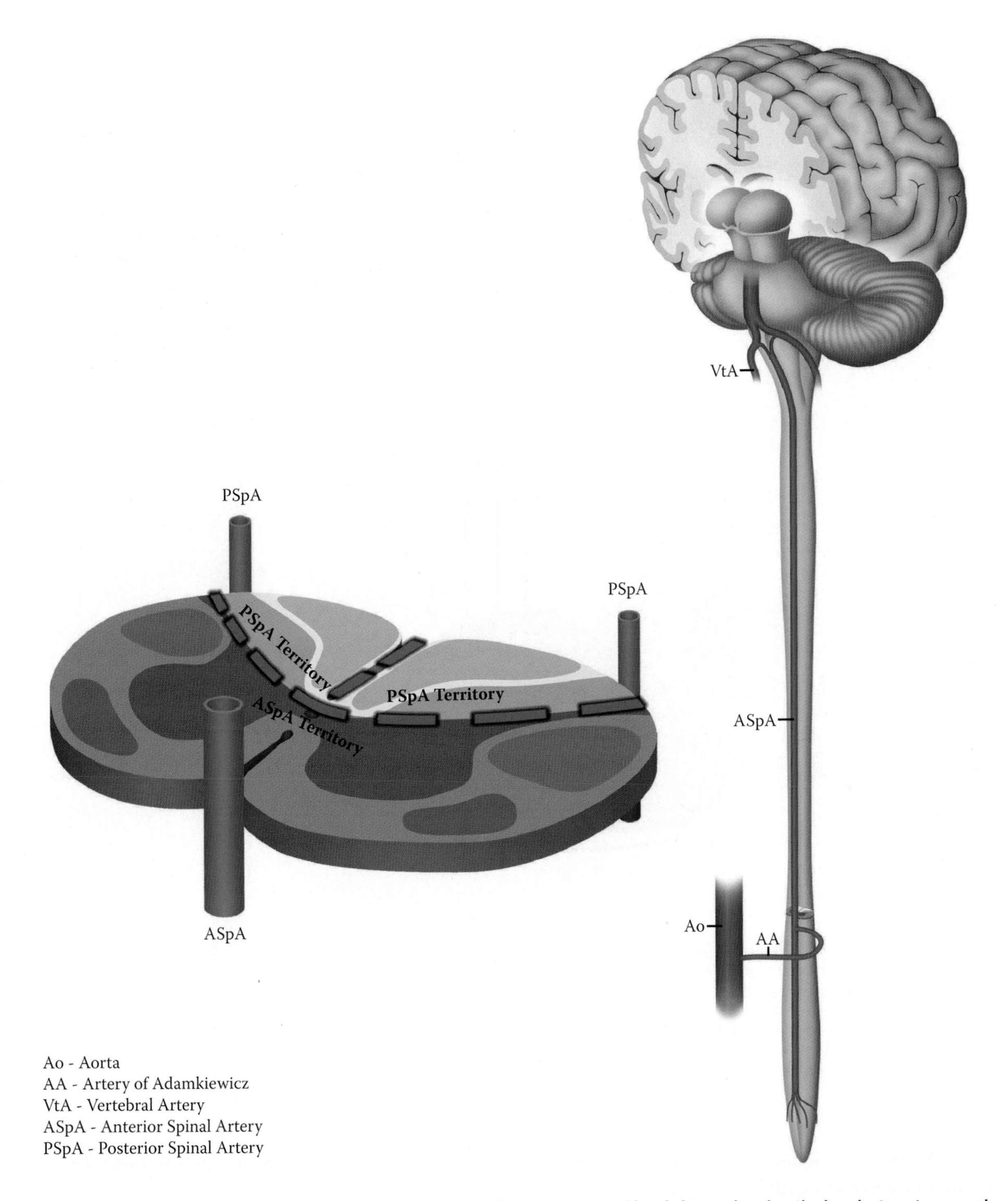

FIGURE 5.4 Anterior spinal artery syndrome: note the interruption of both lateral spinothalamic tracts, sparing of the posterior columns, and interruption of the lateral corticospinal tracts bilaterally.

5.4.1.3 Brown-Séquard Syndrome This syndrome, named after a nineteenth century French-American neurologist, occurs when there is damage to one side of the spinal cord with differential deficits resulting from the crossing of the spinothalamic tract in the spinal cord (Figure 5.5).

For example, if the right side of the spinal cord is destroyed, the corticospinal output to the right side will be damaged as well as the posterior columns on the same side. Therefore, the patient will lose motor power, and sensation for vibration, proprioception and discriminative touch on the

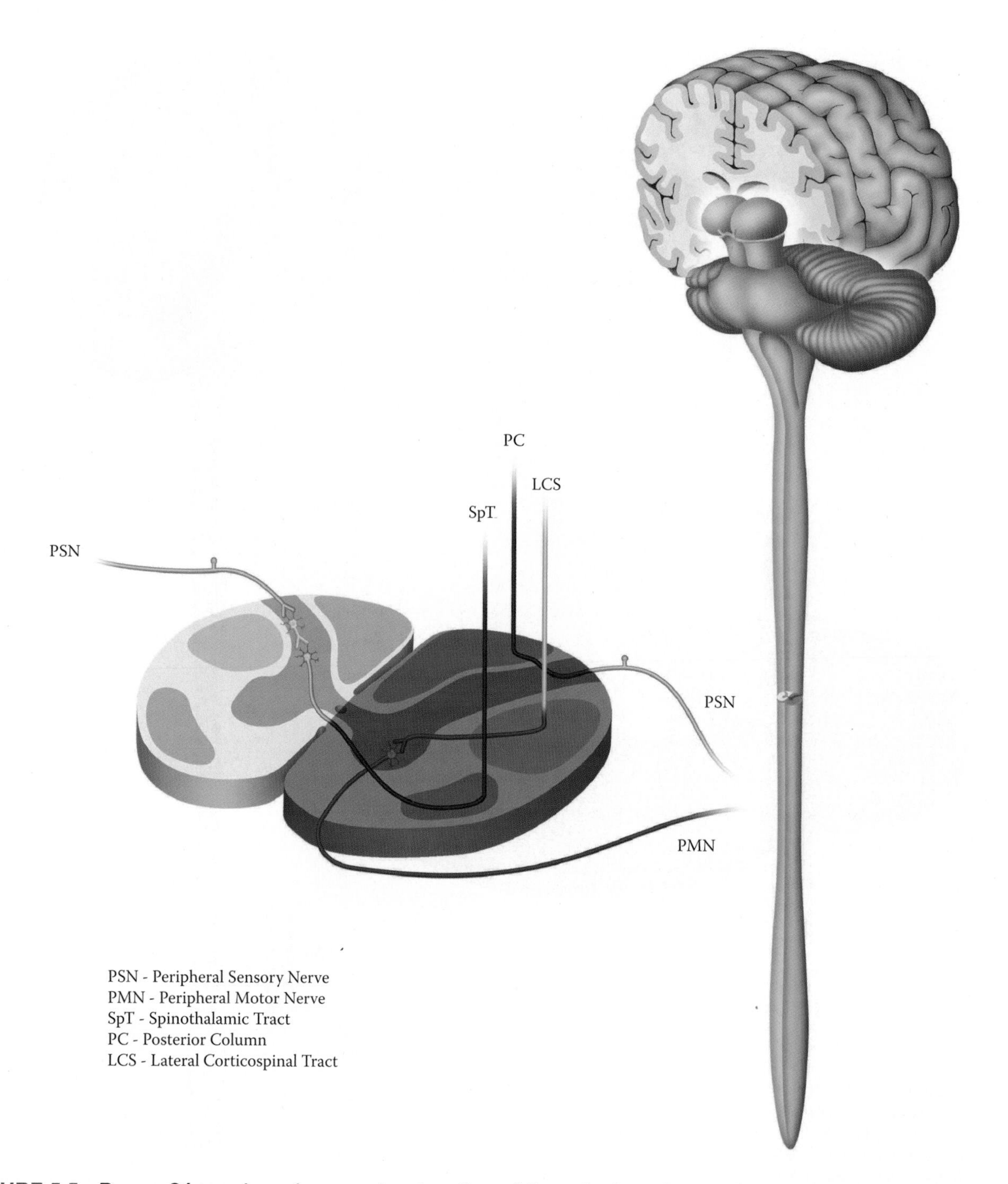

FIGURE 5.5 Brown-Séquard syndrome, a hemisection of the spinal cord: note the tracts interrupted and which side of the body would be affected.

right side below the level of damage. In contrast, the sensation for pinprick and temperature is lost on the left side as the spinothalamic tract serving the opposite side has already crossed in the spinal cord below the level of the damage and therefore would be affected by a right-sided lesion.

The usual cause for this clinical situation is a tumor, usually metastatic, which grows from the lateral side of the spinal canal to compress the spinal cord and the structures within it.

5.4.1.4 Syringomyelia This syndrome results from the expansion of the central canal of the spinal cord, a phenomenon known as a syrinx. The central canal is usually a small conduit for spinal fluid that extends down from the fourth ventricle in the medulla to provide CSF nutrition to the center of the cord. Under circumstances of increased pressure in the spinal canal, such as from abnormal CSF flow at the foramen magnum by a congenital descent of the cerebellar tonsils (a Chiari malformation) or by local trauma or tumors, the canal can undergo expansion and compress the adjacent structures in the spinal cord. The first structure that usually gets compressed is the anterior white matter commissure, which carries the crossing pain and temperature pathway from its second-order neuron cell body to connect to the spinothalamic tract in the anterolateral aspect of the spinal cord on the opposite side. The effect of compression of this commissure (carrying pain and temperature information from both sides, in the process of crossing each other) is to destroy the perception of pain and temperature on both sides of the body over the extent of the dilatation of the syrinx. The posterior columns and corticospinal tracts are usually not affected unless the syrinx becomes very large (Figure 5.6).

Usually syringomyelia occurs in the lower cervical and upper thoracic level; thus the clinical effect is a suspended sensory level of decreased sensation to pain and temperature in a cape-like distribution involving the upper trunk and the arms. Historical descriptions of this condition describe female patients with this condition as having a form of hysteria in that they could not feel pain in the area of the chest and breasts but could feel fine touch, which would be carried by the posterior columns.

Syrinx formation at the level of the lower brainstem is known as syringobulbia. This dilated fluid cavity in the brainstem will interfere with cranial nerve, motor and sensory function depending on the nuclei or tracts that are compressed.

5.5 Etiology: Cletus's Disease Process

The Expanded Localization Matrix, History Worksheet, and Etiology Matrix for this case tell us that this is an acute injury to the spinal cord following severe trauma as a result of the motor vehicle accident.

5.5.1 *Paroxysmal Disorders*

There are several paroxysmal disorders of the spinal cord that cause abnormal movements secondary to uncontrolled discharges from anterior horn cells. A typical example is neuromyotonia in which there is uncontrolled stiffening of the muscles of the arms and legs, resulting in so-called "stiff person syndrome" or Issac syndrome. This is believed to be caused by an autoimmune attack on the spinal cord. These disorders tend to be acute in onset but are recurrent and eventually become chronic.

Spinal myoclonus consists of brief shock-like jerks usually in just a few muscle groups, typically due to segmental damage to the spinal cord.

The history and neurological findings in Cletus's case are in keeping with acute loss of function of two tracts of the spinal cord after an incident of trauma and surgery.

5.5.2 *Traumatic Spinal Cord Injury*

The commonest cause of acute complete spinal cord injury is trauma.

When the spinal cord is disrupted to the point that the various gray matter and white matter tracts are crushed or lose their anatomical proximity to their blood supply, complete destruction of the spinal cord occurs at that level. This leads to a bilateral loss of motor and sensory function below the level of the lesion, as well as loss of bowel and bladder control. Spontaneous recovery is unlikely and the disability is profound and permanent.

In the case of Cletus, there is preservation of function of one spinal cord tract bilaterally, the posterior columns. This could indicate a possible incomplete spinal cord transection due to trauma (see Figure 5.3).

A more common and chronic form of spinal cord trauma is caused by degenerative changes, usually from osteoarthritis in the vertebral bodies and intervertebral disks. These degenerative processes lead to prolapse of intervertebral disks into the spinal canal and remodeling of the bone around joints, which can also lead to impingement on the vertebral foramina (and thus one or more spinal nerves) and also the spinal canal, leading to compression of the spinal cord itself. This condition is known as *spinal stenosis*, which can occur at any level of the spinal axis. If the spinal stenosis is in the cervical or thoracic spine it can lead to spinal cord compression or myelopathy. If this occurs in the lumbar spine below the termination of the spinal cord, it can cause compression of the nerves forming the cauda equina, leading to pain, sensory loss, and weakness of the legs as well as difficulty with bladder control and bowel continence.

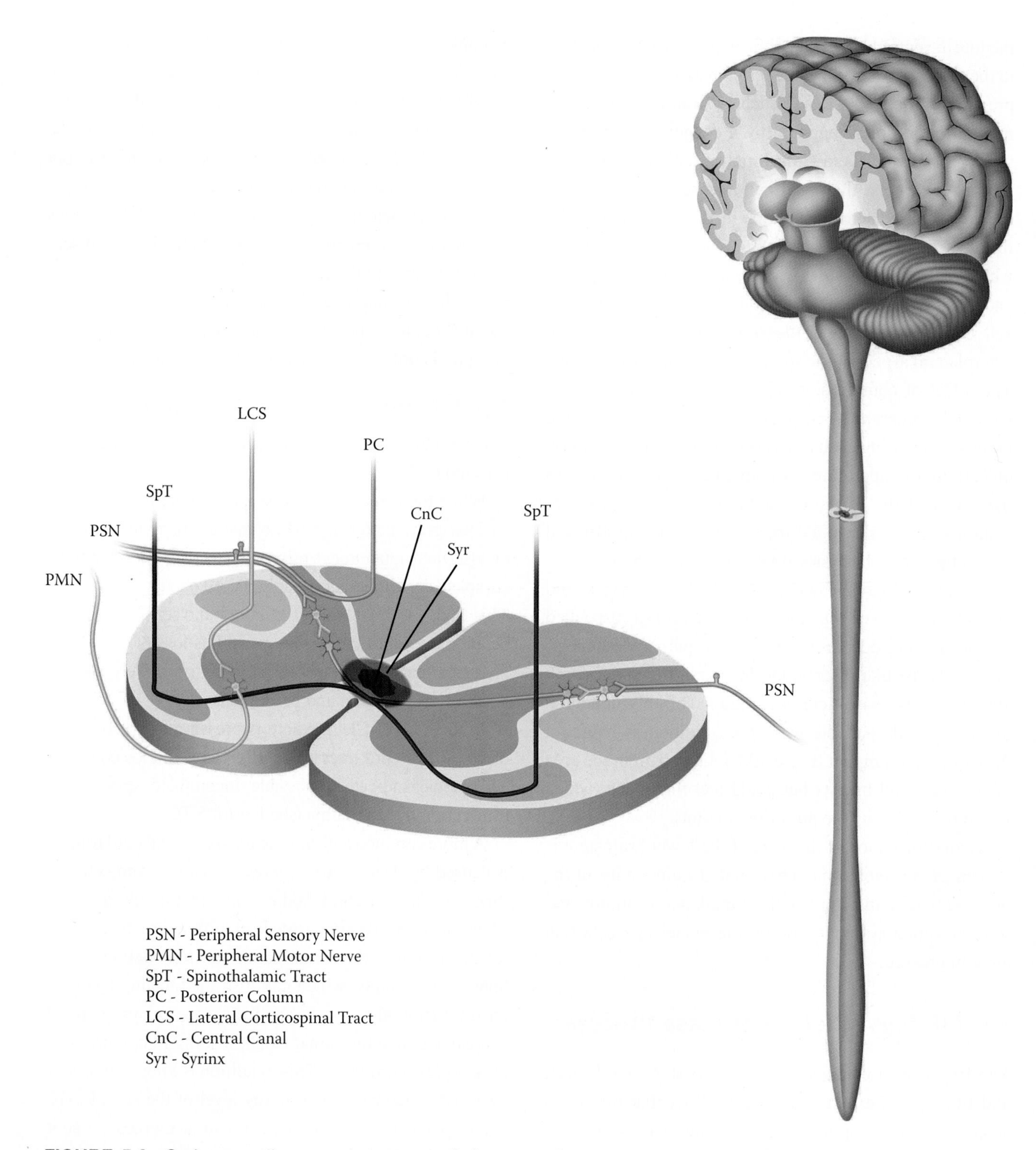

FIGURE 5.6 Syringomyelia, an enlargement of the central canal; note the interruption of the crossing fibers carrying pain and temperature, involving both sides.

5.5.3 Vascular Injury

In the localization exercise, we have already noted that Cletus's deficits fit best with an anterior spinal artery syndrome at or below the level of entry of the artery of Adamkiewicz. Occlusion of the artery of Adamkiewicz leads to infarction of the anterior two thirds of spinal cord at several levels above and below the cord level of the artery of Adamkiewicz. The infarction pattern will include the spinothalamic tracts, the gray matter and the lateral funiculi containing the corticospinal tracts.

If Cletus had been hypotensive and/or hypoxic at the scene of the accident then there may be reason to suspect that his injury might have been due to hypoperfusion of the cord. Under such circumstances there is infarction in the watershed (or arterial border zone) region of the spinal cord. Arterial border zone areas are located where the regional blood supply comes from different arteries and the ischemic injury is in the border zone between those two arteries. In the spinal cord such a border zone is usually located in the cervical level at C6–C7 where the border zone from the vertebral supply of the anterior spinal artery meets the supply from the anterior spinal artery segment which originated from the great radicular artery of Adamkiewcz.

Thus, with hypotension-induced border zone infarcts, the damage is usually confined to the low cervical or high thoracic cord above the entry point of the artery of Adamkiewcz. If the cord had been damaged due to Cletus's hypotensive state, however, he would not have been moving all limbs to pain at the scene of the accident.

Referring back to the conversation between the physician and Cletus's mother at the bedside gives us information that he had to have emergency vascular surgery to repair a partially ruptured thoracic aorta. This was caused by the deceleration trauma when the car hit the rock face and Cletus (unrestrained in the back seat) was subject to significant G forces which pulled his thoracic aorta from its chest wall attachment. His severe systemic hypotension resulted from extensive bleeding from the torn, leaking aorta into the adjacent chest cavity.

During the emergency repair of Cletus's thoracic aorta, the surgeons were obliged to temporarily clamp his thoracic aorta above the artery of Adamkiewcz in order to put a patch on the ruptured area.

The effect of clamping the aorta above the artery of Adamkiewcz caused interruption of blood supply to the spinal cord at the level of entry of this artery. The blood supply to the anterior two-thirds of the spinal cord was interrupted resulting in infarction of this region.

Other vascular causes could include direct trauma causing rupture or occlusion of the artery of Adamkiewcz.

Thus, in terms of localization and etiological diagnosis, Cletus most probably suffered an infarction to the anterior two thirds of his spinal cord at the T10 level, involving several segments, resulting in loss of function of the corticospinal tracts and spinothalamic tracts on both sides.

The prognosis for this type of injury can be guardedly optimistic, in that part of the functional loss may be secondary to reversible ischemic damage rather than complete infarction. As well, the damaged tracts are mainly white matter tissue, which have some capability for remyelination. The recovery, if it does occur, can be very slow—up to 24 months. The outcome would have been significantly worse had he suffered a fracture dislocation of the spine at the same level, with complete cord transection.

5.5.4 Toxic Injury

Various chronic toxic exposures can affect the spinal cord; examples include the heavy metals mercury, lead, and arsenic. These disorders have a chronic course, usually with a combined picture of concurrent progressive cognitive and motor difficulties. A history of exposure is usually related to geographic factors of environmental pollution, such as mercury poisoning in the Minamata region of Japan and in the aboriginal peoples of the Wabigoon-English River area in Northern Ontario.

5.5.5 Spinal Cord Infection

HIV infection causes a vitamin B12-like myelopathy despite normal levels of vitamin B12. This is thought to be due to abnormal metabolism of vitamin B12 and its metabolites within the spinal cord.

Other infections of the spinal cord include such agents as West Nile virus, polio, and Lyme disease.

West Nile is a viral infection transmitted from birds to human by mosquitoes causing a combined viral encephalitis and myelitis. Both gray matter and white matter elements of the spinal cord can be affected.

Poliomyelitis is a viral infection borne through contaminated water, which causes necrosis of the anterior horn cells of the spinal cord giving mainly asymmetrical segmental loss of motor function without sensory findings (see also Chapter 4). It has essentially been eliminated through mass population immunization programs.

Lyme disease is caused by a bacterial spirochete transmitted to humans from deer by a specific tick. This disorder has three distinct phases: an initial cutaneous expanding rash at the site of the bite, followed by a delayed phase of polyarthritis, and then by involvement of the central nervous system.

Infectious myelitis due to herpes zoster can give a diffuse myelitis; this is often associated with disorders that interfere with cell-mediated immunity.

5.5.6 Metabolic Injury

Metabolic disturbances that may affect the spinal cord include loss of the basic nutrients (oxygen and glucose); these disorders usually cause dysfunction in the brain long before the spinal cord.

Low levels of vitamin B12 cause degeneration of the posterior columns, corticospinal tracts and the large fiber peripheral nerves to cause both a neuropathy and myelopathy. This condition is called subacute combined degeneration; the term *combined* refers to the combined degeneration of both the posterior columns and the corticospinal tracts. Replacement of vitamin B12 and folate often results in prompt improvement of signs and symptoms. The clinical manifestations are progressive weakness and loss of sensation usually beginning in the lower limbs, with findings on exam of loss of sensation to vibration and proprioception, weakness, and spasticity. Hyperreflexia is variable depending on the relative contribution of damage from the neuropathy versus the myelopathy.

5.5.7 Spinal Cord Inflammation/Autoimmune

Subacute causes include inflammatory conditions of the spinal cord such as multiple sclerosis and transverse myelitis.

Multiple sclerosis causes spinal cord damage through recurrent attacks, resulting in demyelinating plaques in the cord white matter. Disability depends on the size, location and cumulative number of plaques.

Transverse myelitis is a disorder of subacute demyelination following primary infections—usually in the gastrointestinal tract or respiratory system—by various viruses or by mycoplasma. The mechanism is similar to the peripheral nerve demyelination associated with Guillain-Barré syndrome (see Chapter 4); a similar type of postinfectious demyelination can also affect the spinal cord. This is another example of molecular mimicry in that the immune system is primed to attack an outside invader such as a virus which happens to have an antigenic similarity with a tissue in the nervous system, in this instance the myelin of the spinal cord axon tracts.

Autoimmune or systemic vasculitis can affect the spinal cord. These disorders usually have significant systemic signs and symptoms and also affect the brain at the same time.

5.5.8 Neoplastic Disorders

Tumors of the spinal cord may be intrinsic to the cord (such as astrocytomas) or extra-axial (meningiomas and metastatic tumors). These usually present with subacute or chronic time courses with weakness or sensory loss depending on which tracts or gray matter structures are damaged first.

Paraneoplastic degenerative of the spinal cord can occur in association with a remote carcinoma usually of the lung, resulting in an immune-mediated necrotizing spinal cord degeneration.

5.5.9 Degenerative Disorders

Chronic conditions affecting the spinal cord are usually the result of damage from preexisting acute and subacute conditions. Chronic degenerative conditions are few in number, the most important example being amyotrophic lateral sclerosis (ALS). Degenerative diseases often overlap as late-onset genetic disorders (see below).

ALS is a progressive degenerative disorder affecting both the upper and lower motor neurons in combination and at varying intensities. It is important to understand the term *ALS*. *Amyotrophy* (A) refers to muscle weakness not due to muscle disease. This represents the lower motor neuron component of the damage caused by ALS. Fasciculations are involuntary muscle movements that resemble a "bag of worms" in multiple muscle groups including the tongue. The presence of fasciculations is a cardinal sign that muscles are undergoing denervation secondary to progressive damage to lower motor neurons in the spinal cord or brainstem. The *lateral sclerosis* (LS) part of the name refers to the lateral columns or lateral funiculi in which the corticospinal tracts run. Therefore, the term *ALS* refers to both upper and lower motor neuron damage all in one acronym. The cause of ALS is unknown, although there have been some cases due to genetic deficiencies of enzymes such as superoxide dismutase (SOD), and other cases resulting from a deficiency in an androgen receptor leading to both a predominantly bulbar pattern of ALS and to endocrine abnormalities (testicular atrophy, impotence, erectile dysfunction—Kennedy's disease). So far there is no specific effective treatment for ALS and the mean time from diagnosis to death is 48 months. Most treatment is merely supportive for both physical and mental requirements.

5.5.10 Genetic Disorders

There are many genetic disorders that affect the spinal cord; a good example is spinal muscular atrophy (SMA), which is a specific genetic neurodegenerative disorder affecting the anterior horn cells. Other examples are a variety of genetic variations of ALS, including those noted in the previous section—among others—as well as the various types of spinocerebellar degeneration (SCA; see Table 3.15).

5.6 Case Summary

In summary, Cletus has sustained severe trauma to his chest resulting from a rapid deceleration injury. This led to a partial avulsion and rupture of his thoracic aorta. He was hypoxic and hypotensive at the scene of the accident but noted to be moving all four limbs.

He was transported to the university hospital, where he was found to have absent pulses in the lower limbs, a widened thoracic aorta on chest x-ray and extravasation of contrast from the thoracic aorta on a contrast enhanced CT scan of the chest.

He underwent an emergency procedure during which the surgeon had to clamp his lower thoracic aorta above the artery of Adamkiewcz in order to place a graft at the rupture site.

Following recovery from anesthesia, he was found to have a spinal cord injury involving the spinal gray matter and corticospinal and spinothalamic tracts below T10, with sparing of the posterior columns.

This picture is consistent with an anterior spinal artery syndrome in which the blood supply to the anterior two-thirds of the spinal cord was interrupted, resulting in infarction of the spinal gray matter, spinothalamic, and corticospinal tracts at that level.

After examination by the neurologist, Cletus underwent an MRI of his spine.

The MRI illustrated in Figure 5.7 shows that the anterior aspect of the spinal cord has been infarcted as evidenced by the increased signal on the sagittal and axial T2 images.

The MRI does not show any significant disruption of the vertebral column with subsequent compression of the spinal cord. The injury to the spinal cord therefore has to be due to an indirect cause.

The MRI therefore confirmed the localization and etiological diagnosis of the neurologist.

5.7 Case Evolution

Cletus recovered quickly from the effects of the chest surgery and was anxious to get back home to the bush; he was impatient with the medical, nursing, and allied health staff during his stay in the acute care hospital and in the rehabilitation facility.

He and his family had difficulty understanding why he could not merely have surgery to his back and be permanently "fixed."

Cletus underwent spinal rehabilitation that focused on reestablishing his mobility with artificial devices such as wheelchairs, braces, and crutches. He also underwent training to improve his bowel and bladder function. He received a vocational assessment and had significant psychosocial support during his recovery (see also Chapter 13).

The health care team worked hard to engage Cletus, his family and local social services to help him get back to his home and community and start a new life.

Related Cases

There are e-cases related to this chapter available for review on the DVD and text Web site.

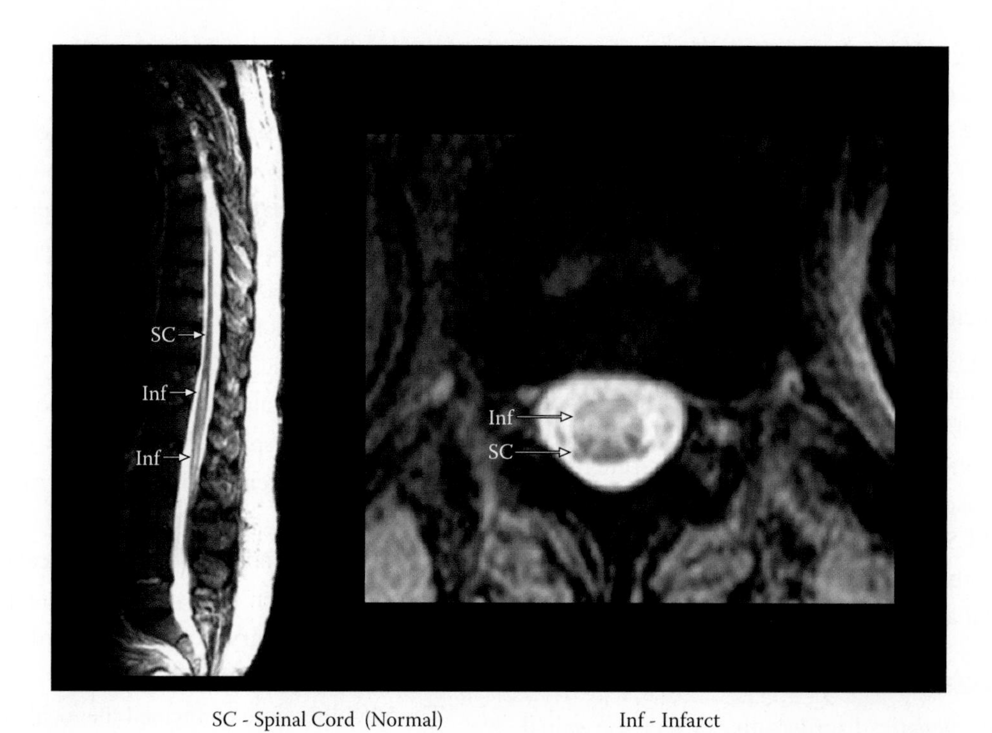

FIGURE 5.7 MRI spinal cord infarction: longitudinal view and axial view. (Courtesy of University Hospitals of Cleveland.)

Suggested Readings

Bottiglieri, T. Folate, vitamin B12 and neuropsychiatric disorders. *Nutr. Rev.* 54 (1996): 382–390. [A review of B12, folate and metabolic effects of deficiencies on the CNS.]

Cheshire, W.P. et al. Spinal cord infarction: Etiology and outcome. *Neurology* 47 (1996): 321–330. [A comprehensive collection of cases and review of spinal cord infarction.]

Fink, J.K. et al. Hereditary spastic paraplegia: Advances in genetic research. *Neurology* 46 (1996): 1507–1514. Hereditary Spastic Paraplegia Working Group. [Guidelines for the classification and diagnosis of hereditary spastic paraplegias.]

Herbert, V. Vitamin B12: An overview. In V. Herbert, Ed., *Vitamin B12 Deficiency*, 1–81. London: Royal Society of Medicine Press, 1999. [A review of vitamin B12 and clinical effects of deficiency.]

Rosenblum, M. et al. Dissociation of AIDS-related vacuolar myelopathy and productive HIV-1 infection of the spinal cord. *Neurology* 39 (1989): 892–896. [A study showing that HIV-associated disease of the spinal cord is due to a metabolic disorder rather than primary infection of the spinal cord.]

Transverse Myelitis Consortium Working Group. Proposed diagnostic criteria and nosology of acute transverse myelitis. *Neurology* 59 (2002): 499–505.

Chapter 6

Bernard

BERNARD

"Uncle" Bernard is the oddball of the family. At times almost noncommunicative and at other times very friendly to all, he is treated by family members with respect. They have learned, however, never to call him "Bernie" to his face. Tough intellectual and philosophical issues are referred to him and often resolved; he is considered "a man of the world." He is also the family guru for any technological questions; he can repair dysfunctional computers and can even program a VCR.

Bernard, now 50, has recently developed a love of pop music and, having purchased a mp3 player (brand name examples include the iPod®), spends many hours a day with his ears plugged in to the latest so-called hits.

Recently, Bernard has mentioned to anyone in earshot that he has a buzz in his left ear that comes and goes; this type of noise, called tinnitus, is sometimes a ringing sound and at other times a hissing sound. Bernard finds this noise very distressing as it seriously interferes with his ability to concentrate and occasionally interferes with his ability to fall asleep. No one, however, except for his six-year-old niece Julie, has had the audacity to suggest that he should perhaps turn down the volume of his music; this advice is, of course, deftly ignored.

After a few months, everyone, including Bernard and even Julie, begins to notice that Uncle Bernie is a little hard of hearing. He starts missing jokes (regardless of whether he laughs or not) and he often changes the telephone headset from his preferred left ear to his right ear. A couple of times recently he has noticed that he is a little unsteady when walking to the bathroom in his house (he lives alone). Now his friends are telling him to get a hearing aid. Others are telling him to get the wax out of his ears.

A little scared and very reluctantly, he decides to talk to "the doc" and finally calls to make an appointment, scheduled for a month later. By the time he arrives in the physician's office, he is a changed man, quite depressed and very worried.

The family physician notes the history, including the absence of any complaint of headache. A thorough physical examination reveals a 50-year-old male somewhat overweight (198 lbs), with normal blood pressure (135/85), a normal chest and abdominal exam, and no abnormality detected on a seven-minute neurological exam except for decreased hearing (using a tuning fork) in the left ear; normal eardrums are clearly visualized on both sides. The fundi are easily visualized and are normal.

The physician arranges for a hearing test at a local hearing clinic that fits hearing aids; Bernard returns a few weeks later to be informed that there is audiologic evidence of a marked hearing loss in his left ear across all parts of the sound frequency range. This time the physician arranges an urgent consultation with a neurologist from a large nearby city where she visits the local hospital one day a month.

Neurological Examination

About three weeks afterward, Bernard arrives (on foot) at the local hospital for his appointment with the neurologist. He presents as a well-groomed middle-aged man who is very tense but fully cooperative.

Cranial nerve examination: CN I is not tested. Eye movements in all directions are intact and no nystagmus is seen on lateral gaze. The pupils are equal in size and react to light symmetrically. Visual fields on confrontation are

Objectives

- To demonstrate how dysfunction of hearing (audition) and balance (vestibular) can affect important aspects of everyday life
- To further the understanding of the brainstem and the cranial nerves, and disease that may occur in that region
- To underscore the point that many people do not have immediate access to the latest technological advances for their medical care

6.1 Clinical Data Extraction

At this point, the student should proceed to the DVD or Web site and complete the history, physical examination, intact. The corneal reflex is tested and found to be intact on both sides. Sensation of the face to touch and pinprick is intact in all three divisions of the trigeminal nerve and equal on both sides; the jaw reflex is normal. Although the facial appearance looks symmetrical (Bernard is a bearded male), testing reveals a slight weakness on forced closure of the left eye, as well as a weakness of the left lower face when the patient is asked to smile or show his teeth. The gag reflex is difficult to elicit but seems intact; voice production also seems unaffected. Shoulder elevation is strong and symmetrical, and the tongue is normal in appearance, with equal movements bilaterally.

Examination with the otoscope shows that each external ear is clear of wax and both eardrums are easily visualized; both have a normal appearance, with a cone of light reflecting in the lower quadrant of the drum. The hearing deficit of the left ear is confirmed using a vibrating tuning fork (at 256 Hz); even Bernard notes that the sound of the tuning fork is heard longer and better on the right side. The neurologist's findings are consistent with a nerve deafness due to loss of hair cells in the cochlea or a lesion affecting the VIIIth nerve, known as a sensorineural deficit; this conclusion cannot be considered reliable without confirmation by a certified audiologist, whose audiographic study would precisely characterize the deficit (see below).

On that particular day the neurologist is working with a fourth-year medical student who asks about the Weber and Rinne tests. The Rinne test compares air and bone conduction for each ear; the patient compares the sound of the tuning fork held close to the ear (air conduction) with the sound heard when it is placed on the mastoid process (bone conduction). Sound is transmitted by the bones of the skull and activates the hair cells of the cochlea. Normally, the sound intensity of the tuning fork is heard better by air conduction. In conductive hearing loss bone conduction is better than air conduction; in sensorineural hearing loss, air conduction is greater than bone conduction, as in normal hearing, but there is a hearing loss in the affected ear. In the Weber test, the vibrating tuning fork is placed on the vertex of the skull in the midline; the sound should be heard equally in both ears. If the person reports that the sound is louder in one ear, then one knows that there is a problem that needs to be sorted out.

Sensation to touch and pinprick is intact and symmetrical in the upper limbs. Muscle bulk and tone are both normal; muscle power at the shoulders, elbows, and hands is graded as 5 and symmetrical (see Table 2.2); biceps, triceps and brachioradialis reflexes are tested as 2+ (normal) on both sides (see Table 2.3). The findings on lower limb testing are also unremarkable; there is no ankle clonus, and the plantar reflex is downgoing on both sides.

Coordination in the upper and lower limbs is intact, as is Bernard's gait. However, Bernard is not able to accomplish tandem walking and is unsteady when asked to turn suddenly, tending to stagger to the left. When asked to close his eyes and stand with his feet together, his balance does not get worse; that is, the Romberg sign is not present.

The mental status examination is difficult to conduct because of the frequent need to repeat questions, but generally appears normal.

The fundi are examined with the room darkened. The discs on both sides demonstrate sharp margins and are assessed as normal. Finally, the blood pressure is taken and recorded as 145/90. His weight is recorded as 195 lbs. Before asking the patient to dress, the neurologist does a focused examination of the skin; no brown patches or small subcutaneous lumps are noted.

The neurologist, having decided upon the localization and presumed etiology, explains to Bernard the probable reason for his deafness, as well as the next steps that need to taken. These include a detailed documentation of his hearing loss in a properly equipped laboratory and a brain scan; all this will have to be done in the "big city."

localization and etiology worksheets, utilizing the information just provided concerning Bernard's problem.

6.2 Review of the Main Clinical Points

The history includes the following:

- Diminished hearing in his left ear, starting a few months ago and increasing over this time
- Intermittent buzzing or hissing sound (tinnitus) in the left ear
- Some unsteadiness when walking, otherwise termed disequilibrium
- No complaint of headache

The neurological examination demonstrates the following:

- A hearing loss in the left ear, of a sensorineural type

- Facial weakness on the left side affecting muscles of both the upper and lower face
- Mild ataxia of gait

6.3 Relevant Neuroanatomy

The history and neurological examination are consistent with a disease process affecting both divisions of CN VIII, the vestibulo-cochlear nerve, but primarily the auditory portion, with the additional involvement of CN VII. Because of the involvement of two cranial nerves, the locus of the disease is likely to be at the level of the brainstem.

The nerves supplying the head and neck, the cranial nerves, have their nuclei in the brainstem (as was introduced in Chapter 1). CN III and IV are found in the midbrain, CN V, VI, VII and VIII in the pons, and the remainder in the medulla (see Figure 1.6B). The three major pathways, two ascending somatosensory and one descending motor, travel through the brainstem (see Figures 1.4A, 1.4B and 1.5A). In addition, the reticular formation with its descending and ascending influences is an important component (see Figure 1.6C). Finally, the brainstem is located in the base of the skull and these cranial nerves have to travel via various foramina and canals in order to reach their destinations peripherally.

6.3.1 *The Facial Nerve (CN VII)*

The nucleus of CN VII is located in the lower pons, in the lateral aspect (see Figure 1.6B), supplying all the muscles of facial expression. The facial nerve has an unusual course within the brainstem, moving inward and upward and then coursing over the nucleus of CN VI before descending to exit at the cerebello-pontine (C-P) angle, adjacent to CN VIII. Both nerves are found initially in the internal auditory canal. The facial nerve then courses through the petrous temporal bone, exiting the skull at the stylomastoid foramen (just anterior and inferior to the ear canal) and distributing its branches (within the parotid gland) to supply the various muscles of facial expression.

The organization of the facial nucleus is unique in that there are two portions, a part supplying the muscles of the upper face (forehead wrinkling and eye closure) and a part supplying the muscles of the lower face (controlling the movements around the mouth and the lips). A lesion affecting the facial nucleus or the exiting fibers of CN VII within the skull will affect both sets of muscles, that is, a lower motor neuron lesion. The portion of the nucleus controlling the upper facial muscles receives its commands from both motor cortices, whereas the portion of the nucleus controlling the lower facial muscles receives commands from only the contralateral motor cortex. Therefore, when a patient has weakness confined to the lower face one looks for a lesion in the contralateral motor cortex or in the descending fibers from the cortex to the pons, that is, an upper motor neuron lesion.

Additional functional parts of the facial nerve are fibers carrying taste from the anterior two thirds of the tongue, as well as parasympathetic fibers to the submandibular and sublingual salivary glands and to the lacrimal gland (which produces tears); both of these constitute a separate branch of CN VII (the nervus intermedius).

6.3.2 *The Vestibular-Cochlear Nerve (CN VIII)*

There are two parts to CN VIII, the auditory component and the vestibular component; the nerve itself is large and divisible into the two major portions described below.

6.3.2.1 The Auditory Component The auditory part of the nerve originates in the region of the hair cells of the cochlea; the nerve's cell bodies, which form the peripheral (sensory) ganglion, are also found within the cochlea. The nerve then courses in the internal auditory canal, a canal within the petrous portion of the temporal bone of the skull, and after leaving the canal the nerve goes through the subarachnoid space. It enters the brainstem at the cerebello-pontine angle, very close to CN VII. Up to this point the nerve is a peripheral nerve with the myelin of the PNS and its associated Schwann cells.

Central Auditory Pathway (Figure 6.1) There are several auditory nuclei, the first of which are found along the nerve itself as it enters the brainstem, with others found at the level of entry, in the lowermost pons and uppermost medulla. At this point the sound is analyzed as to its laterality. After multiple synapses, some of the auditory fibers cross the midline while others remain on the same side. Sound is therefore transmitted upward, toward the cortex, on both sides, in the lateral lemnisci (singular, lemniscus). Hearing is unique among the sensory systems in that the projection is bilateral, although the crossed

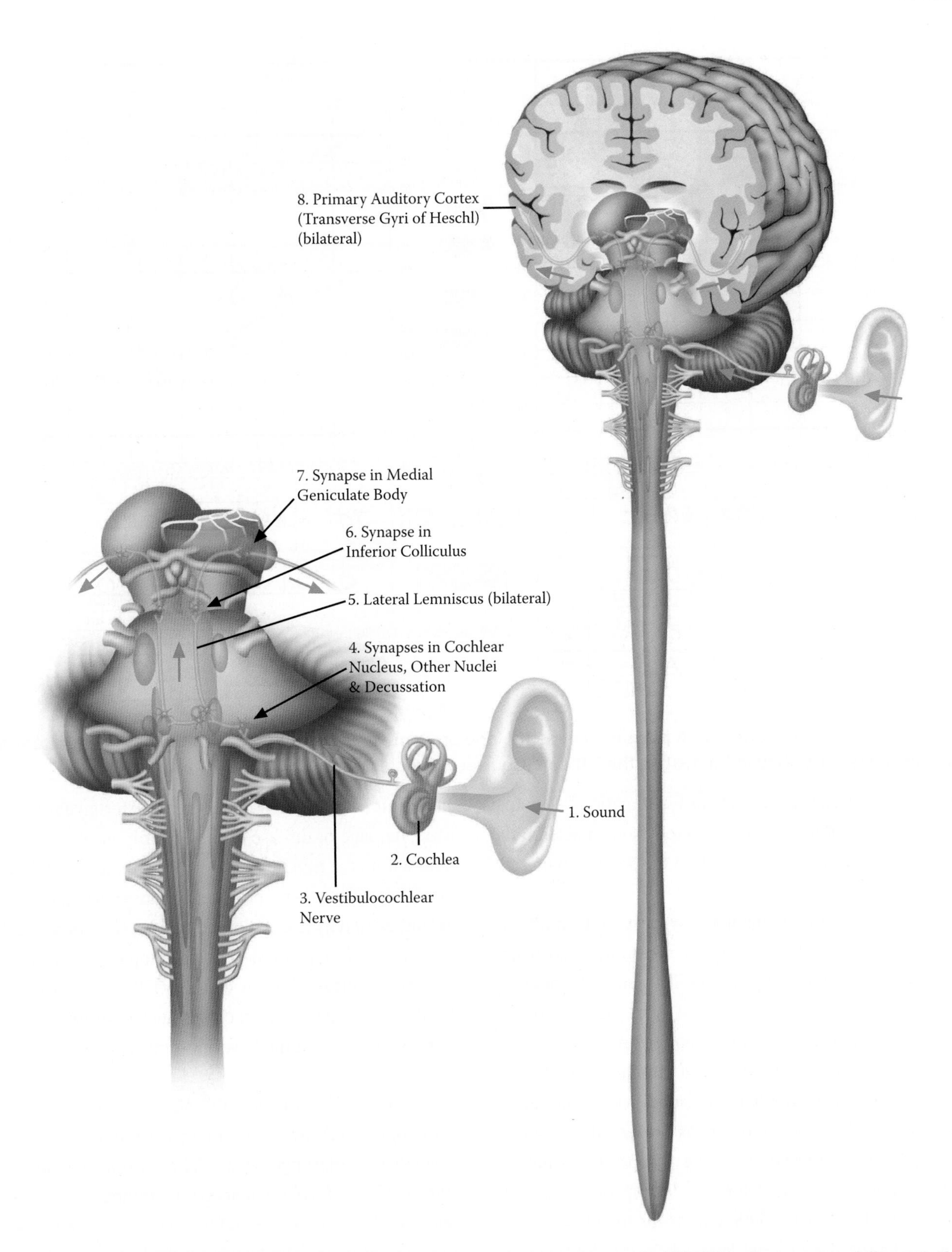

FIGURE 6.1 The pathway for audition (hearing) is shown from the cochlea via CN VIII to the pons and through the brainstem in the lateral lemniscus. The fibers relay in the inferior colliculus, then in the medial geniculate body (MGB) of the thalamus to the primary auditory cortex (transverse gyri of Heschl). Note that the pathway is bilateral. The auditory pathway is animated on the DVD and Web site.

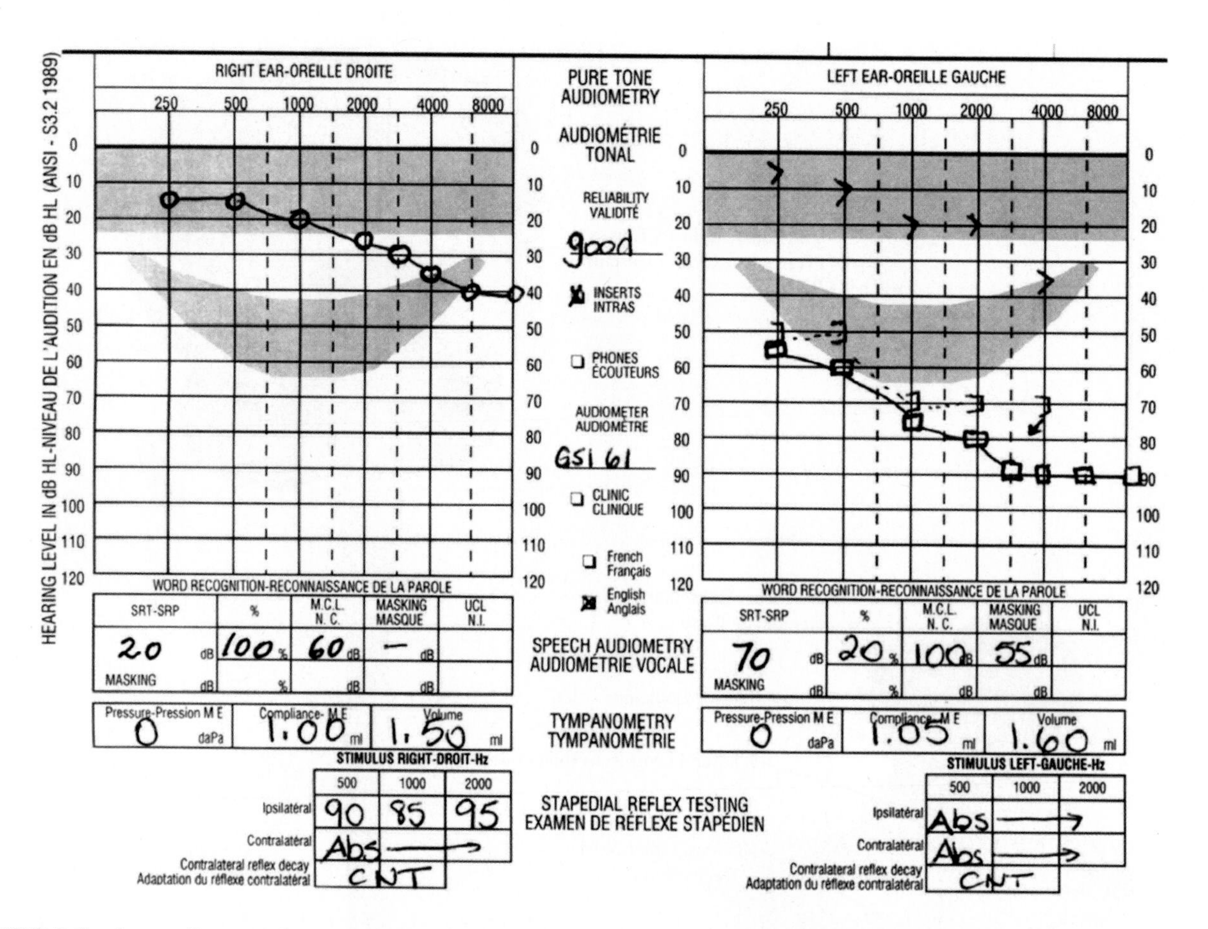

FIGURE 6.2 An audiogram for a patient such as the one presented in this chapter, with the normal right ear and the deficit (as described in the text) in the left ear. (Courtesy of Dr. J. Marsan.)

fibers are more numerous. There is an additional unique feature of the auditory system whereby some nuclei in the pons project from the CNS back to the sensory hair cells of the cochlea.

The upwardly projecting fibers synapse in the midbrain (the inferior colliculi, at the lower midbrain level) where some auditory processing occurs. The auditory pathway then travels upward to a specific relay nucleus of the thalamus (the medial geniculate body), and arrives at the cortex along the transverse gyri of Heschl, located along the superior part of the temporal lobe, within the lateral fissure. Sound frequency (pitch) is preserved throughout the pathway up to and including the primary auditory cortex, where there is tonotopic localization. Adjacent areas of cortex are association areas for sound interpretation, including the adjacent Wernicke's area for receptive speech in the dominant hemisphere.

Deficits of hearing caused by failure of the sound waves to transmit, known as a *conductive* hearing loss, can occur because of blockage in the external ear (e.g. ear wax), damage or disease of the ear drum itself, disease or fluid within the middle ear or damage to the ossicles of the middle ear. The other type of hearing loss is a *sensorineural* deficit, caused by a lesion of the cochlea (loss of hair cells), disruption of transmission of the impulses peripherally (by the cochlear division of the VIIIth nerve within the skull), or lesions of the cochlear nuclei within the brainstem (at the level of the lowermost pons).

Testing Hearing should be tested by a certified audiologist working with an ear, nose and throat (ENT) specialist (otolaryngologist). The testing includes the transmission of sound (ear canal, tympanic membrane and middle ear ossicles) and testing of hearing of sounds of different pitch. The test result for hearing is displayed as an audiogram (Figure 6.2). Particular attention is paid to the discrimination of speech sounds (500 to 3000 Hz, tested at 55 decibels).

A more sophisticated (and complicated) testing of the auditory system is based on the processing of information by the auditory relay nuclei and associated CNS pathways; this technique is known as a brainstem auditory evoked potential test (BAEP, BSEP, BAER). The stimulus is a clicking noise (which consists of all frequencies) of defined intensity presented through an earphone; recording of the response is achieved by electrodes attached to the mastoid processes and the scalp (and averaged by an online computer after hundreds of repetitions). The response consists of a number of well-defined waveforms that can be ascribed to the auditory nerve and its nuclei, the brainstem nuclei, the auditory pathway and the auditory cortex. This testing would be done in order to better define the nature of the lesion of the auditory system. The most common changes of auditory evoked responses are abnormalities of the configuration of the waves, or a delay between the different waves, when comparing one ear to the contralateral ear.

Tinnitus Tinnitus is a sound heard by a person in the absence of an external source for the sound. It is described as a meaningless noise, such as a ringing, buzzing, humming, clicking, roaring or chirping sound, and can be reported in one or both ears; other people cannot hear it. It may be continuous or intermittent, and often depends on the ambient noise level. It is often associated with a hearing loss. Tinnitus can be quite distressing for a person, causing difficulty with concentration and reading and perhaps other tasks; interference with sleep is quite usual. Some will react with emotional distress, which may require active treatment. There is some experimental evidence that tinnitus is a central phenomenon due to a reorganization of the primary auditory cortex, associated with diminished intracortical inhibition presumed to be related to advancing age.

6.3.2.2 The Vestibular Component The vestibular portion of CN VIII originates in other parts of the inner ear: the three semicircular canals (for detection of angular acceleration) and the utricle and saccule (for detection of the head's position in space and hence important for balance and equilibrium). The sensory ganglion is located near these structures and the central processes of these neurons, as peripheral myelinated nerves, with their associated Schwann cells, are also found within the internal auditory canal. The nerve enters the brainstem with its auditory portion at the C-P angle.

Central Vestibular Pathway The information from the receptors is conveyed by the sensory neurons via the vestibulo-cochlear nerve, CN VIII, to the vestibular nuclei in the brainstem. There are four nuclei, located in the uppermost medulla and the lowermost pons, where these fibers synapse (see Figure 1.6B and Figure 1.6C). The information from these nuclei is conveyed to the nuclei controlling the extra-ocular muscles (visuomotor nuclei: CN III, IV and VI), forming the basis for the necessary reflex adjustments of eye movements in response to acceleration and to changes in body position in relation to gravity. The reflex, consisting of conjugate movement of the eyes in the direction opposite to the direction of the head movement in progress, is intended to compensate for the head movement in order to keep the eyes fixed on an object and is called the *vestibular-ocular reflex* (VOR, see below).

Vestibular data are also provided to the reticular formation of the brainstem as well as to the cerebellum so that the movements of the head by the neck muscles and other postural adjustments can be coordinated with those of the eyes; all this is handled by the so-called nonvoluntary (postural) motor system (discussed in Chapter 2). There is also a descending pathway from one of these vestibular nuclei to the spinal cord for postural (nonvoluntary) adjustments to the effects of gravity and alterations in acceleration (as in keeping one's balance hanging onto the strap when the bus starts moving).

Disease can affect the vestibular apparatus itself, the vestibular portion of the VIIIth nerve (within the skull), the vestibular nuclei or their connections to the cerebellum or to the visuomotor nuclei.

Testing Testing of vestibular function can be done best in a controlled setting using the *caloric test*, whereby water (either warm of cold) is introduced into the external ear canal and reflex eye movements are observed (Figure 6.3); the use of refrigerated water at 4°C. is recommended. Both eyes will move at the same time, with a slow phase in the direction of the cold-injected ear; when the eyes reach the limit of how far they can move, there is a rapid (saccadic) compensatory phase in the opposite direction. This rhythmic oscillation movement of the eyes is called nystagmus and its classification is based on the direction of the *rapid*

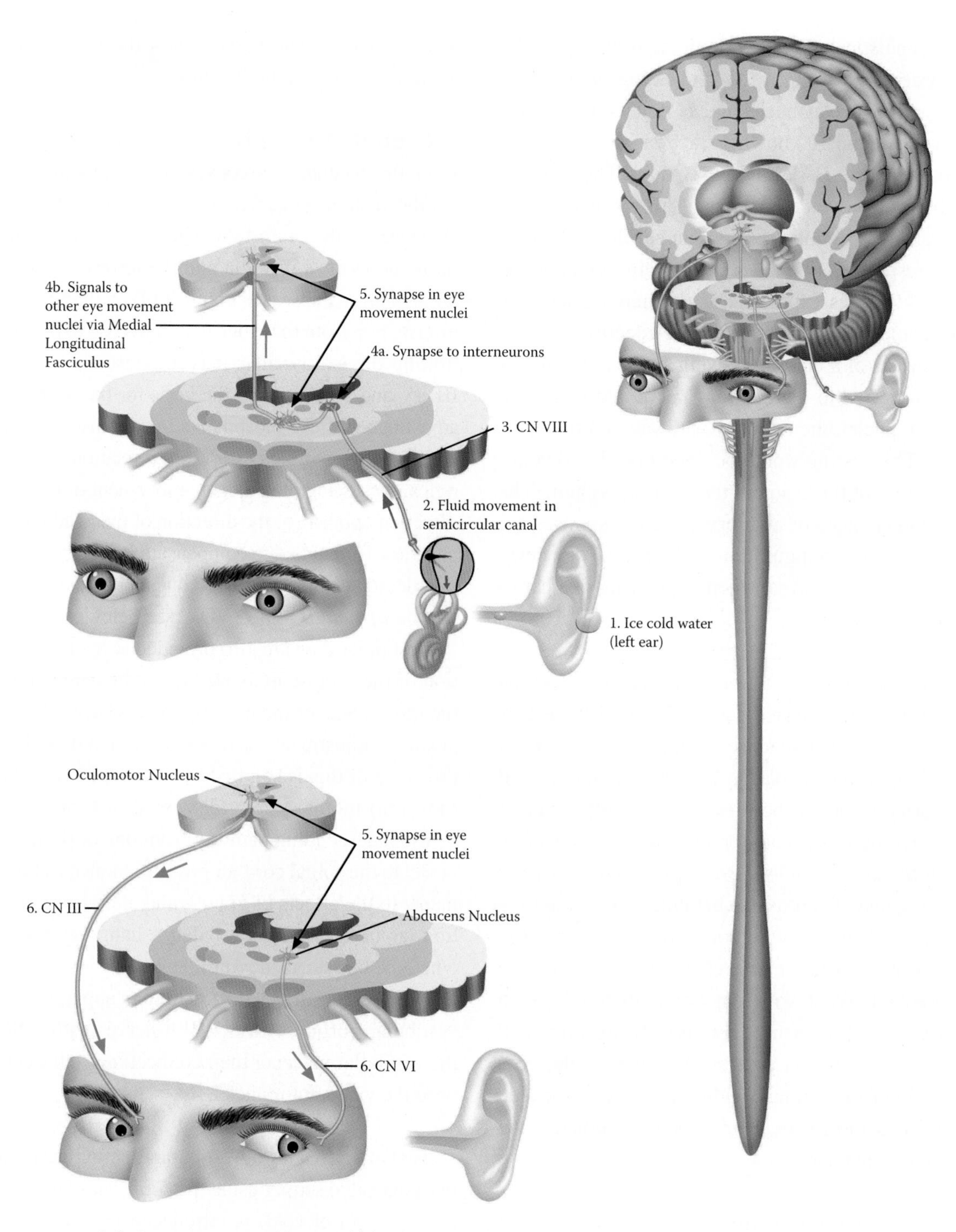

FIGURE 6.3 The vestibular-ocular reflex (VOR). Ice cold water is introduced into the external (left) ear canal, which stimulates the (horizontal) semicircular canal. The stimulus is relayed from the vestibular nuclei to the visuomotor nuclei, the abducens in the pons and also to the oculomotor in the midbrain (via the medial longitudinal fasciculus, the MLF), causing the eyes to move slowly to the same side; the rapid "corrective" return—the direction of the named nystagmus—is to the opposite side and is not shown. In the DVD (and Web site) version of this illustration, the VOR circuit is animated. Note that the ipsilateral slow eye deviation without a fast component is what one sees when the test is performed in a comatose patient.

phase. With the use of warm water, the eye movement slow phase would be away from the ear being tested and rapid nystagmoid movement toward it (see below).

The basis of the test is the stimulation or inhibition of the hair cells of the vestibular apparatus, in this case done artificially by using the water temperature to change the density of the fluid within the semicircular canals. The temperature of the water is transmitted through the tympanic membrane and the air of the middle ear to the bony labyrinth containing the vestibular system. The head is titled back 60 degrees (or elevated 30 degrees) in order to bring one of the vestibular canals (the horizontal) in the proper orientation for this test. It is important to note that a conscious person undergoing the test may experience a rotatory ("dizzy") sensation and this can be very disconcerting. It is also very important to note that the response to the test depends on the integrity of the vestibular part of the VIIIth nerve, the integrity of the pontomedullary region of the brainstem and that of CN III and VI.

As illustrated in Figure 6.3, the ice water is put into the left ear. The stimulus (which in this case is inhibitory) is conveyed via the vestibular division of CN VIII to the vestibular nuclei in the lowermost pons (and uppermost medulla). The basis for the eye movements is the connections to two of the extra-ocular motor nuclei, CN III and CN VI. Two eye muscles are needed for the eyes to move together in the horizontal plane, the medial rectus of one eye (controlled by CN III, at the level of the midbrain), and the lateral rectus of the other eye (controlled by CN VI, at the level of the lower pons). The vestibular nucleus connects with neurons in CN VI on the same side. At the same time there is a synapse with other nearby neurons whose axons cross immediately and ascend in a specific pathway through the brainstem (the MLF, or medial longitudinal fasciculus) to CN III in the midbrain. Activation of these CN nuclei causes the slow phase of the movement of the eyes to the same side, the left; the rapid saccadic return of the eyes is in the opposite direction, to the right, and is the basis for the classification of the nystagmus. The mnemonic used for this test is COWS: cold opposite and warm same, referring to the fast corrective component of the nystagmus (for an awake patient).

A similar test done by manually rotating the head (in order to stimulate the same system) is called the "doll's eye" maneuver, or the oculo-cephalic reflex. The examiner grasps the head (elevated as before) and turns it rapidly to one side, noting the reflexive movement of the eyes in the opposite direction; there is no expectation of a rapid return phase. When the head is turned to the opposite side, the eyes again move in the converse direction. It is important to be aware that oculo-cephalic reflex testing should not be done in a patient with a neck injury.

Dizziness Dizziness is a term used by patients to describe a sensation of turning or spinning, or that the room is moving. The word can describe various abnormal sensations, including spinning, light headedness and feeling faint, unsteady or off balance. *Vertigo* (true vertigo) is the medical term for the spinning sensation, either of the person or of his or her surroundings.

6.3.3 *Sensory and Motor Pathways*

See Figure 6.4. The somatosensory pathway for discriminative touch and proprioceptive sensation in the spinal cord, the posterior column, crosses in the lower brainstem and becomes the medial lemniscus. It ascends through the medulla in the midline area, moving more laterally as it ascends through the pons and into the midbrain. The pathway for pain and temperature, the spinothalamic tract, crosses in the spinal cord and ascends; in the brainstem, it is situated more laterally. Both tracts merge at the level of the uppermost pons and are found together in the midbrain. Lesions affecting the brainstem would result in a contralateral loss of either or both sensory modalities, depending on the location and extent of the lesion.

The voluntary motor pathway, the corticospinal tract, is found in the anterior region of the brainstem, occupying the cerebral peduncles in the midbrain, lying in the anterior bulge of the pons, and found within the pyramids in the medulla anteriorly (hence its name, the pyramidal tract, see Chapter 7). With a lesion of this pathway, neurological testing would reveal a contalateral weakness (paresis) or paralysis and an increase in reflex responsiveness; a Babinski response would indicate involvement of the corticospinal tract at any point along its pathway.

6.4 Localization Process

The history and neurological examination are consistent with the neuroanatomical information that there is a disease process affecting both divisions of CN VIII as well as CN VII. The major question is whether the process is

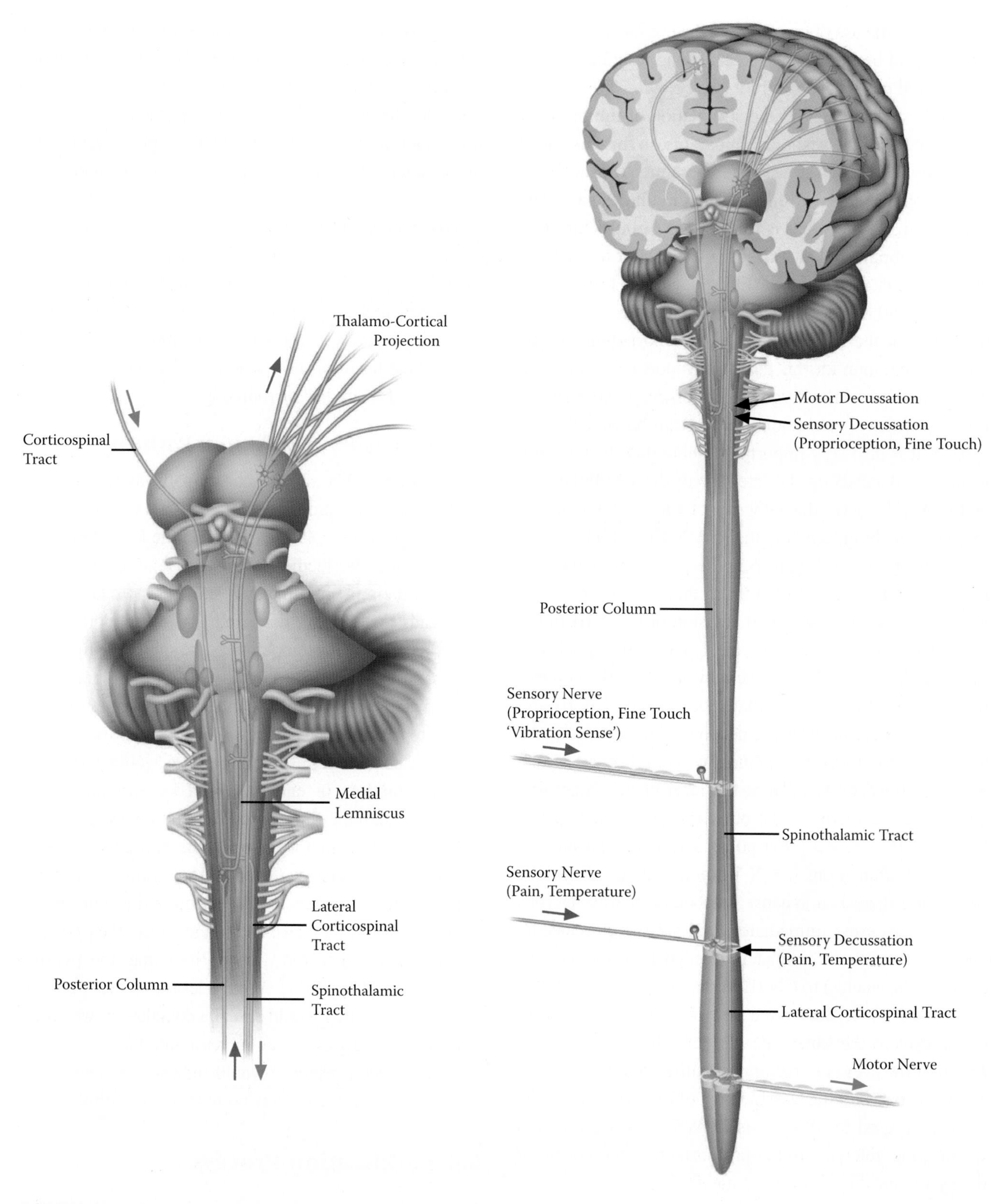

FIGURE 6.4 The three important longitudinal pathways traversing the brainstem are shown, the two ascending, medial lemniscus (for discriminative touch) and spinothalamic (for pain and temperature), along with the major motor descending corticospinal tract.

occurring outside the brainstem (extra-axial) or within the brainstem (intra-axial).

The hearing loss in our case is restricted to one ear (the left one). As the central auditory pathway is bilateral within the brainstem, a loss of hearing due to a lesion within the CNS should cause a hearing deficit in both ears, although not equally. This shifts the locus of the lesion in this case outside the brainstem itself (extra-axial). In addition, if the lesion were within the brainstem, one would anticipate some involvement of the long tracts, sensory or motor, as these tracts course through the brainstem. No somatosensory deficits were noted in our patient and there were no changes in motor strength or in tendon reflex responsiveness, thereby pointing again to a lesion outside the brainstem, but within the skull.

The fact that the facial weakness involves both the upper and lower face is very consistent with involvement of the VIIth nerve nucleus within the lower pons or the nerve after it exits the brainstem, that is, a lower motor neuron lesion. If the lesion were to be within the brain, above the level of the pons, affecting the corticobulbar fibers to the nucleus of CN VII on one side, there would be a deficit of only the lower face (on the opposite side). Again, since there is no involvement of the long tracts (sensory or motor), the locus is again pointing to a lesion of the nerve itself, outside the brainstem.

It should be noted that the neurologist also tested the corneal reflex on the left and found it to be normal. This result certainly indicates that the sensory component of CN V has not, as yet, been affected by the disease process. Involvement of the upper sensory division of this nerve would lead to a decreased corneal reflex in both eyelids when the testing is done on the side of the lesion (see Figure 2.3), but the response would be normal when tested on the other side. A decrease in the response on one side could also be due to the involvement of CN VII, as the palpebral (eyelid) muscles are supplied by CN VII. In Bernard's case, the pathology in CN VII was too mild to interfere with the corneal reflex; with more severe facial nerve pathology, the corneal reflex would be diminished on the side of the weakness, normal on the opposite side, and the patient would report that corneal sensation during testing was the same in both eyes.

The unsteadiness of walking that was seen when the patient was asked to tandem walk or change direction could be caused by a disease affecting the VIIIth nerve within the skull (before it enters the brainstem) or by disease of the brainstem affecting the vestibular nuclei, or it could also reflect cerebellar pathology. The vestibular nuclei project to the central region of the cerebellum (the vermis) to assist in the coordination of gait. Other cerebellar coordination tasks—regulated by the lateral regions of the cerebellum (the cerebellar hemispheres, discussed in Chapter 7)—were tested and found to be within normal limits. The problem with gait could therefore be due to the involvement of the nerve itself peripherally or to a lesion within the CNS.

As both CN VII and CN VIII are found in the internal auditory canal and both enter and exit the brainstem at the C-P angle, the likelihood is that there is a disease process occurring in either of these locations affecting both cranial nerves.

6.5 Etiology: Bernard's Disease Process

Bernard's symptoms have been noted for more than three months, therefore falling into the category of a chronic disease process (see Table 3.13). In fact his symptoms have become worse in this time period, indicating the progressive nature of the disease process.

6.5.1 Hearing Loss

In considering a hearing loss, there are certain parameters that are specific for the auditory system. One of the most common occurrences is an accumulation of earwax in the external ear canal, more often affecting men; this was not the case with Bernard. A hearing loss for the higher frequencies is quite common as people age, but this usually affects both ears (to some degree) and is most commonly found after the age of 60. Another common reason for a hearing loss is exposure to excessively loud sound, leading to a destruction of the hair cells of the cochlea, usually affecting both ears; Bernard is not involved in an industrial job or a hobby with chronic exposure to loud noise, although he has recently been connected to his mp3 player using high volumes. It is not clear whether his hearing problem started before or after beginning to use this device. If the mp3 player were to contribute to the hearing loss, however, both ears would be affected.

6.5.1.1 Paroxysmal Disorder A paroxysmal disorder is an acute phenomenon or an acute event superimposed on a chronic condition. The history in this case does not coincide with such a disease process.

6.5.1.2 Traumatic Injury There has been no accident or trauma to the head in this case.

6.5.1.3 Vascular Disorder Again, vascular events fall under the category of acute, and our patient's history is not compatible with a vascular event, either infarction or hemorrhage.

6.5.1.4 Toxic Injury (Including Drugs) Bernard has not taken any medication that could lead to deafness (e.g., the antibiotics streptomycin and gentamycin; quinine). Aspirin overdose can cause bilateral tinnitus, but nothing in the history suggests that he has been taking pain killers for arthritis or any other pain condition. Chronic lead poisoning has been associated with progressive deafness (e.g., postulated for the composer Ludwig van Beethoven), but Bernard is not making his own wine nor does he store wine in lead-lined containers (such as crystal), or use pottery dishes with lead glazes (one needs to ask these questions specifically). Besides, a toxic type of process would be expected to affect both ears equally.

6.5.1.5 Infectious Diseases Chronic infectious processes are also unlikely as causes of unilateral sensorineural hearing loss and ipsilateral facial weakness, with the single exception of cholesteatoma, a chronic granulomatous disease complicating a long-standing otitis media, primarily in children. In addition, mention should be made of sarcoidosis, a chronic granulomatous process of unknown etiology that may produce gradually evolving multiple cranial nerve pareses, sometimes unilaterally. It is not clear whether sarcoidosis belongs in the infectious disease category or with the inflammatory/autoimmune disorders.

6.5.1.6 Metabolic Diseases Although metabolic diseases (e.g., diabetes or other endocrine disorders) are, by their nature, chronic for the large part, nothing in the history indicates that Bernard is suffering from a chronic systemic disease. He has not complained of fatigue or shortness of breath while walking or biking. Besides, again, a metabolic type of disease process would be expected to affect both ears equally.

6.5.1.7 Inflammatory/Autoimmune Chronic inflammatory/autoimmune diseases of the CNS, with the exception of sarcoidosis (see above), seldom produce sensorineural hearing loss. In addition, were a disease process in this category to involve the acoustic nerves, it would invariably produce a bilateral hearing loss.

6.5.1.8 Neoplastic One must consider the neoplastic disease category in any chronic disease. In the present case, we would be considering a tumor in the region of the petrous temporal bone of the skull or, alternatively, affecting the nerves at the cerebello-pontine angle. There is a particular tumor of the VIIIth nerve, known as a vestibular schwannoma (commonly misnamed an acoustic neuroma, discussed below) which would take into account all the neurological findings and be consistent with the history of Bernard's case. The growth of this Schwann cell tumor occurs at the cerebello-pontine angle and/or within the internal auditory canal, causing pressure on both divisions of the VIIIth nerve and eventually also on the VIIth nerve. Although benign, continued growth of this tumor will lead to involvement of CN V, as well as brainstem compression at that level.

A cerebello-pontine angle meningioma, although somewhat less common, would also be possible. A brainstem glioma is also to be considered, but these are always associated with long tract sensory and motor deficits. Last and perhaps least likely is a metastasis; in this case, there is no known primary.

6.5.1.9 Degenerative Disorders Menière's disease, a degenerative disease of the vestibulo-cochlear apparatus, is also a likely possibility in this case. This disease is characterized by deafness and tinnitus, usually involving only one ear. It affects both men and women, with a disease onset about the age of 50. The hallmark of this disease is the occurrence of acute attacks of dizziness, sometimes described as "explosive vertigo," lasting several minutes, accompanied by nausea and sometimes leading to vomiting. Once the unilateral degenerative process is complete and the sensorineural hearing loss in the involved ear is profound, the vertiginous attacks stop occurring. In a severe attack, the person may fall and there is an intense sensation of rotation of the environment. Our patient Bernard has not reported any such episodes. In addition, Menière's disease is not accompanied by unilateral lower motor neuron facial weakness.

6.5.1.10 Genetic Disorders Tumors involving peripheral nerves, including cranial nerves, are part of the genetic disease neurofibromatosis, especially the type 2 variety. The disease may be unilateral or can occur bilaterally. In this disorder, there are typically also congenital brownish pigmentary skin patches referred to as "café-au-lait" spots, as well as small, rubbery neurofibromas underneath the skin. (Remember that the neurologist did do a cursory examination of the skin at the end of the physical.)

In addition, there are many familial forms of deafness, producing either deafness alone, or deafness along with other disease processes (retinitis pigmentosa, peripheral polyneuropathy and cardiac conduction deficits, among many). In all such diseases, however, the hearing loss is bilateral.

6.5.2 *Dizziness*

As was mentioned, it is important to realize that dizziness is often difficult for a patient to describe. One should inquire as to whether it occurs spontaneously, on movement such as changing position quickly or when getting out of bed in the morning (benign paroxysmal positional vertigo), or while driving in a car. A person with marked anemia may in fact also complain of dizziness; sometimes the person, after questioning, will indicate that the sensation can be better described as a sense of "light-headedness." A viral inflammation of the inner ear may give rise to a vestibulitis, resulting in several days of intense vertigo and vomiting; recovery may be very slow, over weeks or months. None of these conditions would be expected to be accompanied by a hearing loss.

6.6 Case Summary

At this point, with the knowledge of the relevant anatomy, the localization of Bernard's disease is the left vestibulocochlear nerve, likely within the internal auditory canal, as well as the left facial nerve. The most likely etiological diagnosis is a tumor affecting the left CN VIII, a *vestibular schwannoma*, most often called an acoustic neuroma.

6.7 Evolution of the Case

After the expenditure of much energy and time by others, arrangements are finally completed for the additional testing to be done three weeks following the neurological examination.

Bernard's symptoms continue to worsen. He is now almost totally deaf in the one ear, has definite one-sided facial weakness and is having more unsteadiness of gait. The tinnitus is now less of a problem, for some unknown reason.

6.8 Investigations

Bernard's left-sided hearing deficit is found to be significant and is recorded as a moderate to severe sensorineural hearing loss, with loss of speech discrimination (shown in Figure 6.2, right panel, for a similar case). His hearing is completely normal in the unaffected (right) side.

The results of the testing of the auditory evoked responses show a prolonged latency between waves I and III, when compared with the opposite side, consistent with a lesion at the level of the acoustic nerve and/or its nuclei within the brainstem on the affected side.

Caloric testing is also done, much to Bernard's dismay. The left (abnormal) ear is tested first, and there is no reflex response of the eyes (no nystagmus), and hence no subjective response either, indicating a loss of the function of the vestibular portion of CN VIII on that side. The involvement of the vestibular nerve to this degree explains some of the patient's difficulties with gait. Testing of the unaffected right ear, however, elicits the normal response of nystagmus, along with extreme dizziness, accompanied by nausea and vomiting. (Thankfully Bernard had followed the instructions and had not taken any food after midnight.) It takes almost four hours before Bernard is able to get up and walk about without feeling dizzy or almost falling over.

Later that day, Bernard is taken to neuroradiology. A gadolinium-enhanced MRI image reveals the presence of a significant tumor of CN VIII located in the internal auditory canal and extending to the cerebello-pontine angle (see Figure 6.5). The tumor measures 2.5 cm in greatest diameter and is beginning to compress the brainstem in that area. The MRI for cases involving a possible C-P

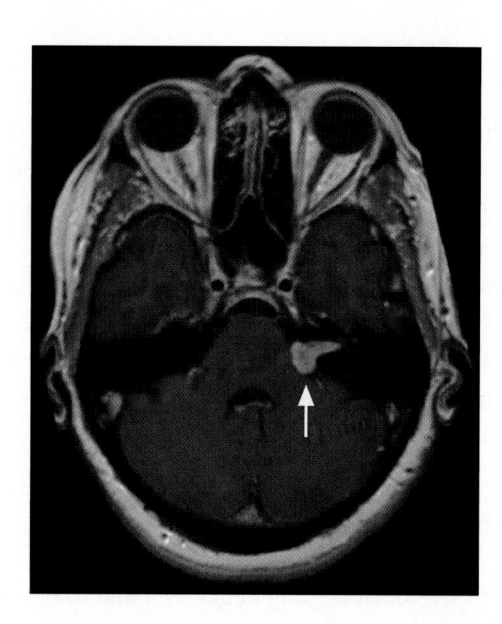

FIGURE 6.5 Radiograph. The contrast-enhanced MRI of the brain shows a typically shaped (ice cream cone-shaped) lesion on the left side, with a narrow "stem" laterally and an expansion medially impinging on the brainstem at the cerebello-pontine (C-P) angle, compatible with a vestibular schwannoma. (Courtesy of Dr. M. Kingstone.)

angle tumor can also be viewed with most newer machines using finer "cuts" of the brainstem and the skull in that region, without enhancement. Being strapped inside this space-age machine, along with experiencing all the clunking and other noises, is a great experience for Bernard, one that he loves to talk about whenever asked.

The results are relayed to the neurologist who wisely decides to fax her report followed by a telephone call to the family practitioner. The neurologist explains the diagnosis of *vestibular schwannoma* (an acoustic neuroma, as mentioned previously) as the most likely cause of Bernard's problems, and the likelihood of requiring neurosurgery.

6.9 Treatment, Management, and Outcome

Bernard is contacted and brought to the physician's office for a special appointment to be informed of the diagnosis. Notwithstanding his initial reluctance to consider surgery "to his brain," he quickly realizes, after explanations utilizing some anatomical diagrams, the actual location of his tumor and what might happen if it continues to grow. Bernard agrees to have a consultation with a neurosurgeon, back in the big city. A few phone calls more and the appointment is confirmed for the following month. It is now eight to nine months since the onset of symptoms and three to four months since Bernard first saw his family physician.

At the time of the consultation, the neurosurgeon explains the two main treatment options: fractionated stereotactic radiotherapy to the tumor bed or surgical removal. Bernard opts for the latter and the surgery is done four weeks later. The tumor is located as expected. This is very delicate and difficult surgery as the neurosurgeon must identify the various cranial nerves and other structures, and attempt to remove the tumor without damaging the acoustic division. As it is now almost one year since Bernard started complaining of his hearing loss, the tumor is now quite large, making the surgery more difficult. The surgery lasts about seven hours and the tumor is finally removed. The post-operative recovery is without complications. The pathology confirms the putative diagnosis of a benign vestibular schwannoma (see below). Ten days after the surgery, Bernard is able to go home, where he receives supportive services while his wound heals and until he is able to look after himself independently.

Bernard is seen in follow-up by his family physician about one month later. He is in a much better frame of mind and almost back to his usual outgoing self. There is some residual facial weakness noted on examination, but most people would not notice this (especially after his beard grows back). There is no longer any complaint of tinnitus, but, unfortunately, there is a complete loss of hearing in his left ear. Bernard slowly learns to live with this deficit and everyone around him learns to speak a little louder, or one at a time, or to his intact side.

A hearing aid is not usually recommended in such cases, as there has been a loss of the nerve itself, a sensorineural deficit, and hearing aids can only be useful if there is partially reduced hearing that can respond to sound amplification. However, newer devices are being tried, one of which is a bone-anchored hearing aid which could transfer sounds via bone conduction from the deaf side to the unaffected side. Bernard is seriously considering getting one of these devices.

6.9.1 Pathology

The term *acoustic neuroma* is somewhat misleading, as the tumor in fact usually originates on the vestibular

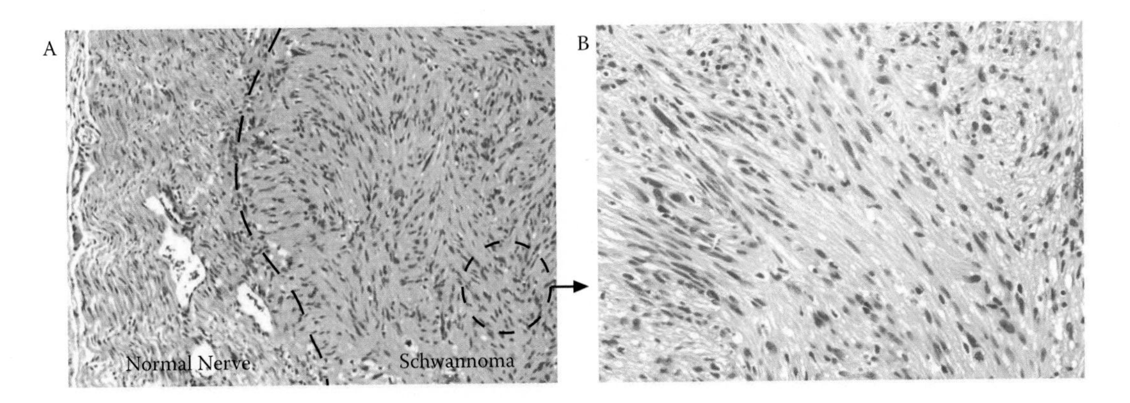

FIGURE 6.6 **Stained sections from a vestibular schwannoma. (A) The normal portion of the nerve with a more linear appearance is shown on the left; the pattern of cells of the schwannoma on the right side is more random. (B) The tumor is shown at higher magnification. Note that the cells in both areas, normal and tumor, are elongated (spindle-shaped). (Courtesy of Dr. J. Woulfe.)**

portion of the nerve (see Figure 6.6). The cells giving rise to this tumor are the Schwann cells, and the term *neuroma* is not appropriate as those tumors are of connective tissue origin. (Another erroneous term that has been used for these tumors is *neurolemmoma*.) The correct nomenclature is *vestibular schwannoma*. Growth of these tumors is slow and leads to pressure on the acoustic division, causing hearing loss and tinnitus. Continued growth of the tumor leads to pressure on CN VII at the C-P angle or within the internal auditory canal, thereby causing the facial motor weakness, of both upper and lower facial muscles, that is, lower motor neuron involvement.

It should be noted that since vestibular schwannomas are benign and often very slow growing, small tumors that are not causing significant symptoms can be monitored for growth; sometimes the decision is made to leave them untouched, without surgical intervention, particularly in older patients.

Related Cases

There are e-cases related to this chapter available for review on the text DVD and Web site.

Suggested Readings

Jackler, R.K. and C.L.W. Driscoll, Eds. *Tumours of the Ear and Temporal Bone.* Philadelphia, PA: Lippincott Williams and Wilkins, 2000.

Musiek, F.E. and J.A. Baron. *The Auditory System.* Upper Saddle River, NJ: Pearson Education, 2007.

Roland, P., B.F. Marple, and R.N. Samy. *Diagnosis and Management of Acoustic Neuroma. A Self Instructional Package*, 2nd ed. Alexandria, VA: American Academy of Otolaryngology, 2003.

Chapter 7

Bobo

BOBO

A few weeks ago Françoise McCool paid a lengthy visit to her older sister Bernadette, having not seen her for three years. Bernadette, aged 32, married Benoit Baguette, a chef, about five years ago. Together they opened a French restaurant in Fort Lauderdale, Florida, primarily catering to the numerous expatriate French Canadians in the area. Affectionately known to the locals as Benny and Bobo, Benoit and Bernadette have become highly regarded fixtures in the dining establishment, offering an eclectic mixture of traditional French-Canadian dishes and Continental cuisine.

After the initial excitement over the long-awaited reunion had dissipated, Fifi became concerned about her sister's changed appearance. Always an exuberant, vivacious (not to say hyperactive) personality, Bobo had become slow moving, almost lethargic. Her face, normally a kaleidoscope of changing emotions, seemed devoid of expression, almost a mask. At times, particularly later in the day, Bobo had trouble keeping her head erect, tending to slouch forward at her desk or table, holding her head with her left hand.

At first, Fifi told herself that Bobo was just very tired, possibly depressed. After all, the restaurant business is unrelenting in its demands on the owner's time and energy, with few possibilities for time off. As the days passed, however, Fifi began to notice several other alarming symptoms in her sister.

The most obvious problems occurred when Bobo walked from station to station in the restaurant kitchen. She had a tendency to keep her right arm held rigidly in front of her body rather than by her side, her elbow and fingers partially flexed. In addition, her right foot had a pronounced tendency to turn inward at the ankle as she walked. Toward the evening she typically walked slightly on her toes, unsteadily, and sometimes stumbled when walking upstairs. At the best of times she walked at a slow pace with short steps. Finally, Bobo's handwriting had deteriorated enormously: her formed letters were very small, sometimes rendering her writing illegible. In the past, Bobo had always been the one who wrote the daily specials on wallboards in the restaurant; about six months ago she had to delegate this responsibility to someone else.

When Fifi at last mentioned her concerns to her sister, Bobo conceded that she was concerned about her state of health, but had concluded that she was just overtired and needed a rest. After some reflection, Fifi replied, "Bobo, this is more than just being tired. I just realized that you are starting to look a bit like Papa did around the time you married Benoit. Papa's doctor thinks he has a mild form of Parkinson's disease. At any rate, Papa has done very well on his medication. Perhaps you need the same thing!"

Objectives

- To learn the main anatomical components of the central nervous system involved in the initiation, cessation and coordination of movement
- To understand the ways in which cerebral cortex, basal ganglia, cerebellum, thalamus, brainstem and spinal cord collaborate in the control of posture, locomotion and complex hand movements
- To understand the symptoms and signs resulting from a disease process affecting each of the three principal central components of the motor control system: the direct and indirect descending motor pathways, the basal ganglia and their connections, and the cerebellum and its connections
- To review the principal neurotransmitter molecules involved in the central movement control circuits

7.1 Solving the Clinical Problem

Before proceeding, as in previous chapters, the student should develop hypotheses concerning the probable localization of Bobo's disease process within the nervous system and the nature of that process, utilizing the history, examination, localization and etiology worksheets in the accompanying DVD and Web site. Once you have completed this process, you are ready to proceed with the remainder of the chapter.

Bobo required some convincing, but eventually she made an appointment with her family physician who—of course—turns out to be *you*.

Upon reviewing Bobo's file, you note that previously she has been in excellent health. Your only contact with her has been for prenatal care during her pregnancy, then for the delivery of her son. She has been too busy at work to come for regular check-ups. According to your records, she takes no medications other than oral contraceptives.

Your assessment of Mme. Baguette takes place at the end of the afternoon, in your "add-on" slot. Bobo comes with her sister, having worked all through the lunch sitting at the restaurant. The general physical examination is normal; all of the abnormal findings concern the nervous system.

Bobo sits on the examining table in a stooped position; she seems to have trouble holding her head and upper trunk erect enough to look at you face-on. You perform a brief mental status exam that is perfectly normal, although the volume of Bobo's voice appears decreased or muffled. In conversation she has virtually no facial expression, even when making an amusing, self-deprecating remark. In addition, you are struck by the fact that she almost never blinks. Her eye movements are full in all directions.

During your examination of her upper limbs you find that she has consistently normal muscle strength, proximal and distal. Passive movement around the wrists, elbows and shoulders, however, reveals a marked increase in muscle tone throughout the range of movement—at the wrist, for example, there is equal resistance to passive stretch whether the wrist is being flexed or extended. At the elbows there is also a jerky, ratchet-like quality to the resistance, more on the right side (a phenomenon known as cog-wheeling). The fingers of the right hand are maintained in flexion at the metacarpophalangeal joints, and extension of the interphalangeal joints. On both sides there is marked slowing in rapid hand and finger movements. A handwriting sample reveals a strikingly small character size.

In the lower extremities muscle strength is also normal, but muscle tone is increased at the knees and ankles. In particular, there is high tone in the invertors of both ankles, worse on the right. When she walks, she is stiff-legged, tending to go up on her toes, her feet turned inward, right more than left; she seems unsteady and takes tiny steps. At the same time, the flexed posture of her right arm becomes more striking; there is no arm swing on either side. Tendon stretch reflexes are uniformly increased while the plantar responses are flexor bilaterally; there is no ankle clonus.

When you comment on your positive findings to your patient, her sister (with her medically trained eye) points out that Bobo's gait and sitting posture, although far from normal, were much better at breakfast time than at present.

7.2 The Main Clinical Points

Bobo's chief complaints are that she has difficulty in walking and in climbing stairs, as well as progressively deteriorating handwriting.

The essential elements of Bobo's medical history and neurological examination are as follows.

7.2.1 *History*

- Insidious onset of symptoms over a period of at least six months, possibly as long as three years, early in the fourth decade of life
- Difficulty in maintaining the normal position of the head and upper trunk
- Increasing difficulty walking and climbing stairs; a tendency to trip over her own feet
- A progressive deterioration in handwriting, with a tendency to form excessively small letters (a finding known as *micrographia*)
- A worsening of symptoms over the course of the day
- A parent with "mild" Parkinson's disease, responding well to medication

7.2.2 *Physical Examination*

- On examination, normal vital signs and systems overview; normal mental status
- A flexed posture of the head and upper trunk
- Low voice volume; muffled speech; lack of facial expression; paucity of spontaneous blinking

- Normal muscle strength, proximal and distal, but a diffuse increase in muscle tone in response to stretch; at the elbows there is a "ratchet" quality to the tone during passive movement
- Abnormal posture of the right arm at rest, with flexion of the elbow, flexion of the metacarpophalangeal joints and extension of the interphalangeal joints
- A tendency for the feet to turn inward at the ankles (exaggerated inversion), with toe-walking, right side worse than left (usually referred to as an equinovarus posture)
- Abnormally small stride length; absent arm swing while walking
- Increased muscle tendon reflexes

As with the previous cases, it is important to recognize some important "negatives":

- Although Bobo is very slow-moving in general and has reduced finger movement speed, she is not weak, either proximally or distally, and does not have muscle wasting.
- She does not have a Babinski sign in either foot (see Chapters 2 and 3).
- Eye movements are normal, with no limitation of up-gaze.
- There is no hand tremor, either at rest or with activity.

In short, Bobo has a gradual-onset disorder of postural control and speed of movement, possibly hereditary. Except for her motor control problem, she seems to be functioning normally.

7.3 Relevant Neuroanatomy and Neurochemistry

As always, our first question concerning Bobo must be: *where* in the motor system is the problem?

Unlike Fifi (in Chapter 4), Bobo is not weak, has increased rather than decreased muscle tone in response to stretch, and has exaggerated rather than absent muscle tendon reflexes. In consequence we can immediately exclude the presence of a lesion involving lower motor neurons. Although she does not have Babinski signs, Bobo's clinical findings otherwise suggest the presence of a problem at the level of the upper motor neuron.

This being the case, we should examine in more detail the organization of the upper part of the motor system.

As a first step, building on material provided in Chapters 1, 2 and 4, let us look at the ways in which movement commands are transmitted from the motor areas of the cerebral cortex to the lower motor neurons (the final common pathway) in the brainstem and spinal cord.

7.3.1 Descending Motor Pathways

As we noted in Chapter 2, there are two principal pathways for the downward transmission of motor messages, the so-called *direct* and *indirect pathways*, both illustrated in Figure 7.1 (see also Figure 2.5). The two pathways work together in (1) carrying out a specific complex movement with hand or foot and (2) achieving the body positions necessary to permit the performance of such a movement.

The *direct pathway* for movement control appears to have evolved later and has been added to the indirect pathway to allow for voluntary precision movements, primarily of the hands and fingers. As might be expected, it is particularly well developed in primates. The neurons transmitting messages required, for example, to employ the right thumb and index finger in holding a piece of chalk while writing a menu, have their cell bodies in the deeper layers of the motor cortex anterior to the left central fissure. The axons originating in these neurons carry their action potentials all the way to the gray matter of the cervical spinal cord in one fell swoop: they pass through the cerebral hemispheric white matter and enter the internal capsule, a large white matter tract adjacent to the basal ganglia. The motor axons in the internal capsule then descend lateral to the thalamus, subsequently passing downward into the anterior midbrain via the cerebral peduncles. From there the axons voyage through the ventral part of the pons into the ventral medulla where they form part of a visible bulge referred to as the pyramid (due to its pyramidal shape in cross-section).

From the left pyramid the motor axons destined for the right cervical cord region pass posteriorly and downward through the lower medulla, crossing the midline and entering the upper cervical spinal cord as part of a distinct bundle known as the (right) *lateral corticospinal tract*. Once they have arrived at the level of the spinal cord involved, let us say, in finger-thumb apposition

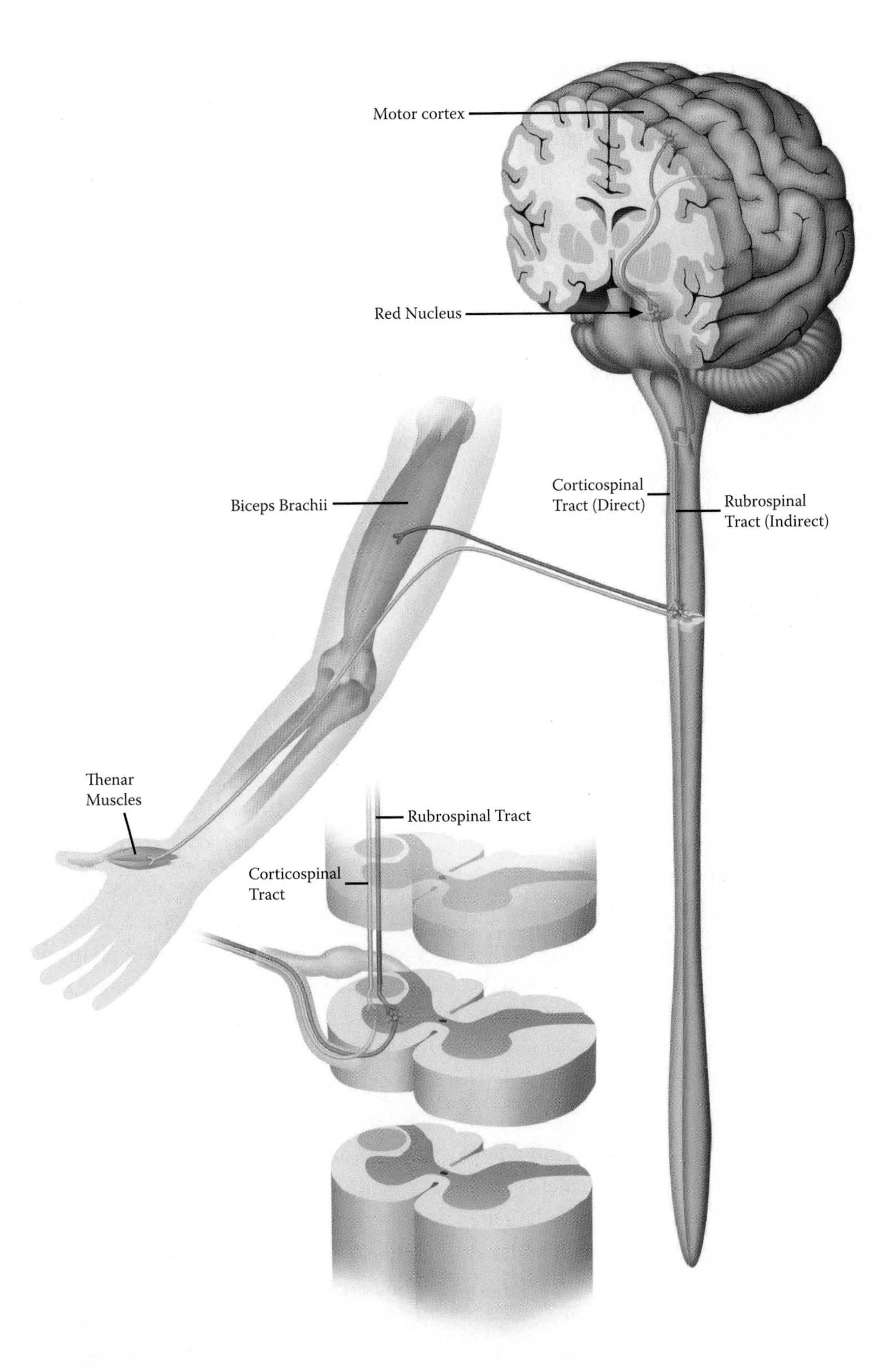

FIGURE 7.1 Direct and indirect descending motor pathways. The light green neurons constitute the direct pathway, the dark green neurons the indirect pathway. Note the synapse in the course of the indirect pathway in the red nucleus in the midbrain.

(C8 and T1), the axons enter the right anterior horn of the gray matter and synapse either directly on the appropriate anterior horn cells (lower motor neurons) or indirectly via an interneuron.

Thus, when Bobo used to write out the menu du jour on the restaurant blackboard, the intricacies of letter and word formation were primarily transmitted to the final common pathway in the cervical cord by the direct motor pathway. Since the pathway passes through the medullary pyramid (while the soon-to-be-described "indirect motor pathway" does not), it has come to be known as the *pyramidal motor system*.

Clearly, in order to write out the lunch menu, Bobo needs the cooperation of much more than the muscles involved in manipulating a stick of chalk. She will need to be able to maintain a stable, standing position, with her head steady, her eyes facing the board. She must keep her shoulder partially abducted (and stable), her elbow partly flexed, and her wrist in a neutral position. All of the body and limb positioning, with respect to the targeted blackboard, is the job of the *indirect motor pathway*, often referred to by clinicians as the "extrapyramidal motor system," a term that is simplistic and somewhat misleading.

Neurons involved in carrying messages to the necessary trunk and postural limb muscles have their cell bodies in more widespread areas of the cerebral cortex: not just the precentral (motor) gyrus, but also the so-called *premotor* cortex (anterior to the primary motor cortex) and even the "sensory" cortex just posterior to the central fissure. These axons also travel downward through the cerebral white matter and internal capsule to the midbrain, pons and medulla where they terminate in a variety of intermediary areas, the most important of which are the red nucleus (midbrain Figure 7.1) and the pontine and medullary reticular formation (central brainstem). These "indirect" motor relays are therefore referred to, respectively, as the corticorubral and corticoreticular pathways.

In turn, motor axons from the intermediary relay centers pass inferiorly to the spinal cord, to the cervical, thoracic and lumbosacral cord regions, as appropriate, to connect via interneurons with anterior horn cells projecting to proximal arm and leg muscles, as well as trunk musculature. The spinal cord tracts containing these motor axons are the rubrospinal tract (lateral column white matter) and the reticulospinal tracts (medullary and pontine—lateral and anterior columns, respectively); see representative examples in Figure 7.2.

All of these indirect pathway tracts, if disconnected from the sensorimotor cortex, exert a tonic influence upon their anterior horn cell targets, the net result of which is an increase in tone in the trunk and proximal limb muscles. Lack of cortical modulation of the rubrospinal and medullary reticulospinal pathways results in increased tone primarily in the flexor muscles, especially of the upper limbs. Similar disinhibition of the pontine reticulospinal tract causes increased tone in the extensor muscles of all four limbs. It is for this reason that "upper motor neuron" lesions are accompanied by increased muscle tone.

We will return to the effects of lesions in one or another of the "direct" and "indirect" motor control pathways when we consider in more detail the localization of Bobo's disease process. First, however, we must describe the components of the motor system involved in the initiation and cessation of movement (the basal ganglia connection) and in the precise coordination of movement, once initiated (the cerebellar connection).

7.3.2 The Basal Ganglia: The "Accelerator" and the "Brake"

The location and anatomical description of the basal ganglia were outlined in Chapter 1. With the exception of the caudate nucleus, components of all the other basal ganglia participate in the overall apparatus of motor control. To these we must add two specific thalamic nuclei located in the antero-lateral part of that structure: the ventral lateral (VL) and ventral anterior (VA) nuclei. Figure 7.3 shows the anatomical locations and relationships of the basal ganglia and thalamus.

With respect to the initiation and cessation of movement circuitry, the key structures are, in what amounts to a functional sequence, the motor cortex, the putamen, the globus pallidus externus (GPe), the subthalamic nucleus, the globus pallidus internus (GPi), and the VA/VL nuclei of the thalamus. The latter nuclei then complete an elaborate loop by projecting back to the motor cortex.

The anatomical circuitry and neurochemical sequences involved in the basal ganglia's role in movement control are complex. Since this is an introductory text, it is necessary, and useful, to over-simplify the situation sufficiently in order to communicate the basic principles required for

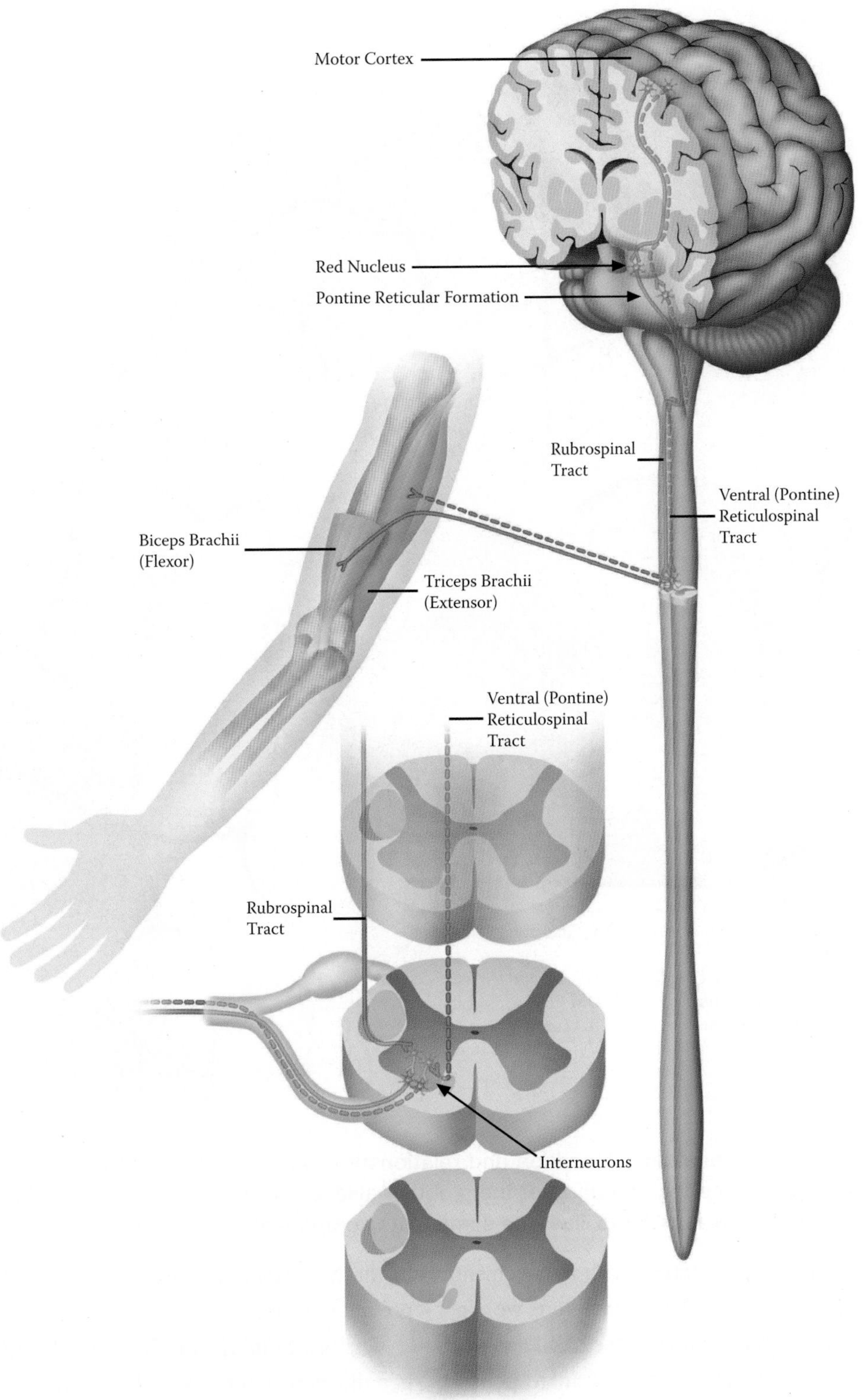

FIGURE 7.2 Indirect descending motor pathways: proximal flexor and extensor activities. The solid green neurons represent the pathway for the flexor muscle, the dashed neurons the pathway for the extensor muscle.

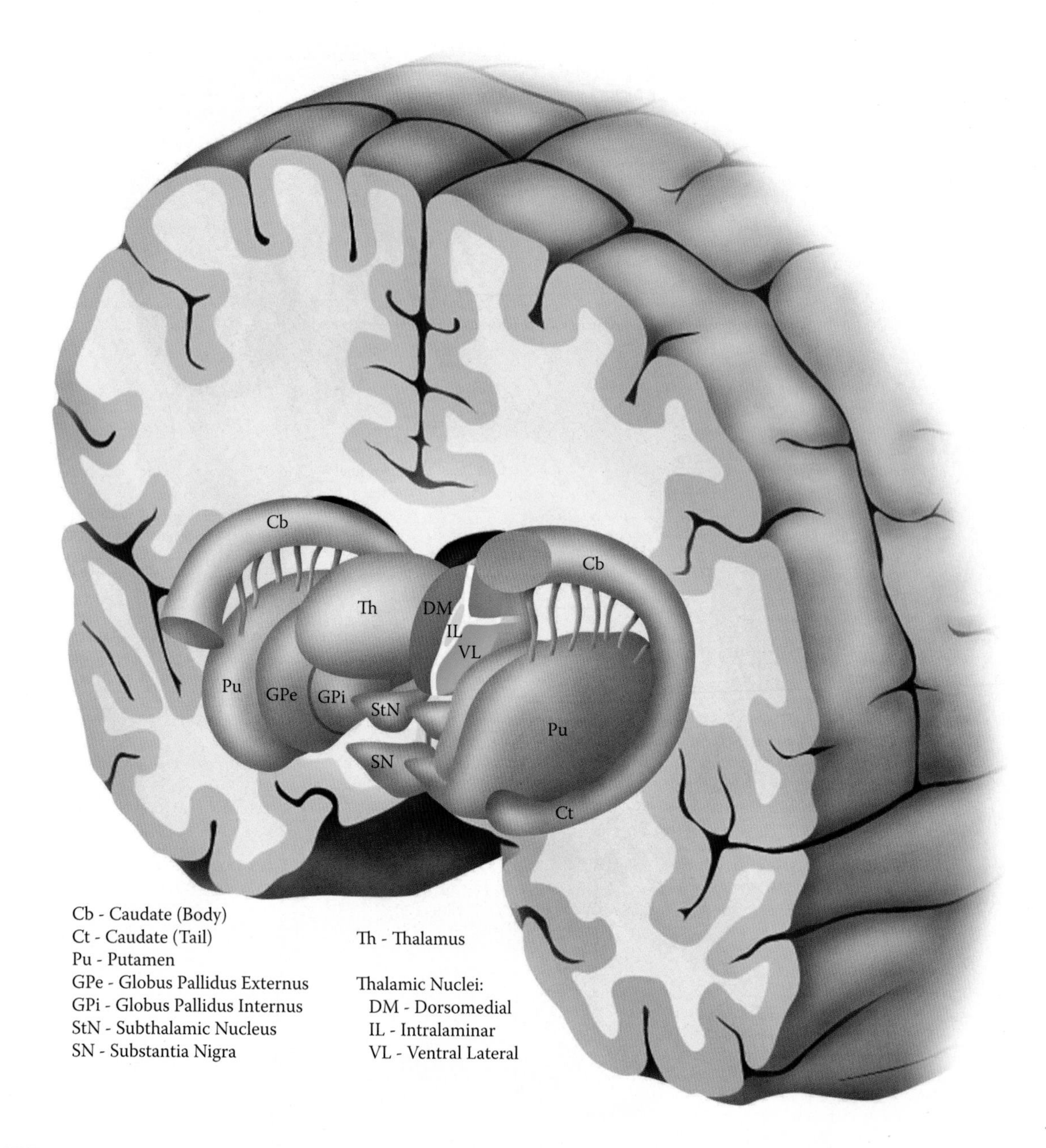

FIGURE 7.3 **Basal ganglia: anatomical location and relationships, including the subthalamic nucleus (StN) and the substantia nigra (SN). Note the two parts of the globus pallidus: externus and internus. The left thalamus is sectioned in the coronal plane in order to illustrate some of its component nuclei.**

an initial clinical approach to a patient with a movement disorder. From the practical point of view, therefore, we can divide the movement control circuitry of the basal ganglia into two components: a so-called "direct" circuit responsible for the initiation of movement and an "indirect" circuit whose function is the cessation of movement. It is important for the student not to confuse the terms *direct* and *indirect* (in the context of basal ganglia circuitry) with the direct and indirect pathways conveying motor messages to the brainstem and spinal cord already described (see previous section).

The basic circuits for the "direct" and "indirect" basal ganglia connections are illustrated in Figure 7.4. The *direct* circuit is so named because there is a direct connection between the putamen and the GPi, the output from the GPi proceeding thence to the VA/VL thalamic nuclei. In contrast, the *indirect* circuit involves a diversionary loop from the putamen to GPe, then subthalamic nucleus,

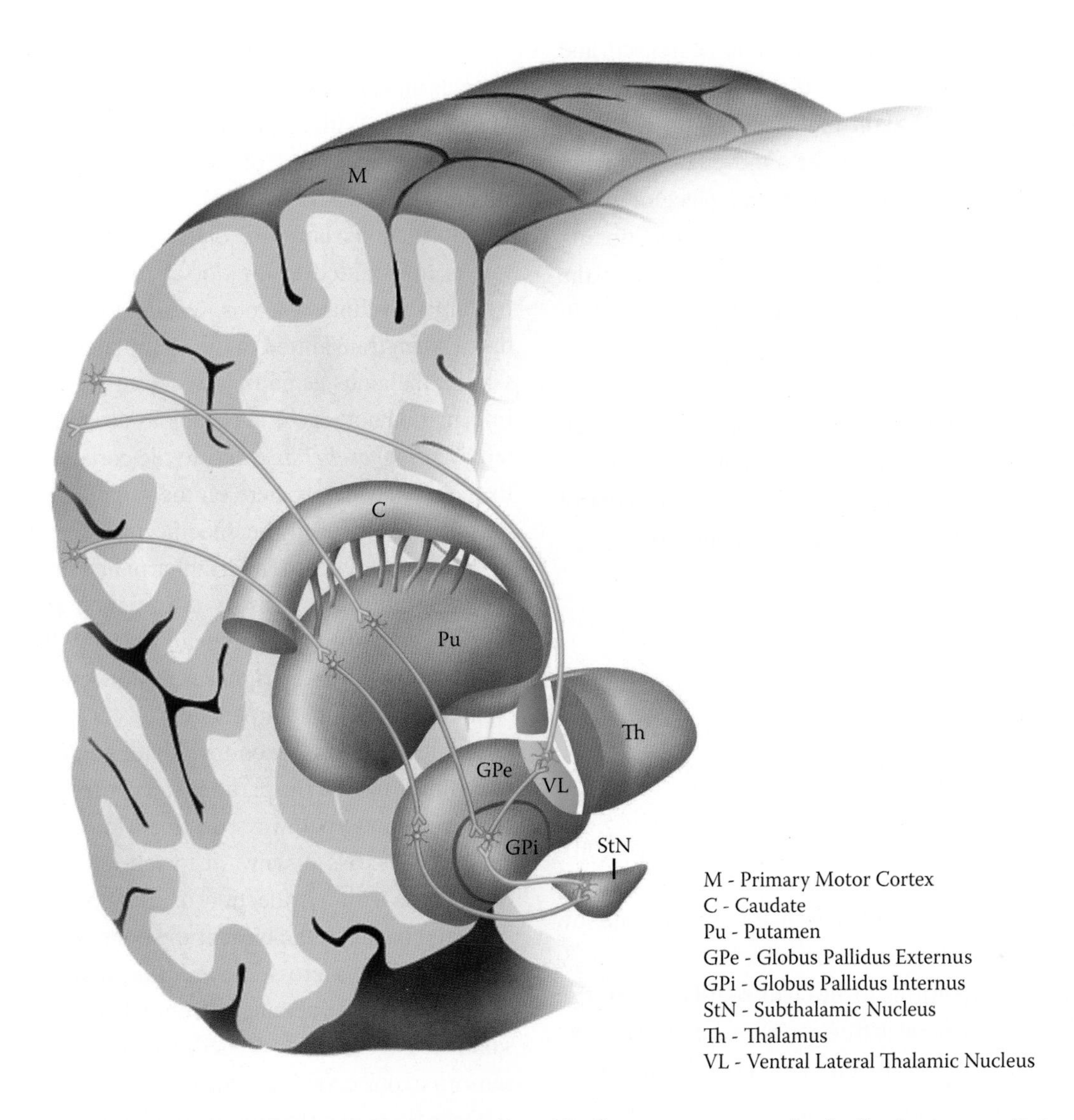

FIGURE 7.4 Basal ganglia motor circuitry, phase 1. For this figure, a green color indicates an excitatory neuron (not a motor pathway neuron as is standard for this text); a red color indicates an inhibitory neuron (not an interneuron as in other illustrations). These circuits are "animated" in the text DVD and Web site.

finally to GPi. In effect, the net role of the direct circuit is to excite the motor cortex (via the thalamus) and thus to initiate movement, while the role of the indirect circuit is to inhibit the motor cortex.

In a sense, the direct circuit can be compared to the accelerator pedal in the automobile we have been intermittently employing to illuminate aspects of human locomotion—push on the gas pedal and our car moves forward. The indirect circuit, on the other hand, is like the brake—press on it and our car stops moving. Using this analogy, and to avoid confusion, we will henceforth refer to the "direct" basal ganglia circuit as the *accelerator circuit*, and the "indirect" circuit as the *brake circuit*.

The normal function of the VA/VL thalamic nuclei is to excite the motor cortex while the normal function of GPi is to inhibit that excitation (see animated version of Figure 7.4). Thus, the "normal" output from the basal ganglia would be having the brake applied; in order to initiate movement the accelerator circuit inhibits GPi, *removes* the brake and permits increased activity in VA/VL. While this concept undoubtedly seems confusing, the student simply has to recall the basic mathematical principle that a "minus" (putamen inhibits GPi) plus a "minus" (GPi inhibits VA/VL) yields a "plus": inhibition of inhibition = excitation.

For the brake circuit, in contrast, the output from the subthalamic nucleus is excitatory: thus GPi's inhibitory function is increased, VA/VL activity is decreased and movement cannot occur. In effect, the subthalamic nucleus represents a foot applied to the brake.

In summary, for Bobo to initiate a movement, whether a spontaneous smile or an upstroke with a piece of chalk, her putamen-GPi accelerator circuit must be functioning properly. In order to stop the chalk upstroke so that the succeeding down-stroke may take place, the GPe-StN-GPi brake circuit must also be doing its job.

Thus, for the menu du jour to appear on the blackboard, both basal ganglia circuits must be in good working order. For the writing to be performed evenly and efficiently and for it to be legible, however, a second set of motor control circuitry will be required: the cerebellar connection.

7.3.3 The Cerebellum: The "Steering Wheel"

Put simply, the function of the cerebellum, as concerns movement control, is to compare information from cerebral cortex about movement commands just being transmitted with information from peripheral sensory receptors concerning the posture and speed of the part of the body being moved. This comparator function is a continuous process occurring, epoch by epoch, or millisecond by millisecond, throughout a given movement.

In other words, returning to the chalkboard analogy, the cerebellum at a given point in time is simultaneously receiving information from Bobo's left motor cortex concerning commands just being sent to her cervical cord (to allow for a further component of an upstroke movement), and information from muscle and joint sensors in her right arm and fingers concerning the part of the upstroke movement that took place a few milliseconds earlier. Her cerebellum, having analyzed this data, then sends a message back to the motor cortex that in essence is saying: "Fine, your previous command had this result, in comparison with what was intended—here is what should happen next to allow the action to continue in the manner desired."

As was the case with the basal ganglia, the anatomy of the cerebellum is extremely complex. In order to address Bobo's clinical problem in a logical fashion, it is simply necessary for us to outline the broad aspects of the cerebellar connections, and how they link with the rest of the apparatus controlling movement.

As for the basal ganglia, the location, anatomical components and attachments of the cerebellum were described in Chapter 1. With respect to its role in movement control, the main connections of the cerebellum are outlined in Figures 7.5 and 7.6.

Information from skin, joints, and muscle spindle receptors is transmitted to the dorsal roots of the spinal cord by large myelinated axons. While some of this sensory data is then transmitted to the contralateral sensory cortex via the thalamus as conscious sensation, parallel sensory information not reaching conscious awareness is transmitted by *spinocerebellar tracts* to the cortical gray matter in the cerebellar hemisphere on the same side as the limb(s) concerned. Since, in the specific instance of Bobo writing on the blackboard, her right arm and hand are primarily doing the work, sensory information about right arm posture and position is being conveyed by the right spinocerebellar tracts to the right cerebellar hemisphere (via the right inferior cerebellar peduncle).

At the same time, axons originating in Bobo's left cerebral motor cortex (in parallel to corticospinal and corticoreticulospinal pathways destined for the right cervical spinal cord) pass downward to the left side of the pons to synapse with a collection of relay neurons located in scattered nuclei in the base of the pons. Axons from these pontine neurons cross the midline and enter the *right* middle cerebellar peduncle, from where they travel to the same area of the right cerebellar hemispheric cortex as the sensory axons carrying information from the right arm.

Once these two streams of converging information have been compared and analyzed by the cerebellar cortex, appropriately consolidated information is then transmitted to neurons located in large nuclei in the central part of the cerebellar hemisphere (see Figure 7.5). For right arm movement this information, originating in the largest neurons of the cerebellar cortex, the *Purkinje cells*, is transmitted to the right *dentate nucleus* in the central region of the cerebellar hemisphere. From there, information is relayed upward in axons located in the right superior cerebellar peduncle. In the midbrain, the axons cross the midline and enter the left red nucleus and either synapse with red nucleus neurons or pass through en route to the left thalamus where they synapse with neurons in the VL thalamic nucleus. These latter neurons then project back to the left sensorimotor cortex to complete the feedback loop.

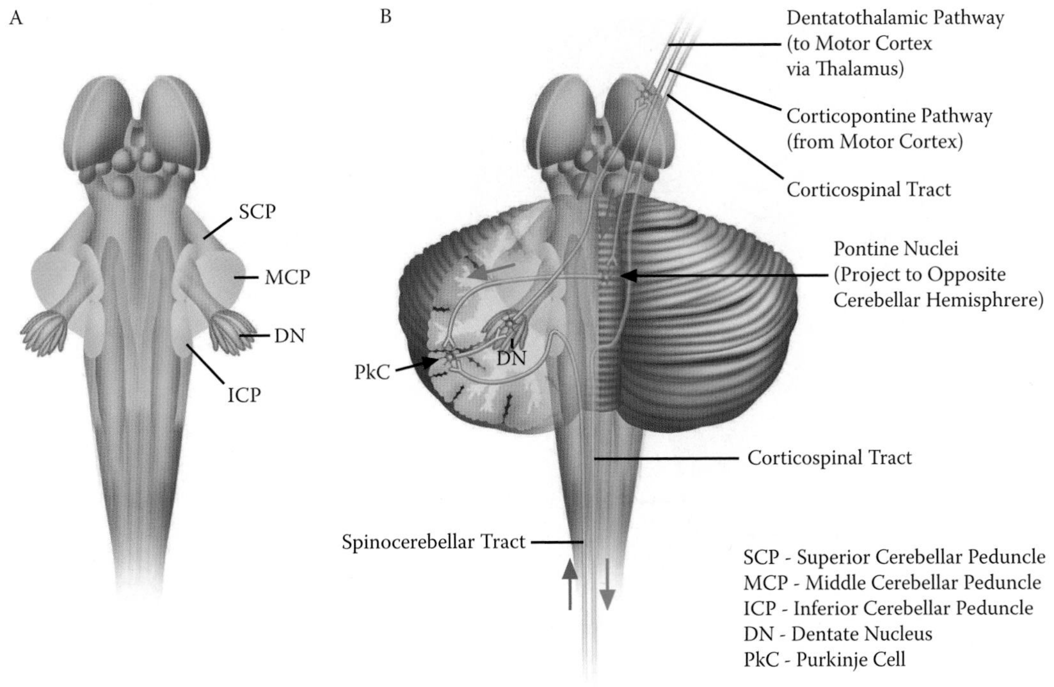

FIGURE 7.5 **Cerebellum connections and pathways from the dorsal perspective: (A) Brainstem with the three cerebellar peduncles, superior, middle and inferior. (B) Cerebellar pathways: inputs from spinal cord (spinocerebellar) and cortex (via pons); output to motor cortex via the thalamus. The descending corticospinal pathway is also included.**

As we have already intimated, there are functional differences between parts of the cerebellum. The midline portion of the cerebellum (the vermis) and the para-midline regions of the cerebellar hemispheres are concerned primarily with integration of trunk postural control and leg movement, that is, the facilitation of sitting, standing and walking. The lateral regions of the cerebellar hemispheres, by far the largest in terms of volume and surface area, are concerned with arm, hand and finger movements.

To return to Bobo and her menu board, the right cerebellar cortex is conveying information rostrally to the left cerebral motor cortex in a just-in-time fashion to help instruct the cortex how to direct the next part of the movement previously planned by the left supplementary motor area. In this way, for example, the chalk is held with consistent (rather than varying) pressure against the board and, in creating the letter D (for D'Hôte), is made to move upward just enough to make the letter the same height as all the other capital letters in the line. The chalk is then made to change direction, proceeding downward and to the right in a curving fashion to eventually arrive at the starting point of the letter D, no more and no less. One can thus imagine that, without the input of the cerebellum, the letter D might be too short or too high, the curve back to the origin poorly formed, and the line perhaps continued past the origin to produce a symbol unrecognizable as a "D".

Clearly the process of successfully producing a letter D will also require constant visual feedback, information relayed from the visual cortex both to the motor cortex and to the cerebellum.

Finally, to complete the automobile analogy, our car may be moving forward, at speed, but will drive into a ditch (or worse!) if, based on constant feedback as to the car's speed and direction, steering wheel adjustments are not constantly being made. Without the basal ganglia circuit, the "automobile" will not leave its parking spot; without the cerebellar circuit, the "automobile" will not stay on the road. Both components are required to get from point A to point B.

Although we have confined our discussion of the cerebellum to its role in movement control, it is important to recognize that the cerebellum also plays a role in the integration of higher brain functions such as language, memory and learning.

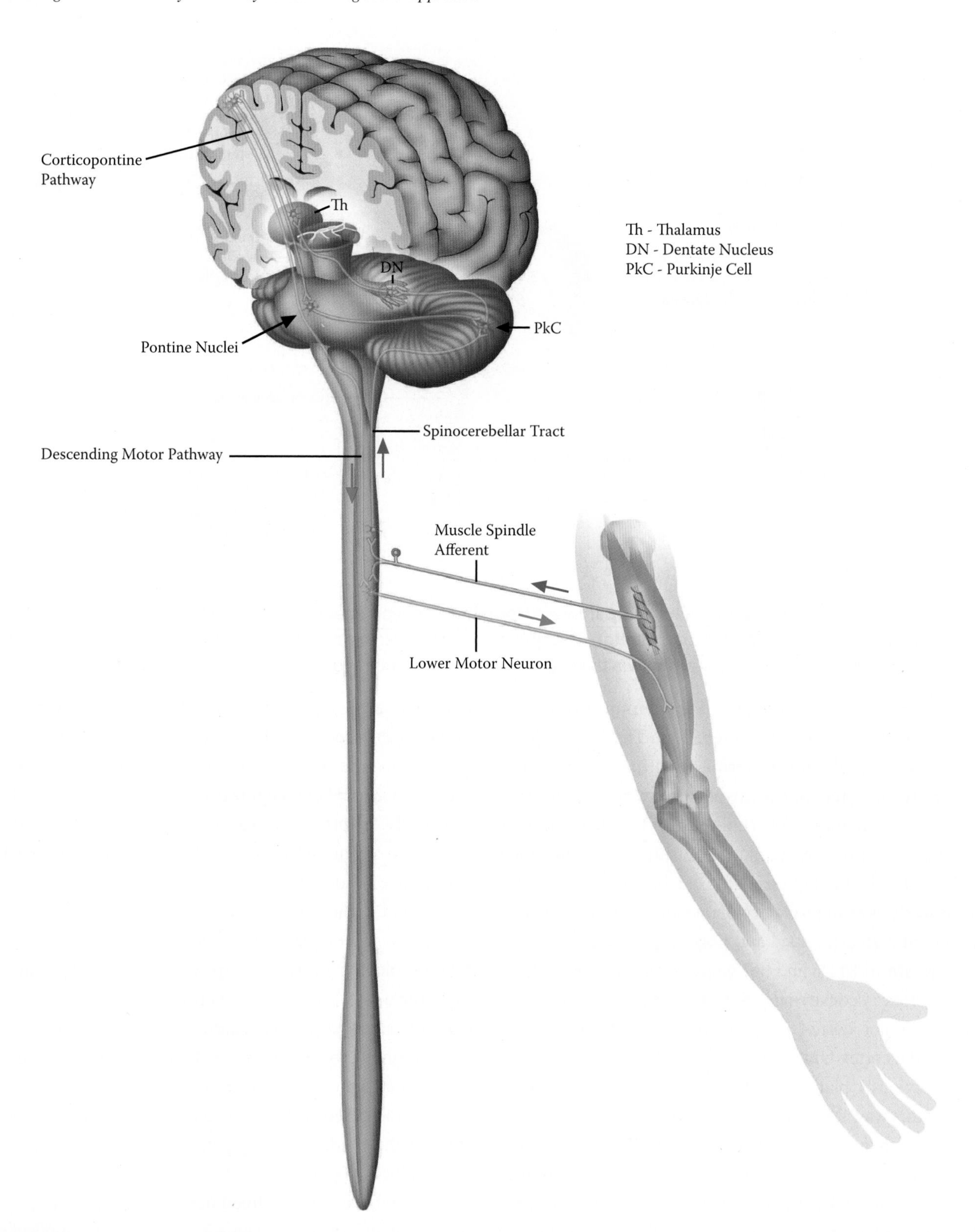

FIGURE 7.6 Cerebellum: basic circuitry for motor function. The color code is the same as for Figure 7.5. Note the "double crossing" of the cerebellar information. These circuits are animated in the text DVD and Web site.

7.3.4 *Neurotransmitters Involved in Movement Control*

The structure and function of neurotransmitter molecules will be considered in some depth in Chapter 10. At this stage in the text it is simply necessary to note that a variety of transmitter molecules are involved in cell-to-cell signaling in the components of the motor control system discussed in this chapter.

The most important transmitters are glutamate (an excitatory compound) and gamma-amino butyrate (also known as GABA, an inhibitory compound). Glutamate is the transmitter operating in the descending motor pathways (the upper motor neuron), as well as the excitatory components of the basal ganglia and cerebellar circuitry; the inhibitory components of the latter circuits are mediated primarily by GABA (see Figure 7.4).

Other transmitters implicated in basal ganglia circuitry are acetylcholine, substance P, enkephalins and the monoamine transmitter dopamine; dopamine will be discussed in more detail later in this chapter.

As is well known, the lower motor neuron transmitter is acetylcholine.

7.4 Localization Process

We have already concluded that, with hyperactive, rather than absent muscle tendon reflexes, and increased, rather than decreased muscle tone, it is highly unlikely that Bobo has a disease process affecting peripheral nerves, neuromuscular junctions or muscles. Since her disorder includes diminished action in facial muscles, it is also improbable that she has a lesion confined to the spinal cord.

Bobo's symptoms and signs point to a disease process affecting the upper levels of the motor control system, but without any associated sensory or cognitive abnormalities. While she may have impaired abilities to walk, talk and write, she has lost none of her capacity to reason, to learn and to remember. This means that a diffuse disease process affecting cerebral cortex, central white matter or entire central gray matter structures such as the thalamus, is not likely.

Based on what we have already learned about the organization of the upper levels of the motor system, we have to consider three main possibilities:

1. The disease could be affecting the descending motor pathways, both direct (e.g., impaired finger movement control) and indirect (e.g., impaired postural control of body and limbs), either in the cerebral hemispheres or in the brainstem.
2. There could be a disorder involving the basal ganglia circuitry.
3. There could be a disorder affecting the cerebellum and its connections.

7.4.1 *Descending Motor Pathways*

Pure disruptions of the direct or "pyramidal" motor control pathways are relatively rare as they require the presence of a lesion confined to the motor cortex (the precentral gyrus) or to the medullary pyramids. Lesions involving the cerebral white matter, internal capsules or cerebral peduncles tend to affect both the direct and indirect pathways as they travel intermixed at these levels. When an isolated dysfunction of the direct pathway does occur, it is characterized by a loss of independent finger movements and fine motor control in the hands (as well as an equivalent loss of rapid movements in the feet and toes) and a positive Babinski sign. There is *no* significant disturbance in muscle tone. Thus, with an isolated direct pathway lesion, one could still pick up an apple and eat it, but would have difficulty paring the skin off the fruit.

In contrast, the effects of an indirect "extrapyramidal" motor pathway lesion are far more dramatic. Disruption of the corticoreticular and corticorubral connections (as, for example, with a lesion of the posterior limb of the internal capsule) results in an inability to move the hands and feet in a purposeful fashion, an increase in muscle tone in the limbs, hyperactive muscle tendon reflexes and Babinski signs. If the lesion is unilateral, the control of the posture of the corresponding arm and leg may be so compromised that the patient may be unable to bear weight on the leg or use the arm to reach for an object. When the lesion is bilateral, the patient cannot walk, or even sit without support, and requires a wheelchair.

In practice, most disorders affecting the descending motor pathways disrupt *both* the direct and indirect pathways, leading to a combination of the deficits outlined above. A lesion of the left internal capsule, for example, will result in a loss of fine finger movement in the right hand, loss of a pincer or even a crude grasp, difficulty extending the fingers and wrist, and inability to dorsiflex

FIGURE 7.7 Patient with left spastic hemiparesis following cerebral vessel thrombosis in the context of a relapse of ulcerative collitis. There is weakness of voluntary facial expression on the left, flexion of the left arm at the elbow, and fisting of the left hand. (Patient's release and with permission; courtesy of Dr. P. Humphreys.)

the right foot and toes. Muscle tone in response to stretch on the right side will be increased, particularly in the finger flexors, wrist flexors and pronators, elbow flexors, knee flexors, plantar flexors and foot invertors. This combination of weakness and increased muscle tone results in a characteristic "hemiparetic" posture, with fisted hand, drop-wrist posture, flexed elbow, and tendency to walk on the toes with the foot inverted (an *equinovarus* posture). An example is shown in Figure 7.7.

With this description in mind, let us examine Bobo's clinical findings for similarities and differences. On the plus side, Bobo has a flexed posture of the forearm on the right, an equinovarus posture of the right foot, increased muscle tone in the arms and legs, and abnormally brisk muscle tendon reflexes. That said, however, there are many important differences. In particular, formal muscle strength testing revealed *no* weakness, not even in intrinsic hand muscles. Rapid hand and finger movements are not abolished: they are just very slowly performed. The posture of her right hand was not characteristic of an indirect motor pathway deficit in that her interphalangeal joints were maintained in extension rather than flexion. Finally Bobo did not have positive Babinski reflexes, a sine qua non of lesions involving the descending motor pathways.

Thus, the result of our comparison suggests that it is unlikely that Bobo has a pathological process primarily affecting the descending motor pathways.

7.4.2 Basal Ganglia Circuitry

If we review the information already given about the motor control system, and compare this with what we found on Bobo's neurological examination, we find a number of similarities. A major difficulty for Bobo was the spontaneous and voluntary initiation of movement: she did not blink; her face was expressionless; fine finger movements were slow and her handwriting abnormally small; she was able to walk, but awkwardly and at a slow rate.

There were also several things, previously mentioned, that we did *not* find in our examination of Bobo, and that also support the possibility of a basal ganglia disturbance: the lack of muscle weakness per se, and the flexor, rather than extensor, plantar responses.

Bobo also had two other findings consistent with basal ganglia dysfunction that we have not yet addressed in our consideration of this component of the motor control system: she had difficulty maintaining an upright posture of her head and upper trunk; there was a diffuse increase in muscle tone affecting both agonist and antagonist muscles, sometimes with a "ratchet"-like quality.

That the basal ganglia play a role in postural control is based largely on evidence from clinicopathological correlations. Individuals with acquired bilateral injury to the basal ganglia (for example, children with dystonic cerebral palsy following perinatal anoxic damage to the putamen; adults with carbon monoxide poisoning and injury to the globus pallidus) characteristically have difficulty maintaining head control and a normal sitting/standing posture. This postural control mechanism is presumed to be mediated via the sensorimotor cortex and indirect motor pathway projections to the brainstem reticular formation.

Similarly, basal ganglia pathology, whether unilateral or bilateral, is characteristically accompanied by an abnormal tone pattern quite distinct from that resulting from lesions in the indirect descending motor pathways. Instead of tone being primarily increased in proximal and distal flexor muscles (see previous section), muscle tone in basal ganglia pathology typically involves opposing flexor and extensor muscles simultaneously. If the examiner were to briskly flex and extend the elbow joint(s) in someone with an indirect motor pathway lesion (e.g., internal capsule infarct), there would be a marked resistance from the biceps brachii but little resistance from the triceps. In addition, in this scenario, continued stretching of the biceps would typically result in an abrupt reduction in resistance as the muscle lengthened: the so-called "clasp-knife" phenomenon. In contrast, a similar maneuver in an individual with a basal ganglia (e.g., globus pallidus) lesion would result in an equal degree of difficulty extending *and* flexing the elbow, without any sudden-release phenomenon toward the end of the range of movement. We have clearly documented this latter phenomenon, also known as lead-pipe rigidity, in our examination of Bobo.

In certain specific basal ganglia disorders (to be described later), the rigid tone may be accompanied by a rhythmic series of brief tone reductions throughout the range of movement. This rhythmic tone fluctuation, present in muscles acting around Bobo's elbow joints, is known as *cog-wheel rigidity*. Some individuals with cog-wheel rigidity (although not Bobo) also have a coarse *tremor* of the hands and fingers when the hands are not in use; this *resting tremor* is often referred to as a "pill-rolling" tremor because it appears as if the person is rolling a small object between the thumb and fingers.

Finally, for reasons that are unclear but presumably relate to relative brain immaturity, children and young adults with basal ganglia dysfunction often demonstrate twisted trunk and limb postures. The patient's head may be uncontrollably turned to one side (torticollis); one arm may be held in extension across the front of the body, the wrist flexed and the fingers extended; one leg may be flexed at the hip joint while the other is extended. These various twisted postures are collectively referred to as *dystonia*.

On the whole, therefore, Bobo has demonstrated many characteristic features of a disorder of basal ganglia circuitry.

7.4.3 The Cerebellum and Its Connections

Since, as we have seen, the cerebellum is involved primarily (with respect to movement control) in the ongoing smooth coordination of a specific movement throughout its performance, a disturbance in cerebellar function would be expected to result in poorly coordinated movement. Such is, in fact, the case. Movements in a person with cerebellar dysfunction, whether he or she is walking across a room or writing on a blackboard, are not so much slow as they are erratic.

We have already noted that the midline and paramedian portions of the cerebellum are largely concerned with control of trunk movements. Thus, it would not be surprising to learn that patients with midline cerebellar dysfunction tend to stagger when walking and to veer sideways. With severe disruption of function they may be unable to maintain a standing or sitting posture without lurching over, a phenomenon known as *truncal ataxia*. It is often said that patients with midline (vermis) cerebellar lesions walk as if they are drunk: this is no coincidence as midline cerebellar function is easily suppressed by ethanol.

On the other hand, impaired function in the cerebellar hemispheres results in incoordination of limb movements, both of arms and legs (see Chapter 3). In reaching for an object, for example, the patient is unable to gauge exactly how far to extend the arm and may either undershoot or overshoot the object (a phenomenon referred to as *dysmetria*). At the same time, given that antagonistic muscles cannot be respectively contracted and relaxed in a smooth, graded fashion, the antagonistic muscles tend to shorten or lengthen in an irregular, step-wise fashion—this results in a coarse, oscillating tremor during the intended activity, known as an *intention tremor*.

As was the case with basal ganglia dysfunction, muscle strength is fairly normal in cerebellar disorders. When someone with cerebellar dysfunction is requested to contract a muscle as strongly as possible (for example, to resist a powerful pull against the muscle), there will be difficulty contracting the muscle in a consistent fashion such that the tension may appear to vary from moment to moment. Likewise there may be difficulty reducing the amount of muscle contraction in a regular, timely fashion (as, for example, when the pulling force against the limb has been suddenly reduced), resulting in the limb overshooting—the *rebound* phenomenon.

As might be anticipated, handwriting in an individual with cerebellar dysfunction is irregular in size and spacing of letters. Rather than being uniformly small, the letters tend to be large and poorly formed, often unrecognizable.

Rather than being increased, muscle tone in cerebellar disorders is typically low, and muscle tendon reflexes are reduced, rather than hyperactive.

Taken individually and together, the characteristic motor abnormalities in an individual with cerebellar pathology appear completely different from what we have noted on Bobo's neurological examination.

Thus, of the three main components of the upper echelon of the motor control system, the *basal ganglia circuit* seems the likely target of Bobo's disease process.

7.5 Etiology: Bobo's Disease Process

Given that Bobo has had her symptoms for months, possibly years, a review of Table 3.13 would lead one to conclude that the time frame is too long for the paroxysmal, traumatic and vascular disease categories, but that all of the other categories are theoretically possible. As in other chapters, we will briefly consider each disease category, focusing most of our attention on the degenerative and genetic categories as these appear by far the most likely.

7.5.1 Paroxysmal

Paroxysmal disorders of the basal ganglia are rare, but well described. The most important example is paroxysmal kinesigenic dystonia, an autosomal dominant hereditary disorder. Affected individuals, appearing otherwise normal, suddenly develop unilateral or bilateral limb dystonic posturing while in the process of standing up or starting to walk or run. As a result, the patient is unable to continue with the activity and may fall to the ground. Even though this is not an epileptic disorder, the symptoms usually respond well to antiepileptic drugs such as carbamazepine.

7.5.2 Traumatic

Although rather unusual, one of the lentiform nuclei may undergo hemorrhagic contusion with a closed head injury. Initially the patient is typically in a deep coma with a flaccid hemiparesis; the latter eventually evolves into a severe dystonic hemiparesis. Clearly neither the history nor the time frame supports the idea of a traumatic etiology.

7.5.3 Vascular

Both ischemic and hemorrhagic strokes may selectively damage the basal ganglia, usually the caudate head, putamen and globus pallidus in combination, typically in one cerebral hemisphere. If the patient survives the event—less likely in the case of a basal ganglia hemorrhage—there will again be a dense dystonic hemiparesis. Vascular disorders of the cerebral hemispheres will be considered in detail in the next chapter.

7.5.4 Toxic

Both basal ganglia may be selectively damaged by a number of neurotoxic agents, of which the most common examples are carbon monoxide and manganese. With carbon monoxide poisoning, the onset of symptoms is acute; they appear shortly after the patient awakes from an initial comatose state. With manganese poisoning—typically following inhalation of the metal during the process of arc welding—the symptoms evolve over weeks to months.

Many antipsychotic drugs (neuroleptics such as chlorpromazine, stelazine, fluphenazine and risperidone), especially in high doses, produce a reversible basal ganglia disorder much like Bobo's.

7.5.5 Infectious

Some infectious agents, especially neurotropic viruses, may affect the basal ganglia more or less exclusively and produce an isolated motor disability, typically with severe impairment of the bulbar (facial, lingual and pharyngeal) musculature. Such localized encephalitides, however, are invariably of acute (days) or subacute (weeks) length, and do not evolve gradually over many months.

Bacterial meningitis—especially tuberculous meningitis—may produce thromboses of small vessels entering the brain from the circle of Willis (see Chapter 8). The result is a putaminal infarction and a characteristic form of dystonic hemiparesis in which the involved arm is extended at the elbow, hand fisted, and partly abducted or flexed at the shoulder, as if the person were trying to push open an imaginary door.

Finally, a condition very much like Bobo's, if usually more severe, developed in many individuals following the 1918 to 1919 influenza pandemic. Known as Von Economo's encephalitis, the disorder largely disappeared as the original patients died, but is still occasionally encountered.

In Bobo's case there is no history of any relevant infection.

7.5.6 Metabolic

Most metabolic diseases affect the cerebral hemispheres as a whole, rather than specifically targeting the basal ganglia. A characteristic gliosis with large protoplasmic astrocytes develops in the basal ganglia, as elsewhere in the brain, in some patients with chronic hepatic failure; this phenomenon may help explain the flapping tremor (or *asterixis*) and cog-wheel rigidity sometimes seen in such patients. Although this suggests some similarity with Bobo's clinical picture, the presence of hepatic failure would be known well in advance of the development of neurologic symptoms.

7.5.7 Inflammatory/Autoimmune

Some postinfectious autoimmune disorders may produce basal ganglia dysfunction, usually in a subacute fashion over several weeks. An excellent example is post-streptococcal choreoathetosis, or *Sydenham's chorea. Chorea* refers to an uncontrollable, rapid, dance-like movement of the fingers, hands, toes, feet, or even the face and the vocal apparatus. *Athetosis* refers to a slower, more proximal, twisting movement, usually of the upper limbs, when an attempt is made to reach for an object. Quite apart from the fact that the time course of Bobo's neurological disorder is much more protracted than for Sydenham's chorea, her movement disorder is a paucity or absence of movement, rather than an excess.

Collagen-vascular diseases such as lupus erythematosus may also produce chorea, usually in a subacute or chronic, insidious fashion, but seldom dystonia or akinesia.

7.5.8 Neoplastic

Primary CNS tumors (gliomas, germinomas) and metastatic malignancies may selectively involve the basal ganglia. A neoplastic disease is highly unlikely in Bobo's case, however, for the simple reason that the disease process seems to involve the basal ganglia bilaterally. A tumor would likely affect one cerebral hemisphere, but not both sets of basal ganglia simultaneously without involving other structures such as the internal capsules and thalami.

7.5.9 Degenerative/Genetic

There is a wide variety of degenerative disorders affecting the basal ganglia; many, but not all are of genetic origin. Before considering the most important of these disorders—and the most probable in Bobo's case—it would be helpful at this point to consider in more detail the specific basal ganglia syndrome Bobo has developed.

7.5.9.1 The Parkinsonian Syndrome That Bobo might have a form of parkinsonism was recognized by Fifi, who had already witnessed similar symptoms appearing at a somewhat later age in her father. Parkinsonism refers to a specific collection of symptoms and signs stemming from a common sporadic degenerative disease of a component of the basal ganglia, *Parkinson's disease*. This disease, first described in the early nineteenth century by James Parkinson, is one of the most common causes of progressive, severe motor disability in the sixth to eighth decades of life.

The principal features of idiopathic Parkinson's disease include a resting tremor of the pill-rolling type, increased muscle tone of a rigid type, slowness of movement (bradykinesia) or complete inability to initiate movement (akinesia), and postural control problems. These four main features are most readily recalled by the use of the acronym *TRAP*, standing, respectively, for tremor, rigidity, akinesia and posture. Affected individuals typically have an expressionless face with a fixed stare, and reduced blinking; drooling is common (Figure 7.8A). Movements are performed slowly and with effort; handwriting is micrographic (Figure 7.8B). In standing, parkinsonian individuals tend to be stooped forward, with neck and elbows semiflexed (Figure 7.8C). The gait is slow and shuffling, with reduced arm swing and a continuous resting tremor of the fingers. Postural control is poor and balance easily lost, particularly if the ground is uneven. The rigid tone is obviously more apparent on physical examination, with cog-wheeling at the elbows particularly prominent.

Later in the illness, many affected persons may develop features of a progressive dementia (see Chapter 9). Postural hypotension is also a common problem, as are other autonomic instabilities.

The cardinal pathological abnormality in idiopathic Parkinson's disease is degeneration of the large, pigmented neurons in the substantia nigra, the most inferiorly placed

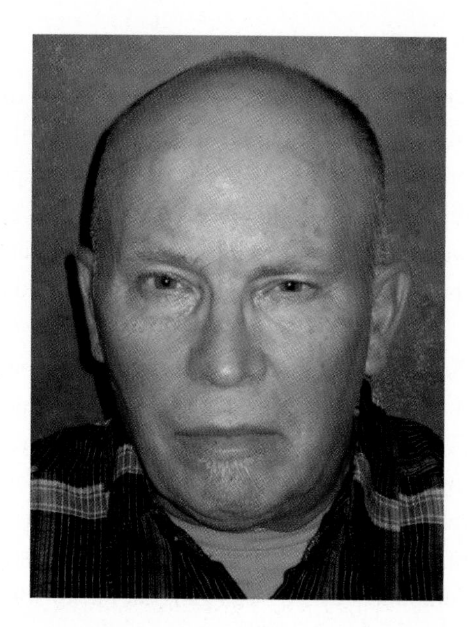

A

B

C

FIGURE 7.8 **Clinical manifestations of Parkinson's disease. (A) Mask-like facies; (B) micrographia; (C) lateral view of patient showing typical trunk posture. (Patient's release and with permission; courtesy of Dr. D. Grimes.)**

component of the basal ganglia located in the midbrain posterior to the cerebral peduncles (Figures 7.3, 7.9A and 7.10B). Some degenerating neurons have prominent intranuclear inclusions, so-called *Lewy bodies* (Figure 7.9B); these are accumulations of a compound known as *ubiquitin*.

Although Bobo obviously has many features of sporadic idiopathic Parkinson's disease, at 32 she is well below the typical age range for the disorder. In addition, her father also has parkinsonism, apparently of a mild, less aggressive form than would normally be seen in someone who is presumably around 60 years of age.

Familial forms of parkinsonism do exist, typically beginning in the third to fifth decades of life, and are pathologically distinct from sporadic Parkinson's disease; the clinical picture is often accompanied by some elements of *generalized dystonia*. Dementia and severe postural instability do not typically develop in early-onset, familial parkinsonism. A number of genes have been implicated in

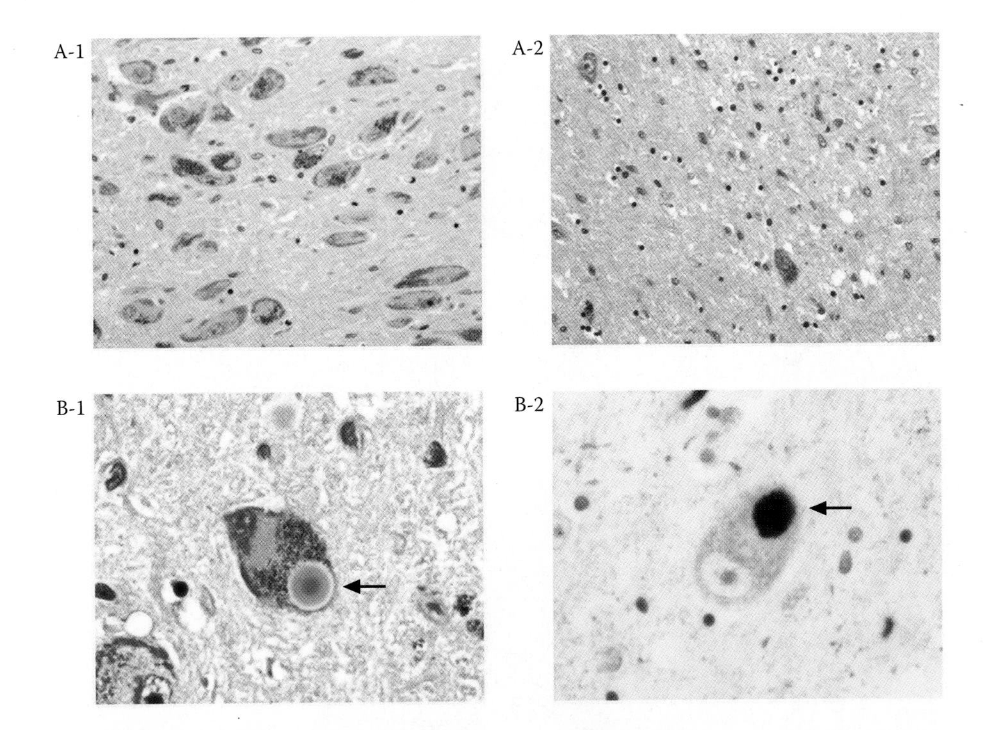

FIGURE 7.9 Neuropathological features of Parkinson's disease. (A) Panel 1, photomicrograph of normal substantia nigra neurons; panel 2, substantia nigra neurons in Parkinson's disease showing depigmentation and neuronal loss. (B) Panel 1, high power photomicrograph showing typical Lewy body (arrow); panel 2, Lewy body, ubiquitin stain (a different neuron from the one shown in panel B-1). (Courtesy of Dr. J. Woulfe.)

the production of familial parkinsonism, the most common mutations being in the parkin gene at 6q25.2. Parkin mutations do not usually produce significant nigral neuronal degeneration unless homozygous and thus appear as autosomal recessive disorders, sometimes as early as age ten.

Not all parkinsonian syndromes are neurodegenerative in nature, however. One of the most striking exceptions is *dopa-responsive dystonia*, usually an autosomal dominant disorder with variable expressivity whose mechanism is a defect in dopamine synthesis. In its most aggressive form, the disorder becomes manifest around the middle of the first decade of life, with progressive limb dystonia, legs more than arms, and impaired control of head and trunk posture. Milder forms of the disease produce a mixture of dystonic and parkinsonian features beginning in early to mid-adulthood. For both the early and later forms of the disorder there is a characteristic diurnal variation in symptom severity, with symptoms becoming progressively more severe over the course of the day and improved after a nap or rest. You will recall that Bobo had a significant diurnal variation in symptom severity, as witnessed by her sister.

From what we have learned in this section, it appears most likely that Bobo has either a form of familial, early-onset parkinsonism/dystonia or a later-onset form of dopa-responsive dystonia. Before considering other potential diagnoses, let alone what might be done to help Bobo, however, we must focus on the substantia nigra and its role in parkinsonian-type movement disorders. As well, as suggested by the self-explanatory term *dopa-responsive dystonia*, we must consider the role played by the monoamine transmitter dopamine in the control of posture and movement.

7.5.9.2 The Substantia Nigra In considering the contribution of the basal ganglia to the control of movement, we have presented, for the sake of clarity and simplicity, only the basic elements of basal ganglia circuitry: the accelerator and brake mechanisms. We must now review the subsidiary and crucial role played by the substantia nigra.

There are reciprocal connections between the putamen and substantia nigra that complement the previously described circuits involving putamen, globus pallidus and subthalamic nucleus. With respect to the observed deficits

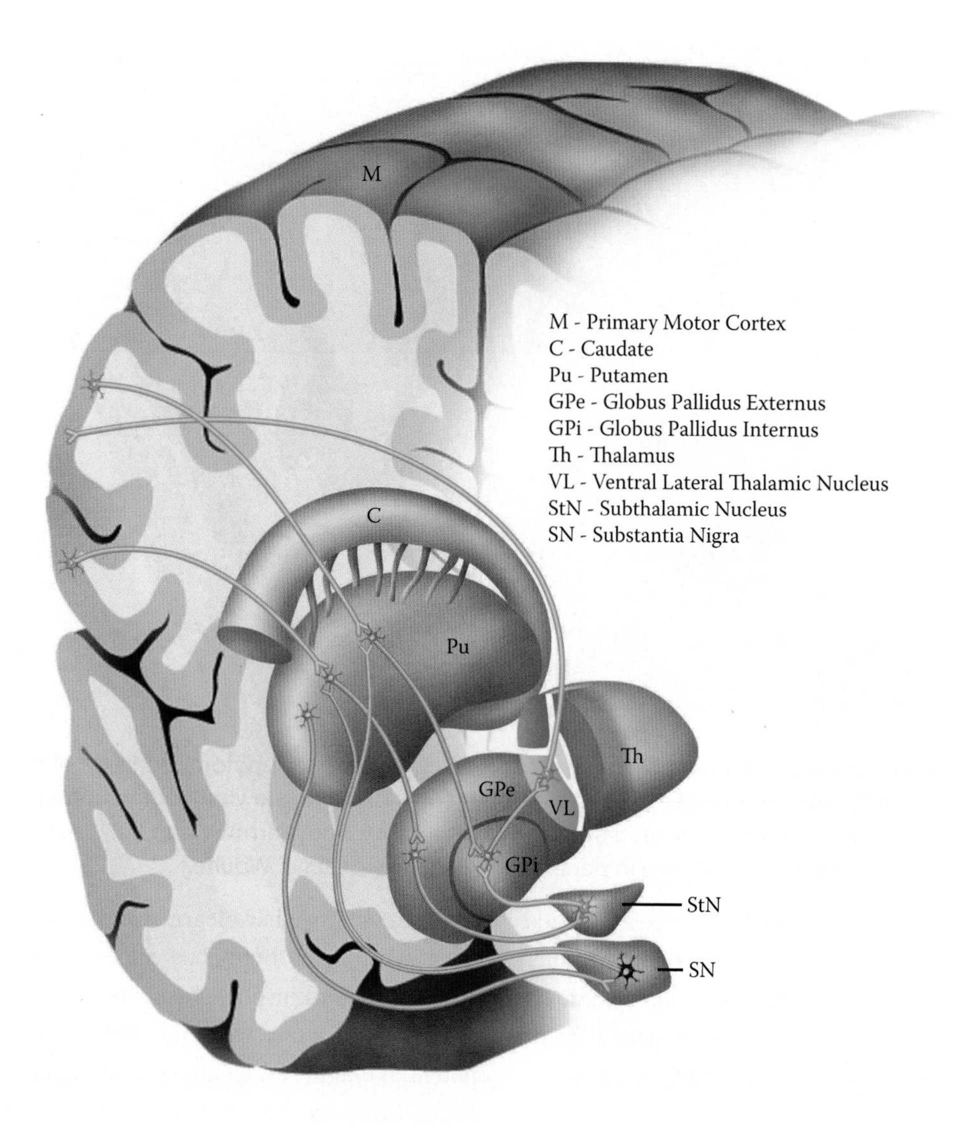

FIGURE 7.10 (A) Basal ganglia motor circuitry, phase 2. For reasons explained in the text, nigrostriatal neurons have both a red and a green color.

in the parkinsonian syndrome, the key component in the "dialogue" between putamen and substantia nigra is the projection returning from substantia nigra to putamen: the *nigrostriatal pathway* (see Figure 7.10A).

As we noted earlier, the substantia nigra contains large pigmented neurons whose axons project to the corpus striatum (caudate, putamen and nucleus accumbens), as well as to the basal forebrain and frontal lobe cortex (see Chapter 12). Axons destined for the corpus striatum (in the case of movement control, the putamen) originate in the compact portion of the substantia nigra (pars compacta). The main neurotransmitter elaborated by the synaptic terminals of the substantia nigra neurons, as you may already have guessed, is *dopamine*.

In the putamen, the dopaminergic terminals impinge on neurons involved in *both* the accelerator and brake circuits (Figure 7.10A). Even though the substantia nigra axons project the same chemical "message" to both pathways, the effects are *opposite*. The explanation for this apparent contradiction can be found in the dopamine receptor subtypes present on the respective putaminal neurons for the two pathways. Putaminal neurons participating in the accelerator circuit (direct to GPi) elaborate a class of dopamine receptors that respond by exciting the cell (D1 receptors). The end result is a facilitation of the accelerator circuit and initiation of movement. On the other hand, the putaminal neurons participating in the brake circuit (via the subthalamic nucleus) contain receptors that respond by inhibiting the

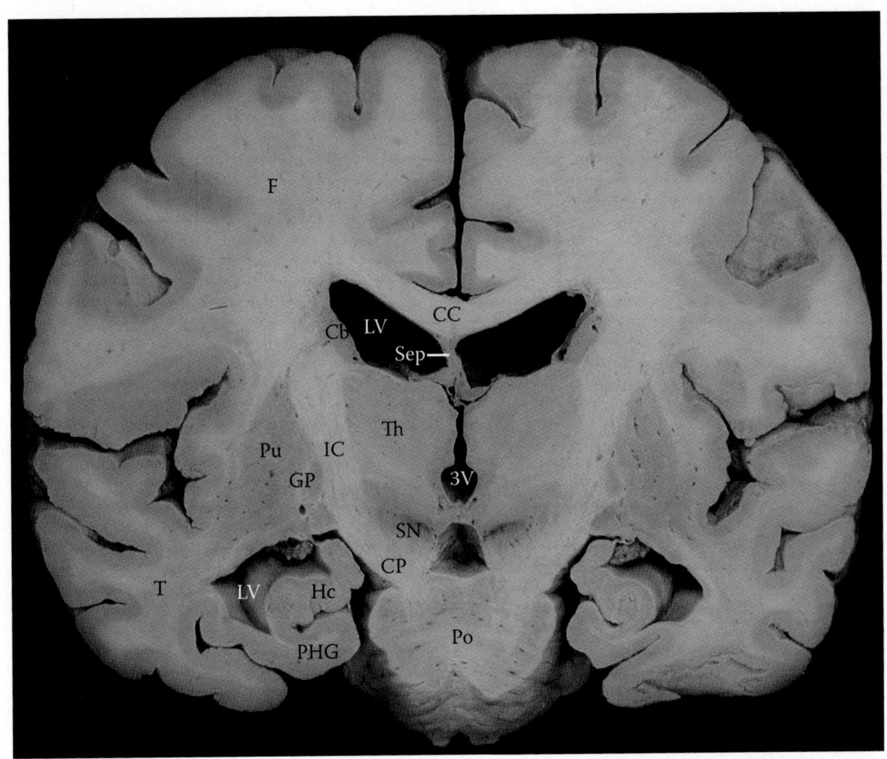

F - Frontal Lobe
T - Temporal Lobe
CC - Corpus Callosum
LV - Lateral Ventricle
3V - IIIrd Ventricle
Sep - Septum Pellucidum
Cb - Caudate (Body)
Pu - Putamen
GP - Globus Pallidus
Th - Thalamus
SN - Substantia Nigra
IC - Internal Capsule
CP - Cerebral Peduncle
Po - Pons
Hc - Hippocampus
PHG - Parahippocampal Gyrus

FIGURE 7.10 (B) Coronal section of cerebrum showing the anatomical location of the substantia nigra and other central gray matter structures.

cell (D2 receptors). In this case the result is an inhibition of the brake circuit, or again, a facilitation of movement. Thus, through both accelerator and brake circuits, the main contribution of the nigrostriatal dopaminergic projection is, normally, to facilitate the initiation of movement.

You will immediately recognize, therefore, that a degeneration of pars compacta neurons, or a metabolic "failure" of the nigrostriatal connection, will result in the opposite: a profound inability to initiate movement, one of the cardinal features of parkinsonism.

For sporadic Parkinson's disease, and for early-onset familial parkinsonism/dystonia, as we have seen, the pathological substrate is progressive degeneration of dopaminergic neurons in the pars compacta of the substantia nigra. In the case of dopa-responsive dystonia, the nigral neurons remain intact, but are depigmented; they are lacking an enzyme, usually GTP cyclohydrolase 1 (GCH1), whose function is to facilitate the first step in the synthesis of tetrahydrobiopterin (BH4), a key factor in the synthesis of dopamine (see Figure 7.11). BH4 is a necessary cofactor for conversion of the essential amino acid phenylalanine to tyrosine and, in turn, the conversion of tyrosine to dihydroxy-phenylalanine (dopa), the immediate precursor of dopamine. Thus, a deficiency in the production of BH4 will result in a deceleration in the production of dopa/dopamine, particularly in situations where the nigrostriatal pathway is active for long periods. This inability of dopamine synthesis to keep up with "demand" may explain why patients with dopa-responsive dystonia become more incapacitated as the day goes on and partially recover overnight.

7.5.9.3 Other Causes of Progressive Dystonia

Since Bobo's clinical picture includes dystonic elements, we would also have to briefly consider other causes of progressive dystonia that are not typically associated with the

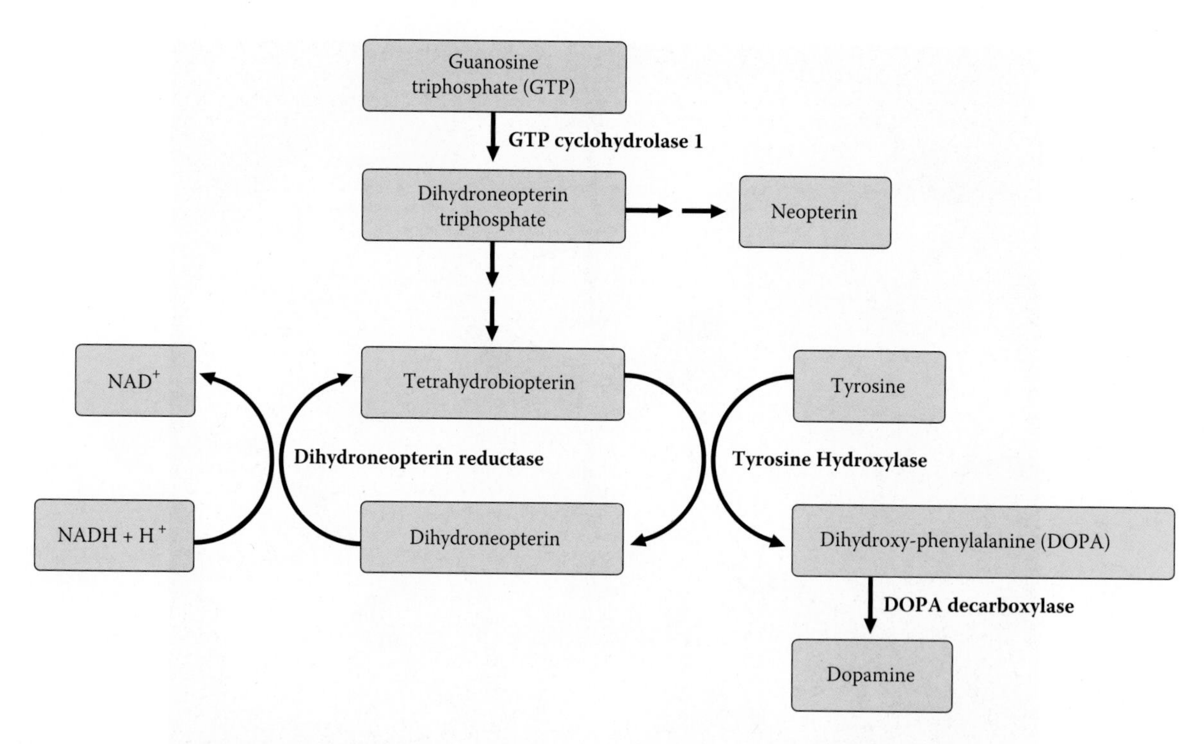

FIGURE 7.11 The dopamine synthesis biochemical pathway.

parkinsonian characteristics of tremor, rigidity, akinesia and postural instability. There are, for example, a variety of hereditary dystonias of which the most common is *idiopathic torsion dystonia*, or hereditary dystonia type 1, due to a mutation in the torsin gene. *Wilson's disease*, an autosomal recessive disorder of copper transport, produces a progressive dystonia, dysarthria and dysphagia typically beginning in the teenage years, and is accompanied by visible evidence on MRI of degeneration in the corpus striatum. Rigidity and dystonia can also be seen in the early-onset forms of *Huntington's chorea* (discussed in Chapter 14), an autosomal dominant neurodegenerative disorder of the corpus striatum due to a mutation in the huntingtin gene (further discussed in the appendix to Chapter 14 on the DVD and Web site). At Bobo's age, however, chorea would be the predominant symptom. Finally, progressive rigidity and dystonia can be seen in *familial strionigral degeneration*, an autosomal dominant disorder usually becoming manifest in the second to fourth decades, and leading to a much more profound degree of disability than that seen in Bobo and her father.

7.6 Case Summary

To recapitulate, Bobo has a slowly progressive disease characterized by increasing difficulty maintaining postural control, walking, and writing. The main clinical findings of a mask-like facies, muscle tone rigidity, cog-wheeling, and abnormal twisted limb postures suggest a parkinsonism/dystonia disorder; this, in turn, places the likely focus of disease in the basal ganglia, in particular the substantia nigra. While idiopathic Parkinson's disease is the most common degenerative disease affecting these structures, Bobo's relatively young age and positive family history suggest that she has either familial parkinsonism/dystonia syndrome or dopa-responsive dystonia.

7.7 Approach to the Investigation of Patients with Parkinsonism/Dystonia

The initial work-up should include a cranial imaging study, preferably MRI. While no abnormality would be expected in sporadic Parkinson's disease, hereditary early-onset parkinsonism, dopa-responsive dystonia or idiopathic torsion dystonia, evidence of basal ganglia atrophy and/or signal abnormalities may be found in disorders such as Wilson's or Huntington's disease. Wilson's disease, if suspected, can be confirmed by the finding of a low serum level of the copper transport protein ceruloplasmin.

In both familial and sporadic parkinsonian disorders, whether or not accompanied by nigral degeneration, there is a marked decrease in output of dopamine from the substantia nigra. This phenomenon can be documented by the finding of an abnormally low level of homovanillic acid (HVA), the main metabolite of dopamine, in the cerebrospinal fluid. CSF studies can distinguish between juvenile parkinsonism and autosomal dominant dopa-responsive dystonia/parkinsonism in that, while HVA levels are low in both disorders, neopterin and tetrahydrobiopterin levels are also low in the latter disorder (see Figure 7.11) and normal in the former.

For Bobo, the most specific investigative tool is leukocyte DNA analysis for possible mutations in relevant genes. Thus, for early-onset autosomal recessive parkinsonism/dystonia, characterization of the parkin gene would likely detect a mutation. Likewise, for dopa-responsive dystonia one would search for mutations in the GTP cyclohydrolase 1 gene; for idiopathic torsion dystonia, defects in the torsin gene, and so on.

7.8 Investigations

Bobo was referred to a local neurologist who, after arranging for a series of investigations, gave Bobo a trial of low-dose levodopa/carbidopa taken three times daily (for explanation, see next section). Within 48 hours, there was a dramatic improvement in Bobo's gait, upper trunk posture and handwriting. This result eliminated the possibility of idiopathic torsion dystonia as it does not respond to dopamine precursor medications.

An MRI study of the head was normal, as was a serum ceruloplasmin.

A lumbar puncture (done with Bobo temporarily off levodopa/carbidopa) revealed low levels of HVA, neopterin and tetrahydrobiopterin. These results strongly suggested the diagnosis of dopa-responsive dystonia; the diagnosis was confirmed by the finding of a mutation in the GTP cyclohydrolase 1 gene. A subsequent extended investigation of the family revealed that Bobo's father had the same mutation, whereas Fifi's study was normal.

7.9 Treatment/Outcome

After her LP, Bobo was restarted on the same small dose of levodopa/carbidopa. She continued to improve to such an extent that, after three months on medication, her neurological examination was completely normal. Bobo has had no side effects of the medication and, a year later, has delivered a second child—a daughter, who has eventually been found to have her mother's GCH1 mutation and who will be treated promptly if and when she becomes symptomatic.

Why the use of the levodopa/carbidopa combination?

Orally administered dopamine does not cross the blood–brain barrier in significant amounts. Its immediate precursor dopa does cross the barrier, particularly if combined with an agent that inhibits the peripheral degrader of dopa, dopa decarboxylase. Such combinations (known as levodopa/carbidopa) readily enter the brain, where the dopa is taken up by dopaminergic nerve terminals, thus bypassing the metabolic defect in dopamine synthesis. As might be expected, given that dopa-responsive dystonia is not a degenerative disease, the end result is a sustained, dramatic improvement in parkinsonian symptoms. Sustained but gradually waning improvement (reflecting the progressive loss of substantia nigra neurons in these disorders) is also seen in familial early-onset parkinsonism/dystonia, as well as in the early stages of sporadic Parkinson's disease.

Related Cases

There are e-cases related to this chapter available for review on the text DVD and Web site.

Suggested Readings

Clarke, C.E. Parkinson's disease. *BMJ* 335 (2007): 441–445.

Hermanowicz, N. Drug therapy for Parkinson's disease. *Semin. Neurol.* 27 (2007): 97–105.

Segawa, M., Y. Nomura, and N. Nishiyama. Autosomal dominant guanosine triphosphate cyclohydrolase 1 deficiency (Segawa disease). *Ann. Neurol.* 54, Suppl. 6 (2003): S32–S45.

Tuite, P.J., and K. Krawczewski. Parkinsonism: A review-of-systems approach to diagnosis. *Semin. Neurol.* 27 (2007): 113–122.

Web Sites

National Institute for Neurological Disorders and Stroke: http://www.ninds.nih.gov/disorders/parkinsons_disease/detail_parkinsons_disease.htm

Medline Plus: http://www.nlm.nih.gov/medlineplus/movementdisorders.html

Chapter 8

Etienne

ETIENNE

Etienne is Crash's uncle on his father's side. He is a used-car salesman with little regard for his lifestyle. He works long hours and frequently eats greasy take-out food at work.

At age 57, his weight is 240 pounds; at 5 feet 6 inches, this gives him a body mass index (BMI) of 39 kg/m^2. The normal range is 18.5 to 25 kg/m^2. His doctor has told him on several occasions that his weight, blood pressure and cholesterol are too high and that he should quit smoking. His wife continually complains of his snoring; he frequently has spells during which he temporarily stops breathing when asleep. His father died of a stroke at age 55.

About a month ago, while at work, he suddenly noticed a disturbance in his speech, lasting five minutes, during which he had difficulty expressing himself to a customer. He shook it off as just being tired and overworked.

Recently he has been under considerable stress because of his manager's complaints about his poor sales over the last month. Today while he was having a heated exchange with his manager, he began having trouble expressing himself, then developed weakness of his right arm, face and leg to the point where he collapsed to the ground. His boss called 911; the Emergency Services arrived within 15 minutes and called the Regional Stroke Center, declaring a potential Stroke Code; this made Etienne a possible candidate for intravenous tPA (tissue plasminogen activating factor).

Objectives

- To learn the anatomy of the vascular supply of the brain
- To learn the mechanisms of brain cell death from ischemic damage
- To learn the effects of vascular damage to affected regions of the brain and their clinical manifestations
- To learn the major stroke syndromes and their clinical manifestations
- To become sensitized to the medical, physical and psychosocial implications of damage caused by cerebrovascular disease

8.1 Clinical Data Extraction

Before proceeding with this chapter, you should first carefully review Etienne's story in order to extract key information concerning the history of the illness and the physical examination. In the DVD accompanying this text you will find a sequence of worksheets to assist you in this process.

8.2 The Main Clinical Points

Etienne's principal clinical findings were as follows:

- One episode of transient speech disturbance one month before the major event
- Significant risk factors for vascular disease, including obesity, smoking, male, middle age, hypertension, hyperlipidemia, probable obstructive sleep apnea, and a family history of vascular disease
- Sudden onset of global aphasia, dense weakness of the right lower face, arm and leg, hyperreflexia, Babinski sign, and right-sided sensory loss
- Conjugate eye deviation to the left side
- Lack of visual attention to the right visual field and loss of sensation on the right side of the body

8.3 Relevant Neuroanatomy and Physiology

As the clinical history suggests, Etienne's predisposing features and clinical deficits suggest a disorder of the cerebral circulation. Subsequent discussion with respect to ischemic cerebrovascular disease must be based on an understanding of the mechanisms of cell death due to ischemia and a detailed knowledge of the blood supply to the brain. The effects of disturbances of that blood supply on different areas of the brain and their clinical effects will be examined.

The sequence of events that occur when the blood supply is interrupted to a given area of the brain and spinal cord needs to be understood in order to facilitate the process of rapid diagnosis and treatment.

His examination on arrival in the emergency department showed a blood pressure of 170/100 and a heart rate of 84 which was regular.

He was alert but was unable to speak, with inability to understand or to produce any meaningful speech. Both eyes were deviated to the left but could be moved across the midline to the right side with rapid passive movement of the head to the left. The right lower face was weak and there was decreased movement of the right side of the palate. He responded to visual threat on the left side but not on the right.

His motor exam showed 0/5 motor power on the right in the upper and lower limbs, with hyperreflexia at 3+ in the biceps, triceps, brachioradialis, knee and ankle jerks on the right; all reflexes on the left were graded as 2+. The plantar responses were tested: on the right, the reflex response was abnormal, showing an extensor plantar response (positive Babinski sign); on the left, the response was normal, showing a flexor plantar response (negative Babinski sign).

A sensory exam was difficult to interpret due to Etienne's speech difficulties but it was clear that he could feel pinprick sensation on the left over the face, trunk, arm and leg, but not on the right. Coordination testing and gait testing could not be performed due to the speech difficulty and profound right-sided weakness.

The neurologist on call performed a Stroke Code assessment and determined that his NIH Stroke Scale score was 20 (see NIH stroke scale Web site listed at the end of the chapter). A right carotid bruit could be heard but there was no bruit on the left side.

8.3.1 Cerebral Circulation

The cerebral circulation is unique in several ways. The major cerebral arteries and their branches serve specific areas of the brain, as indicated in Table 8.1. Unfortunately, there is little collateral flow between the various cerebral arteries *after* their major branching points. The consequence of this pattern of vascular anatomy is that occlusion of a major cerebral artery or one of its branches will lead to ischemic damage in the area served by that artery. This is the bad news. The good news is that the cerebral circulation has been designed with redundancy between the major cerebral arteries. This is achieved through the supplementary arterial branches known as the communicating arteries. For instance, the posterior communicating arteries (PCom) join the posterior cerebral arteries on each side to the internal carotid arteries on the same side. In addition the anterior communicating artery (ACom) joins the two anterior cerebral arteries together. This arterial network, known as the *circle of Willis*, allows for collateral flow from one carotid to the opposite hemisphere and supply from the basilar artery to the cerebral hemispheres through the posterior communicating arteries. This system of redundancy—similar to an aircraft fuel system—allows for blockage of one or more of the major cerebral arteries with flow being maintained through the circle of Willis. Figures 8.1, 8.2, 8.3, and 8.4 show the relationships of the major cerebral arteries and the territories of the brain that are supplied by these arteries.

8.3.2 Cerebral Artery Occlusion

The mechanisms of how the arteries become blocked will now be considered. Any process that occludes the lumen of one of the cerebral arteries to a critical level (usually over 90 percent) will lead to ischemic injury to the brain tissue that the artery serves. The mechanism by which

Table 8.1 Summary of the Territories of Cerebral Artery Supply Showing the Areas of Overlap (Occurring Mainly in the Deep Structures)*

Artery	Lobe	Function
Anterior cerebral (ACA)	Frontal, parasagittal	Contralateral motor leg, bladder control, executive function
Middle cerebral (MCA)	Anterior, lateral frontotemporal, parietal, anterior thalamus, internal capsule	Contralateral motor, sensory of face, arm, language, memory
Posterior cerebral (PCA)	Parietal, occipital, posterior thalamus, inferior and mesial temporal	Visual fields, contralateral sensory, memory, visual recognition
Basilar	Brainstem, cerebellum, thalamus	Cranial nerve III–XII, sensory, motor, balance, coordination

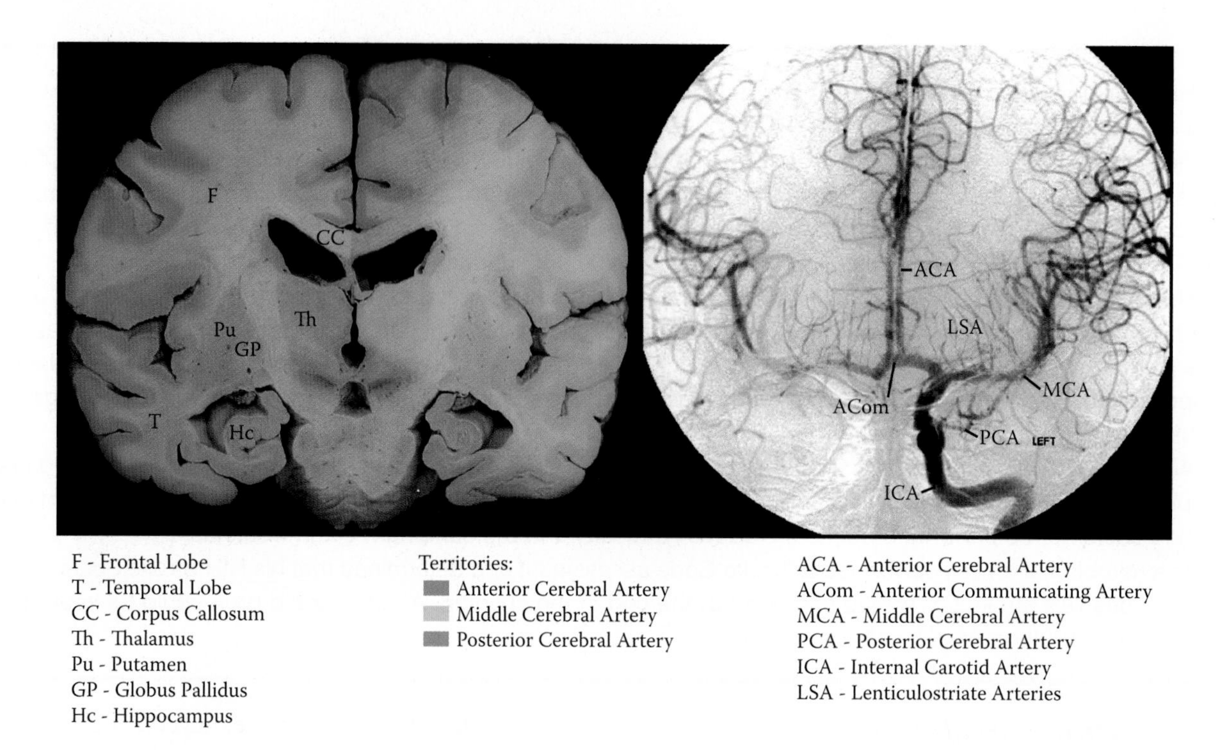

FIGURE 8.1 Blood supply of the brain: the basic pattern of the areas of the cerebral hemispheres of the brain that are served by the anterior (ACA), middle (MCA) and posterior (PCA) cerebral arteries in a coronal brain section, and an accompanying left carotid angiogram viewed in the same plane. The MCA serves a large area of frontal, temporal and parietal cortex, as well as white matter and the lateral thalamus. The ACA serves the medial part of the frontal and parietal lobes, including the cingulate gyrus. The PCA serves the medial temporal lobe, the posterior aspect of the parietal lobe, the occipital lobes and medial thalamus.

the arteries become blocked is usually thromboemboli that have been formed upstream and have floated down to occlude the narrower lumen in the more distal portion of the vessel. Atherosclerosis intrinsic to the large branches of the cerebral arteries in the brain does occur but, in general, is not the commonest cause of arterial occlusion. The commonest source of arterial occlusion is usually from thromboemboli formed at the carotid bifurcation or from emboli from the heart or aortic arch (also called central source emboli).

8.3.2.1 Carotid At the carotid bifurcation, there is turbulent flow due to the change in the hydrodynamics of blood flow at the split of the common carotid into the external and internal carotid arteries. The energy dissipated by this turbulent flow causes endothelial stress and injury at the origin of the internal carotid artery. The combination of this shear stress and factors that accelerate atherosclerosis (such as hypertension, smoking, hyperlipidemia, diabetes and genetic factors) leads to the build up of atherosclerotic material in the subintimal region of the artery wall. The evolution of atherosclerosis in this area includes the build up of fatty material in the affected vessel, with secondary involvement of inflammatory cells that increase the volume of the plaque. The progressive accumulation of this fatty material causes gradual narrowing of the internal carotid artery. The atherosclerotic material can also reach a point where there is damage to the overlying endothelium, leading to a so-called plaque accident. This phenomenon occurs when there is a denuded area of the vascular lining which then leads to build up of thrombus over the surface of the plaque. These thrombi grow in size and, depending on the underlying substrate, produce platelet emboli consisting of either clot or clot mixed with cholesterol and, occasionally, with calcium deposits. Thrombi formed at the origin of the internal carotid artery can then detach and migrate downstream to become lodged in smaller arteries, producing arterial occlusion and neurological deficits. The neurological consequences of the release of this thrombus (which has now become an embolus) depends on which arterial territory the embolus occludes. The middle cerebral artery (MCA) tends to be the preferential end point for emboli originating from the internal carotid artery.

8.3.2.2 Central Source Emboli Central source emboli are formed in the heart or aorta and then travel from the heart to the brain through the carotid arteries, the vertebral arteries, or both. Any patient presenting with an acute focal deficit with a CT or MRI showing ischemic lesions old or new, in multiple territories or on both sides of the brain, should trigger the search for a central source of emboli.

The most common cause of central source emboli is a thrombus in the left atrium due to blood flow stasis associated with atrial fibrillation. Other sources include prosthetic valves, acute anterior wall myocardial infarction, endocarditis, patent foramen ovale (PFO), ventriculoseptal defect (VSD), atrial myxoma, and atherosclerosis with visible thrombus on the aortic arch. Transthoracic (TTE) and transesophageal echocardiography (TEE) are usually able to identify and localize the origin of the central source emboli.

8.3.3 Cerebral Ischemia

Interruption of blood supply to any area of the brain or spinal cord—known as ischemia—will lead to a series of increasingly serious physiological consequences depending on the location, intensity and duration of the interruption of blood flow. In addition, different areas of the brain have different susceptibilities to localized or generalized ischemia. For instance, in adults, the hippocampal cortical areas and the Purkinje cells in the cerebellar cortex are most susceptible to ischemia, whereas the white matter areas are less susceptible.

Acute ischemia is due to a sudden loss of supply of oxygen, glucose and other critical metabolic substrates by blockage of one of the cerebral arteries or its branches. The effect of this acute interruption of cellular metabolism initially leads to disturbances in the electrical function of the neurons, glial cells and transmitting axons (also discussed in Chapter 10). Following this, there is biochemical breakdown of the integrity of the neurons and axons that, if not corrected, is followed by irreversible cell death. In the face of an acute ischemic event, there are generally three affected populations of neurons or axons in the area of ischemia to consider. The first group of cells are those that have been most severely affected and have undergone irreversible changes; these are usually at the center of the area of ischemia and are effectively dead cells. The second group are less severely affected and are still alive but not properly functioning due to a suspension of their electrical membrane functions. This second area surrounds the region of most severely affected neurons and axons and is known as the *ischemic penumbra.* This is the group of damaged cells that can potentially be salvaged from permanent damage by restoring blood flow to the area. The third group of neurons and axons is outside the penumbra, and usually fed by other branch arteries proximal to the occluded artery or supplied by one of the adjacent cerebral arteries.

For the clinician, these concepts are important in that, when a patient arrives in the emergency department with acute ischemic damage, it is important to know the duration and extent of the ischemia as well as to determine the magnitude of the ischemic penumbra; this information will help to decide the choice of treatments to restore blood flow to the damaged areas. There are various neuroradiological imaging techniques available to assist the clinician to determine the extent of damage and the amount of brain tissue that might be salvageable. For CT scanning, these are known as perfusion maps; for MRI scanning, they are known as diffusion weighted images (DWI) and apparent diffusion coefficient (ADC) maps.

8.3.3.1 Middle Cerebral Artery The most common and devastating of all strokes is an occlusion of the left middle cerebral artery (MCA) in a left hemisphere-dominant (usually right-handed) person. This type of stroke is most damaging because of its effect on the centers that support language and motor function on the dominant side. Both Broca's area (responsible for the fluent output of spoken content of language) and Wernicke's area (responsible for the recognition of language content) are served by the middle cerebral artery through two different branches (Figure 8.2).

The middle cerebral artery originates from the internal carotid artery in the middle cranial fossa. It runs along the outer surface of the thalamus and internal capsule giving off small branches which supply these structures. The

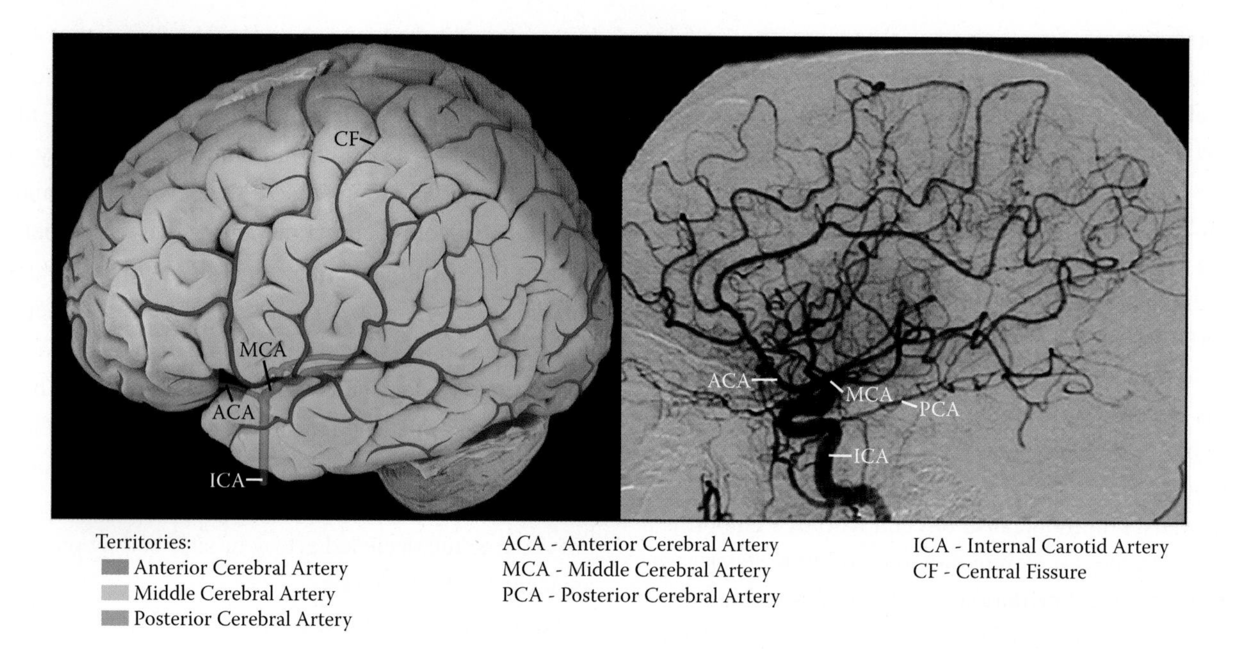

FIGURE 8.2 Blood supply of the brain: a lateral view of the brain with the territories of the ACA, MCA and PCA color-coded, and a corresponding left carotid angiogram.

MCA then breaks up into branches that resemble a candelabra; of these there are two larger branches: one anterior whose branches serve the lateral frontal lobe including Broca's area, and one posterior whose branches serve the lateral temporal and parietal lobes, including Wernicke's area. Thus, the amount of damage to the brain and the associated neurological deficits will depend on the location of the occlusion in the MCA and its branches.

An occlusion of the MCA close to its origin in a dominant hemisphere will give severe deficits of motor and sensory function in the face, arm and leg on the contralateral side, as well as loss of receptive and expressive language function.

Damage to the frontal lobes also affects the frontal eye fields which control the upper motor neurons mediating contralateral conjugate eye movement. If there is damage to the frontal eye fields in one hemisphere, the eyes will respond to the output of the undamaged hemisphere, resulting in conjugate deviation of the eyes toward the damaged hemisphere.

Damage to the optic tract, thalamus and optic radiations can lead to hemianopia. This deficit relates to damage to the white matter and relay structures that transmit information concerning the visual fields from the retina. The visual system will be discussed in detail in Chapter 11. Damage to these areas causes loss of visual perception in the contralateral visual field in both eyes.

If the occlusion is distal to the perforating lenticulostriate branches, thus sparing the internal capsule, then there will be less motor deficit in the contralateral leg; nevertheless, language and movement of the face and arm will be affected.

More distal occlusion of either the anterior or posterior branches will give selective damage. Occlusion of the anterior division results in weakness of the face and arm with a Broca's aphasia. Occlusion of the posterior division causes mild upper motor neuron weakness on the contralateral side, with Wernicke's aphasia and contralateral cortical sensory loss (e.g. loss of stereognosis, see Chapter 2).

Isolated occlusions of some branch vessels will produce deficits specific to the function of the location damaged.

An occlusion of small blood vessels serving the internal capsule or lateral thalamus will result in pure motor weakness or sensory loss on the contralateral side. A small round area of damage can be seen on CT or MRI scan; this is sometimes called a *lacune* or small lake. Lacunes which occur in the lateral thalamus cause pure sensory deficits on the contralateral side.

Branch occlusions to arteries that primarily supply areas of cortex lead to wedge-shaped areas of damage visible on CT scan or MRI, with deficits specific to the location (review Table 3.9). For instance, a branch occlusion to the arterial supply to the left angular gyrus will lead to very specific syndrome of deficits called Gerstmann syndrome: the patient is not able to calculate simple arithmetic, loses the ability to perceive left from right and is unable to identify which finger is being touched by the examiner while the patient's eyes are shut.

8.3.3.2 Anterior Cerebral Artery The anterior cerebral artery (ACA) originates from the internal carotid artery in the middle cranial fossa; it runs up and over the corpus callosum, serving the corpus callosum and the cerebral cortex of the medial frontal gyri, cingulate gyrus and part of the medial parietal lobe.

Occlusion of the ACA close to its origin will cause damage to a strip of cortex that it serves running longitudinally back along the parasagittal area. Damage to this area leads to motor weakness, spasticity and hyperreflexia in the leg on the opposite side (see the motor homunculus, Figure 4.1). Occlusion of this vessel is uncommon, accounting for less than 10 percent of all strokes.

8.3.3.3 Posterior Cerebral Artery The two posterior cerebral arteries (PCA) originate at the termination of the basilar artery; the PCA on each side runs back to the parietal and occipital lobes but also serves the medial and inferior portions of the temporal lobe. Close to its origin, the PCA gives off small branches to the medial thalamus close. There is a forked artery close to the origin of the PCAs, called the artery of Percheron, which has a single origin and serves the medial surface of both thalami adjacent to the third ventricle. Occlusion of this artery can lead to a medial thalamic syndrome characterized by apathy and anhedonia (inability to experience emotional pleasure), similar to a frontal lobe syndrome (see Figure 8.3 and Figure 8.4).

Occlusion of the PCA close to its origin will result in destruction of all the regions that it serves (including the thalamus, parietal and occipital cortices) leading to hemianopia (loss of perception of the visual fields on the opposite side), contralateral cortical sensory loss and thalamic dysfunction.

Occlusion of branches distal to the section of the artery serving the medial temporal and parietal lobes will result in pure occipital lobe damage and isolated hemianopia without other sensory changes.

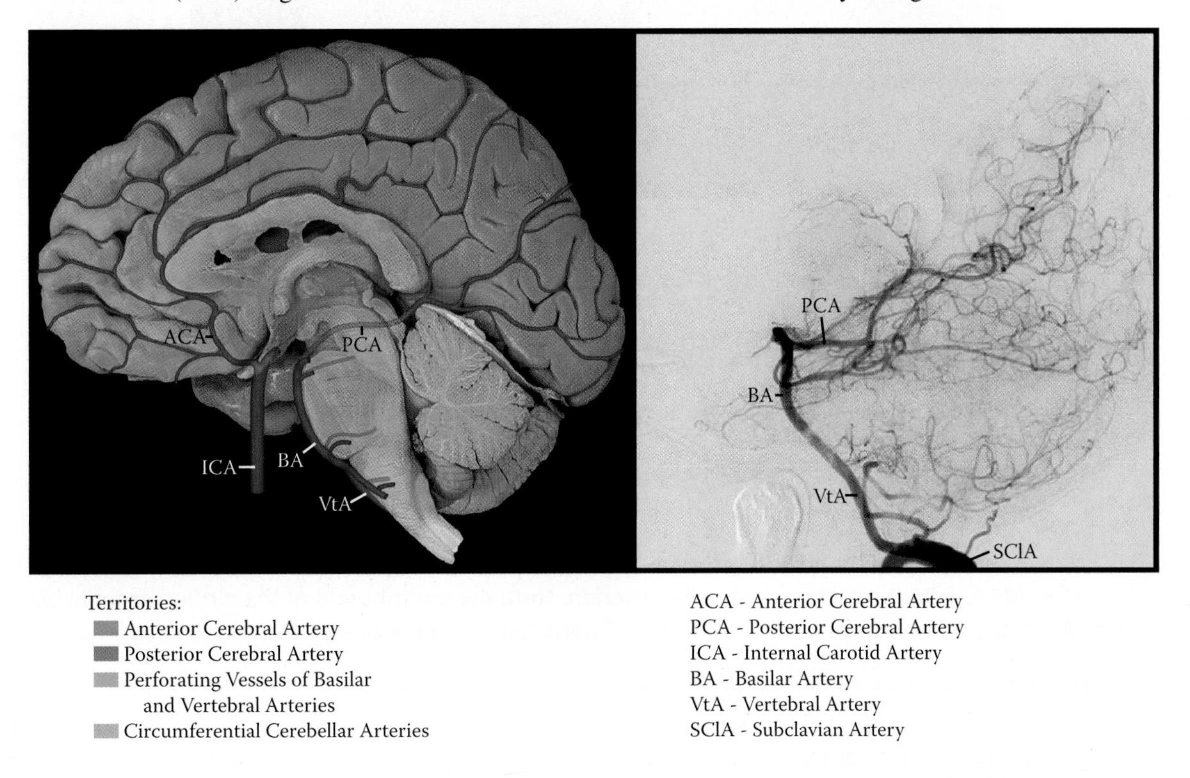

FIGURE 8.3 Blood supply of the brain: a mid-sagittal view of the brain and brainstem showing the territories of the vertebral (VtA), basilar (BA) and posterior cerebral (PCA) arteries, which are color-coded, and a left vertebral angiogram from the same perspective.

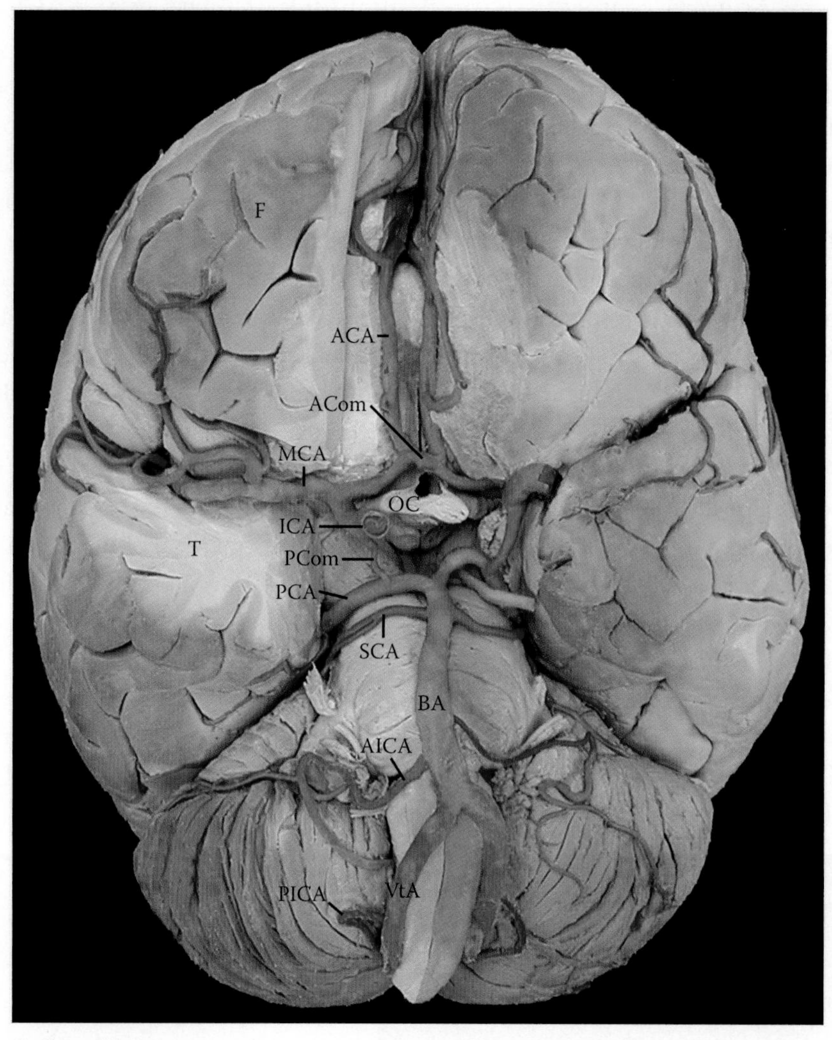

F - Frontal Lobe
T - Temporal Lobe (cut)
OC - Optic Chiasm (cut)
ACA - Anterior Cerebral Artery
ACom - Anterior Communicating Artery
MCA - Middle Cerebral Artery
ICA - Internal Carotid Artery (cut)
PCom - Posterior Communicating Artery
PCA - Posterior Cerebral Artery
SCA - Superior Cerebellar Artery
BA - Basilar Artery
AICA - Anterior Inferior Cerebellar Artery
PICA - Posterior Inferior Cerebellar Artery
VtA - Vertebral Artery

Territories:
Anterior Cerebral Artery
Middle Cerebral Artery
Posterior Cerebral Artery
Perforating Vessels of Basilar and Vertebral Arteries
Circumferential Cerebellar Arteries

FIGURE 8.4 Blood supply of the brain: a basal view of the cerebral hemispheres and brainstem showing the territories of the ACA, MCA, PCA, vertebral and basilar arteries. Note the contribution of the circumferential cerebellar arteries (PICA, AICA and SCA) to the blood supply of the lateral aspect of the brainstem, en route to the cerebellum.

Interruption of blood flow to the medial temporal lobes can lead to partial visual field defects and to memory loss. The hippocampus is situated right at the border zone of the arterial supply from the MCA and PCA. Occlusion of branches of either of these arteries can lead to hippocampal dysfunction and loss of short-term memory.

The two posterior cerebral arteries have a communicating branch between them and the internal carotid arteries at their origin; these are called the posterior communicating arteries (PCom) and are also part of the circle of Willis. They join the PCAs to the ICAs in combination with the AComs, allowing for redundancy of blood flow from one hemisphere to another in the event of obstruction of a major vessel below the level of the circle of Willis (e.g. the internal carotid on one side). The circle of Willis has significant variability and its ability to provide collateral circulation depends on the state of vessels involved.

8.3.3.4 Basilar Artery The basilar artery originates from the joining of the two vertebral arteries at the base of the skull. The basilar artery runs up the brainstem from the medulla to the pons and ends at the top of the midbrain to form the two PCAs (see Figure 8.4).

The basilar artery gives out arcuate branches along its path which serve the medial and lateral aspects of the brainstem and cerebellum at each level. Occlusions of the basilar artery are usually branch occlusions which produce one of the seven brainstem syndromes described below.

Brainstem syndromes depend on the level and the laterality. As was noted in Chapter 1, there are basically three levels of the brainstem: midbrain, pons and medulla, with a medial and lateral syndrome for all three levels.

Arcuate branch occlusions of the basilar artery lead to the medial and lateral pontine and midbrain syndromes. Occlusion of these branches by intrinsic atherosclerosis is common in patients with diabetes and hypertension.

Midbrain lesions involve distal sensory and motor deficits in the face and limbs as well as abnormal function of CN III and sometimes CN IV.

Weber's syndrome is the medial midbrain syndrome; it involves CN III and the cerebral peduncle, leading to a unilateral CN III palsy and contralateral hemiparesis of face and arm.

Claude's syndrome is the lateral midbrain syndrome. It involves the medial and lateral lemnisci, the spinothalamic tracts and the superior cerebellar peduncle. The clinical findings include contralateral sensory loss and ataxia of the ipsilateral limbs.

Lesions of the pons affect distal sensory and motor function as well as producing dysfunction of CN V, VI, VII and VIII.

The lateral pontine syndrome (Foville's syndrome) includes loss of hearing, balance and facial sensation as well as limb ataxia. The nuclei of CN V, VII and VIII are affected as well as the middle cerebellar peduncle.

The medial pontine syndrome (Millard-Gubler syndrome) includes loss of function of CN VI and VII and contralateral hemiparesis of arm and leg. The medial longitudinal fasciculus (MLF) may also be involved; it connects the CN VI to III nuclei to coordinate eye movements. An MLF lesion leads to an internuclear ophthalmoplegia in which the patient can abduct the ipsilateral eye (CN VI) but not adduct the contralateral eye (CN III).

Lesions of the medulla affect distal sensory and motor function as well as producing dysfunction of CN IX, X, XI, and XII.

The two vertebral arteries give off branches that serve the medulla and inferior surface of the cerebellum; these are known as the posterior inferior cerebellar arteries (PICA). Occlusion of a PICA will lead to the lateral medullary syndrome.

The lateral medullary syndrome (or Wallenberg's syndrome) consists of damage to the lateral structures of the medulla and often the inferior surface of the cerebellum. The anatomical components damaged include the inferior cerebellar peduncle, the descending sympathetic tract, the crossed spinothalamic tract, the descending tract of CN V and the lower end of the vestibular nucleus. Depending on the size of the damage to this area, the patient experiences ipsilateral facial sensory loss, ataxia, Horner's syndrome (see Glossary), and loss of pain and temperature sensation on the opposite side of the body.

The medial medullary syndrome (which is rather rare) includes loss of the pyramids and both medial lemnisci leading to bilateral extremity weakness and loss of the sensation for vibration and proprioception.

The most serious brainstem syndrome occurs with a complete occlusion of the basilar artery, leading to extensive ischemic necrosis that is most concentrated in the ventral pons (see Figure 8.5). This leads to a catastrophic condition called *locked-in syndrome* in which the patient has lost all motor function to the rest of the body below the CN III but with preserved sensory, auditory, visual and cognitive function. These individuals are usually ventilator dependent and require total care. Unable to communicate except through eye blinks, they have a poor quality of life; this situation leads to difficult ethical questions with respect to end-of-life decisions.

Central source emboli can originate from the heart or aortic arch and travel through a vertebral artery to the basilar artery. Often these emboli will fragment as they travel up the basilar artery, sending small emboli to its branches and giving multifocal brainstem and cerebel-

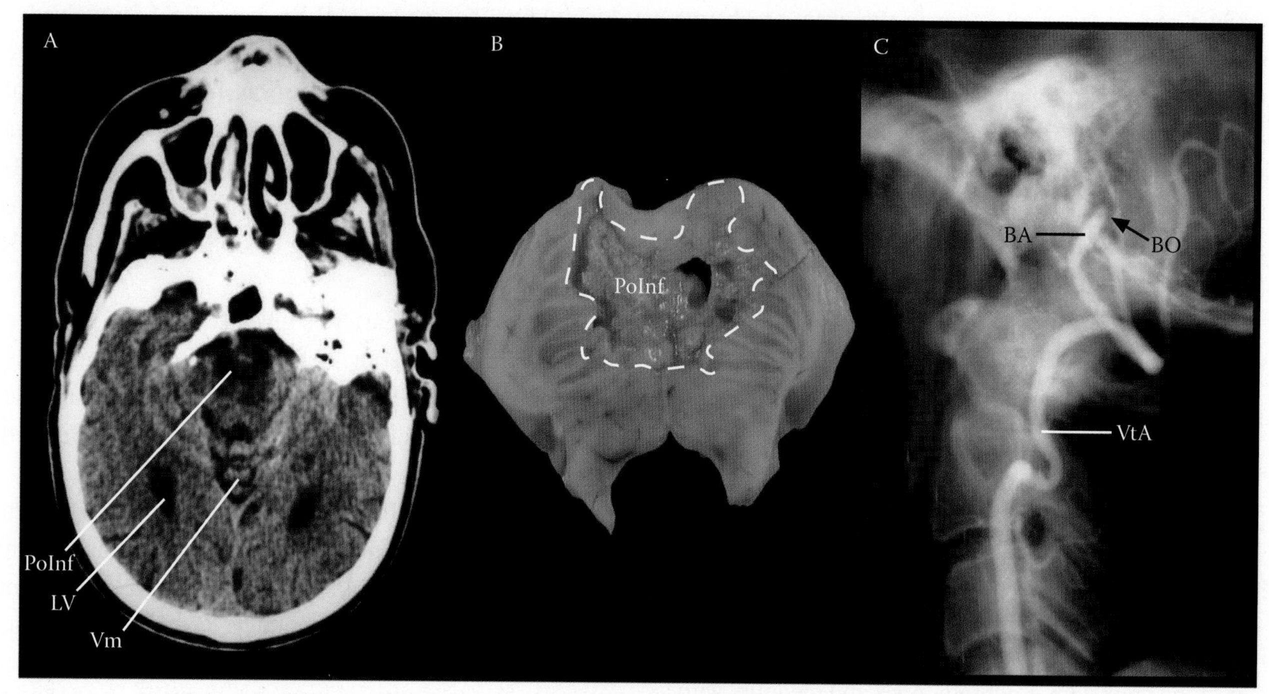

FIGURE 8.5 An ischemic infarct in the pons: (A) CT scan; (B) the pathology specimen of the same area as seen in the CT scan; (C) the angiogram shows no flow in the basilar artery with injection of the left vertebral artery.

lar deficits. If the emboli reach the top end of the basilar artery, the embolus often will fragment into several pieces going into the PCAs on both sides and resulting in infarction in the territories of the PCAs. This is the so-called top of the basilar syndrome. The clinical effect of this condition is to cause multiple infarctions in the brainstem and often in both occipital lobes, depending on where the fragments of the embolus interrupt arterial flow.

8.4 Localization Process

Following the defining event, the completed Expanded Localization Matrix would appear as shown in Table 8.2. As can be seen from the matrix, muscle, neuromuscular junction and peripheral nerve disorders may be eliminated due to the presence of cortical dysfunction and hyperreflexia. The major, more likely localizations target the left frontal, temporal and parietal areas. Brainstem and spinal cord localizations are also ruled out due to the cognitive impairment. The cerebellum as a primary localization is ruled out due to the presence of upper motor neuron weakness.

In terms of localization, prior to the major crisis, Etienne suffered an initial event in Broca's area of the left frontal lobe, affecting his speech, a TIA (discussed in Chapter 3). The presenting symptoms at that time, localizing to the left frontal area, suggest that there was a transient disturbance of flow in a branch of the left middle cerebral artery.

The localization suggested by a sudden transient speech disturbance, had Etienne mentioned it to anyone, should have triggered a search for the source of this event. The vascular territories supplying the left hemisphere can be retraced sequentially starting with the middle cerebral artery to the internal carotid artery, to the common carotid artery, the aorta, and then the heart.

8.5 Etiology: Etienne's Disease Process

In terms of the History Worksheet and Etiology matrix, the two events can be categorized as acute in nature. Since the localization exercise clearly points to a cerebral blood vessel disorder, this section will focus on acute disease processes affecting cerebral vessels.

8.5.1 Paroxysmal Disorders

Acute hemiplegia can present as a manifestation of a paroxysmal disorder. Epileptic seizures (see Chapter 10) can cause prolonged unilateral weakness, Todd's paralysis. This condition usually clears after 12 to 24 hours. There is usually a history of a preceding seizure but sometimes the seizure can be unwitnessed or can occur during sleep.

Migraine headache can be associated with hemiplegia and aphasia in a condition called hemiplegic migraine. These spells are often recurrent and there may be a family history. When an episode occurs for the first time, however, it is essential to perform vascular imaging so as not to miss a serious vascular problem. In Etienne's case, it is unlikely that a hemiplegic migraine attack would occur for the first time at age 57; the disorder is seen primarily in children and young adults.

8.5.2 Traumatic Vascular Injury

Traumatic injury to cerebral blood vessels can occur with sports injuries causing direct trauma to the carotid artery under the angle of the jaw: a typical example is a blow from a flying object such as a baseball or golf ball. Dissections of either carotid or vertebral vessels have been caused by falls, swimming or therapeutic neck manipulation.

Direct trauma to the top of the head can injure vertebral arteries in the neck, especially if there are degenerative changes in and around the course of the vertebral arteries in the transverse processes of the vertebrae.

In military environments with flying debris such as shrapnel, one may encounter direct rupture of blood vessels or vascular spasm related to blast injury.

Clearly none of these conditions would apply to Etienne.

8.5.3 Vascular Disorders

Vascular disorders include infarction or hemorrhage secondary to large or small vessel occlusion or rupture. This is the disease category that best fits Etienne's story. In the preliminary incident, there was a transient small vessel occlusion due to a thromboembolus which formed in the internal carotid artery more proximally. The presence of a single small vessel event should trigger a search for the offending large vessel. An alternative would be small vessel disease due to diabetes mellitus or vasculitis; in such patients the disease is usually chronic and the patient often has other systemic manifestations.

The most common cause of occlusive vascular disease is atherosclerosis, in this case of the left internal carotid artery distal to its origin. As previously explained, the bifurcation of the common carotid into the internal and external carotid arteries leads to turbulent flow of blood; this then causes shear injury to the wall of the internal carotid artery at its origin. This regional increased risk for damage, coupled with Etienne's multiple risk factors for atherosclerosis, led to the formation of a complex plaque at the origin of his left internal carotid artery.

In Etienne's case, the thrombus causing the first embolus probably formed over complex plaque at the origin of the left internal carotid artery and travelled to the middle cerebral artery where it transiently blocked a small branch artery serving Broca's area. Because of the short duration and neurological recovery from the event, it would be classified as a TIA (see Chapter 3) involving Broca's area of the left hemisphere.

The second event, which caused the right sided weakness, expressive and receptive aphasia, eye deviation, and visual and sensory loss involved damage to the frontal, temporal and parietal lobe areas served by the middle cerebral and anterior cerebral arteries. This would suggest an occlusion of the left internal carotid artery.

Therefore, if Etienne had sought medical advice after his first speech disturbance, given his multiple risk factors for cerebrovascular disease, he might have been worked up for a vascular cause of his symptoms. A Doppler or CT angiogram would have revealed a high grade stenosis of the left internal carotid artery, probably of 90 percent, with a complex plaque. He could have then been referred for neurosurgical or neuroradiologic interventional therapy before his stroke occurred one month later, and medical treatment to reduce his risk of stroke could have been instituted.

As was mentioned in Section 8.3, central source emboli refer to embolic material formed in the heart or great vessels either directly or indirectly. These sources include the left atrium due to stasis from atrial fibrillation, diseased or prosthetic heart valves, or infected thrombi forming on heart valves (endocarditis).

Sources of emboli outside the heart that subsequently pass through it include deep vein thrombosis (e.g. in the legs or pelvis), tumor or air bubbles from open fractures; the emboli then pass through right to left shunts such as PFO, atrial septal defects (ASD) and VSD, or through intrapulmonary vascular shunts.

Table 8.2 The Expanded Localization Matrix for Etienne

	Findings		Expanded Localization Matrix											
									Cortex					
			Muscle	NMJ	Peripheral Nerve	Spinal Cord	Brain-stem	White Matter	Frontal	Temporal	Parietal	Occipital	Basal Ganglia	Cere-bellum
Mental Status Exam														
Attention	Normal													
Short-Term Memory	Unable to test													
Long-Term Memory	Unable to test													
Calculation	Unable to test													
Verbal Reception	Receptive aphasia							×	×	×				
Verbal Output	Expressive aphasia							×	×					
Written Input	Unable to test													
Written Output	Unable to test													
Sequential Processing	Unable to test													
Visuospatial	Unable to test													
Integration	Unable to test													
Cranial Nerves	**Right**	**Left**												
Smell CN1	Unable to test	Unable to test												
Visual Acuity CN II	Unable to test	Unable to test												
Color Vision CN II	Unable to test	Unable to test												
Visual Fields CN II	Decreased threat from right	Decreased threat from right						×		×	×	×		
Pupillary Reflex CN II–III	Normal	Normal												
Extra-Ocular Movements CN III, IV, VI	Eyes deviated to left	Eyes deviated to left					×	×	×					
Fundoscopic CN II	Normal	Normal												
Facial Sensation CN V	Decreased	Normal			×		×	×			×			
Jaw Movement CN V	Unable to test	Unable to test												
Corneal Reflex V–VII	Normal	Normal												
Facial Movement CN VII	Decreased lower	Normal	×		×		×	×	×					
Hearing Acuity CN VIII	Normal	Normal												
Doll's Eye/Caloric CN VIII–III, IV, VI	Unable to test	Unable to test												
Gag Reflex CN IX–X	Decreased	Decreased	×		×		×	×						
Sternocleidomastoid Trapezius CN XI	Weak	Normal	×		×		×	×	×					
Tongue Movement CN XII	Weak	Normal	×		×		×	×	×					
Apnea Test Medulla														

	Findings		Expanded Localization Matrix											
									Cortex					
			Muscle	NMJ	Peripheral Nerve	Spinal Cord	Brain-stem	White Matter	Frontal	Temporal	Parietal	Occipital	Basal Ganglia	Cere-bellum
Motor Exam	**Right**	**Left**												
Involuntary	None	None												
Tone	Increased arm and leg	Normal	×		×	×	×	×	×					
Bulk	Normal	Normal												
Power	Decreased arm and leg	Normal	×		×	×	×	×	×					
Reflexes	Increased arm and leg	Normal	×		×	×	×	×	×					
Fatiguability	Unable to test	Unable to test												
Sensory Exam	**Right**	**Left**												
Pin/Temp	Decreased	Normal			×	×	×	×			×			
Vibr/Propio	Decreased	Normal			×	×	×	×			×			
Stereognosis	Unable to test	Unable to test												
Graphesthesia	Unable to test	Unable to test												
Two Point	Unable to test	Unable to test												
Coordination	**Right**	**Left**												
Upper limb	Unable to test	Unable to test												
Lower limb	Unable to test	Unable to test												
Trunk	Unable to test	Unable to test												
	Right	**Left**												
Gait	Unable to test	Unable to test												
Walk	Unable to test	Unable to test												
Stand	Unable to test	Unable to test												
Tandem	Unable to test	Unable to test												
Romberg	Unable to test	Unable to test												

Vascular disorders causing hemorrhage include conditions such as arterial berry aneurysms, arteriovenous malformations, and cavernous venous angiomas (also known as cavernomas). These vascular abnormalities are generally present at birth but expand to rupture and bleed later in life; whether they eventually rupture is influenced by risk factors such as age, hypertension and smoking.

The small branches of the MCA, known as the lenticulostriate arteries, undergo degenerative changes with age and with poorly controlled blood pressure. This may lead to rupture of these arteries leading to hemorrhages involving the internal capsule, lateral thalamus and basal ganglia. These are referred to as hypertensive gangliothalamic hemorrhages.

Blood vessel degenerative diseases that tend to cause hemorrhage include congophilic angiopathy, in which there is infiltration of vessel walls with amyloid material, resulting in weakening of the arterial walls. This condition leads to recurrent cortical hemorrhages with focal neurological deficits according to location, and to progressive dementia. This condition is also part of the larger process of amyloid accumulation elsewhere in the brain in Alzheimer's disease.

Arterial dissection refers to a situation where there is a weakness or injury to the intima of a vessel. If the weakness leads to structural failure of the vessel wall, then blood under arterial pressure creates a false passage between the intima and adventitia through which there is blood flow. This false channel then terminates by usually reentering the main lumen. Complications of this situation in cerebral vessels are restriction of flow of the affected artery and formation of thrombus at the distal end of the false lumen created by the dissection.

8.5.4 Toxic Injury

In general, medications or recreational drugs do not cause focal neurological deficits, although drug addicts who mainline substances that are contaminated with particulate matter may develop emboli that can lodge in any of the cerebral arteries. Progressive visual loss, often starting in just one eye, may develop in the context of an individual who smokes heavily and has poor nutrition, a disorder known as tobacco amblyopia. The damage to the retina is thought to be due to the combined effects of vitamin deficiencies and toxic chemicals in cigarette smoke.

With some exceptions, drugs and medications rarely directly lead to focal ischemic stroke. Any medication that might have a prothrombotic potential must be considered; examples include l-asparaginase, a chemotherapeutic agent, and intravenous immunoglobulin (IVIG). More commonly the withdrawal of medication, in particular anticoagulants for central source emboli, can result in acute infarction. Patients on anticoagulants for established central source emboli are thus at high risk of recurrence if these medications are stopped or mismanaged.

Life-threatening gastrointestinal and genitourinary hemorrhages are obvious indications for stopping systemic anticoagulation. Careful consideration, however, must be given to the risks versus the benefits of stopping, as against resuming, the anticoagulants once the cause of the hemorrhage has been found and resolved.

Medications that cause a precipitous rise in blood pressure can lead to both hemorrhagic and ischemic strokes. These substances can be simple over-the-counter cold medications; in patients with severe hypertension, the use of cocaine, amphetamines, and other vasopressor agents may lead to infarction or hemorrhage. Medications that cause a precipitous lowering of blood pressure (such as medications for hypertension, erectile dysfunction, and angina) can lead to cerebral and ocular hypoperfusion with ischemic damage in border zone areas.

Since the advent of MRI scanning, a new syndrome has emerged involving the dysregulation of the posterior cerebral circulation resulting in ischemic changes and, on occasion, infarction in the occipital lobes. The posterior reversible encephalopathy syndrome (PRES) has been associated with the use of medications that induce or aggravate hypertension. Sometimes the damage from this injury becomes permanent—PERMAPRES.

8.5.5 Vascular Infection

Infections of blood vessels are usually caused by metastatic infection from material released by infected heart valves (endocarditis). These infections are usually bacterial, with the most common organisms being staphylococcus and enterococcus. Infection of a cerebral blood vessel can cause a mycotic aneurysm. The source of the infection of the heart valve itself must be determined; possibilities include dental infections or occult malignancy in sites such as the colon or cecum.

Tertiary syphilis is an example of a chronic infection of blood vessels of the brain resulting in the formation of micro-abscesses or gumma adjacent to the infected blood vessels.

8.5.6 Metabolic Injury to Blood Vessels

Most metabolic disorders cause diffuse cerebral dysfunction and damage in a global rather than a focal fashion. However, sometimes extremes of metabolic disturbance can present with hard focal findings indistinguishable from those of stroke. Extreme hypoglycemia associated with insulin overdose or insulinoma may present in this way.

8.5.7 Inflammatory/Autoimmune

Inflammatory diseases of blood vessels such as systemic lupus erythematosus, temporal arteritis, and primary cerebral arteritis are all associated with multifocal vessel wall inflammation.

These disorders can come on quickly but would not resolve as quickly and completely as did Etienne's original attack. Vasculitis is often associated with systemic symptoms such as headache, arthritis, skin changes, and liver, lung, and, renal dysfunction.

Vascular etiologies not previously mentioned that may also lead to complete occlusion of the carotid, anterior, middle, posterior or basilar arteries include inflammatory diseases such as Kawasaki disease, primarily seen in children. Moyamoya disease refers to an occult occlusion of the intracranial carotid arteries accompanied by the development of extensive regional collateral vessels that appear on cerebral angiograms as a "puff of smoke"—the latter term in Japanese is *moyamoya*. The cause of this disorder is not known but may be related to qualitative abnormalities of collagen in the cerebral vessels.

Other noninfectious inflammatory disorders that can cause ischemic stroke include rheumatoid arthritis, sarcoidosis, Sjogren's syndrome, and Reiter's syndrome. These are often associated with systemic symptoms.

8.5.8 Neoplastic Disorders

Neoplasms of blood vessels are exceedingly rare. So-called glomus tumors of jugular veins or carotid arteries are probably embryonic rests of specialized neural crest cells that are located in these vessels and subsequently undergo neoplastic change.

Neoplasms of the heart, such as atrial myxoma, may be the source of recurrent emboli to the brain. These can cause characteristic murmurs and are best identified with transesophageal echocardiography.

8.5.9 Degenerative Disorders

Degenerative diseases of blood vessels other than atherosclerosis are being increasingly recognized: examples include vascular leukoencephalopathy, a disorder in which there is premature degeneration of small blood vessels in the white matter. CADASIL (cerebral autosomal dominant arteriopathy with subcortical infarcts and leukoencephalopathy) is an example of a progressive small vessel disease of genetic origin also affecting white matter.

8.5.10 Genetic Disorders

Genetic abnormalities of blood vessels include abnormalities of formation of blood vessels such as Sturge-Weber syndrome, a neurocutaneous disorder in which there is a congenital facial vascular nevus accompanied by abnormally developed cerebral vessels on the same side and by early-onset unilateral cerebrocortical infarction resulting in hemiparesis and epileptic seizures.

Fibromuscular dysplasia refers to a disorder of blood vessels, usually involving the renal and cerebral arteries. There are fibrotic changes in the blood vessel walls that cause tortuosity and stenosis. Other complications include aneurysm formation and dissection.

Another familial disorder of blood vessels is the familial cavernoma syndrome in which affected members have multiple cavernomas throughout the brain that can bleed and cause seizures.

Other genetic disorders that can lead to either spontaneous vascular occlusion or dissection include homocystinuria and Marfan's disease.

8.6 Case Summary

In summary, Etienne presented with acute right hemiplegia, global aphasia, right-sided sensory loss and inattention following a previous transient episode of expressive aphasia.

In terms of localization at the time of his stroke, the combination of acute profound weakness of the right side (including face, arm, and leg), gaze preference to the left, visual inattention to the right and global aphasia is indicative of widespread cortical and white matter dysfunction in the left frontal, temporal and parietal regions due to lack of blood flow in both the anterior and middle cerebral arteries. The common source vessel for all of these areas is the internal carotid artery (or, proceeding even further upstream, the common carotid artery).

The thrombus, built up over a plaque occupying approximately 90 percent of the cross-section of his left internal carotid artery, gradually increased in size to the point that it completely occluded the small amount of lumen left. Occlusion of the left internal carotid artery then interrupted flow to the left anterior and middle cerebral arteries—serving the frontal, temporal and anterior parietal lobes—and leading to the neurological deficits. In Etienne's case, the anterior communicating and posterior communicating arteries in the circle of Willis were too small or diseased to allow for significant shunting of arterial blood from the right internal carotid and basilar arteries.

The acute occlusion of his left internal carotid artery by fresh thrombus provided the opportunity for his health care team to give thrombolytic therapy to reestablish blood flow and to prevent and even reverse neurological damage. In this case, he had complete large vessel occlusion which is more amenable to intra-arterial therapy than to intravenous therapy. The common practice, as in this case, is to use an initial dose of intravenous tPA and, if there is no immediate clinical improvement, to then use intra-arterial tPA. tPA catalyzes the conversion of plasminogen to plasmin, resulting in the lysis of newly formed clots. If given within three hours of the onset of symptoms, this treatment is now considered the standard of care for acute ischemic stroke.

8.7 Case Evolution

Etienne was seen in the emergency department within 29 minutes and had a rapid assessment, blood work, CT scan and CT angiogram within 1 hour of the onset of his stroke.

The CT scan showed an evolving infarct involving the left middle and anterior cerebral artery territories without evidence of hemorrhage; the CT angiogram showed an acute internal carotid occlusion.

The neurologist on call notified the neuroradiologist on call that this represented a case of a large vessel occlusion and that Etienne was a possible candidate for intra-arterial tPA and clot removal. It was agreed that IV tPA should be started (up to two thirds of the normal full dosage) and a decision with respect to intra-arterial tPA then be made based on the patient's response to the initial therapy.

Etienne had no past history of gastrointestinal or genitourinary bleeding, no recent surgery or myocardial infarction. His blood work showed a normal platelet count, INR and PTT. It is important to check for abnormalities in platelet quantity and in the clotting system as these could increase the effect of tPA and cause bleeding.

Due to his dense aphasia, Etienne was not able to adequately communicate whether he understood the risks and benefits of both IV and IA tPA treatment; the risks and benefits were discussed with his wife Genevieve who signed the consent for tPA.

At 1 hour 45 minutes from the onset of his symptoms, IV tPA was started. After 35 minutes and two-thirds of the total dose given, he was reassessed and found not to have recovered from any of his deficits. He was then taken to the angiogram suite where a selective left carotid angiogram showed no flow in the left internal carotid artery. IA tPA was infused in conjunction with maneuvers to disrupt the clot. After several attempts, flow in the artery was reestablished. An angioplasty was performed on the stenotic vessel followed by the deployment of a self-eluting stent.

His neurological examination showed gradual improvement over the next 24 hours, during which he regained most of the strength on his right side and much of his speech. A follow-up CT scan at 24 hours showed a medium-sized left frontal infarct.

After 24 hours, he was started on medical treatment to prevent recurrence of stroke, including Clopidogrel (a platelet inhibitory agent), Ramipril (for hypertension) and Atorvastatin (to treat his elevated cholesterol). He was also referred for physiotherapy, occupational therapy and speech therapy assessments.

When seen in the neurologist's office one month later, he had some weakness and clumsiness of his right arm, as well as some difficulty finding words; his speech was still not completely smooth. His follow-up CT scan showed the infarct seen at 4 weeks to have become somewhat smaller but more radiolucent (indicating permanent tissue loss). The follow-up CT angiogram showed that the stent was patent.

With medication, Etienne's blood pressure became controlled, and his cholesterol returned to the normal range. He was referred for a sleep study, which showed severe obstructive sleep apnea with a respiratory disturbance index (RDI) of 65 and desaturations down to 79 percent. He responded to nasal continuous positive airway pressure (CPAP) at a level of 15 cm of water; this completely controlled the obstructive sleep apnea events and associated desaturations.

He continued to follow his rehab program, lost 30 pounds and returned to work with a reduced schedule three months after his stroke.

This scenario represents one of the most serious clinical presentations of stroke, in this case with a gratifying outcome.

Related Cases

There are e-cases related to this chapter for review on the DVD and Web site.

Suggested Readings

Adams, H. P. Jr. et al. Guidelines for the early management of adults with ischemic stroke. *Stroke* 38 (2007): 1655–1711. [Principal guideline statement for the use of thrombolytics in acute stroke.]

Chalela, J.A. et al. Magnetic resonance imaging and computed tomography in emergency assessment of patients with suspected acute stroke: A prospective comparison. *Lancet* 369, no. 9958 (2007): 293–298. [A review of methods of DWI, ADC maps for MRI and CT perfusion mapping.]

Feldmann, E. et al. The Stroke Outcomes and Neuroimaging of Intracranial Atherosclerosis (SONIA) trial. *Neurology* 68, no. 24 (2007): 2099–2106.

Golledge, J., R.M. Greenhalgh, and A.H. Davies. The symptomatic carotid plaque. *Stroke* 31 (2000): 774–781. [A description of the clinical symptoms associated with carotid stenosis.]

Hacke, W M. et al. Thrombolysis with alteplase 3 to 4.5 hours after acute ischemic stroke. *N. Engl. J. Med.* 359, no. 13 (2008): 1317–1329.

Lee, J.M., G.J. Zippfel, and D.W. Chio. The changing landscape of ischemic brain injury mechanisms. *Nature Suppl.* 399 (1999): A7–A14. [A review of the sequence of ischemic damage and ischemic cascade.]

Lyden, P. Thrombolytic therapy for acute stroke: Not a moment to lose. *N. Engl. J. Med* 359, no. 13 (2008): 1393–1395.

Skinner, C. Neurological complications of endocarditis in *Endocarditis: Diagnosis and Management*, K-L Chan and J. Embil, Eds. New York: Springer, 2006: 241–251.

Web Sites

Barthel Index: http://www.strokecenter.org/trials/scales/barthel.pdf. [The most common index to score disability after stroke.]

Modified Rankin Scale: http://www.strokecenter.org/trials/scales/modified_rankin.pdf. [Another commonly used scale to score disability.]

NIH Stroke Scale: http://www.ninds.nih.gov/doctors/NIH_Stroke_Scale_Booklet.pdf. [The most commonly used scale to score stroke severity in acute ischemia to determine use of thrombolytics. The scale has its limitations in that deficits have unequal weighting, for example, aphasia can score equally to facial weakness yet its contribution to disability is significantly greater.]

Chapter 9

Armand

ARMAND

Armand is Crash's rich uncle. He had been a successful businessman in real estate over the years, building up a local firm consisting of five partners of whom he is the most senior. Two of his sons have gone into the business and are also partners.

Armand served in the army during WWII, married his high school sweetheart when he returned from the war, had then settled down to a comfortable life and had raised his family.

Armand is now 82; nevertheless, he still gets dressed in his suit and tie and drives himself to work every day. Two months ago, however, he had an accident with his car—he had missed a red light and hit another car. No serious damage was done but the police had charged him with dangerous driving. His excuse was that it was a rainy day and he had not seen the light change. The Ministry of Transport had sent him a notice that he needed a medical assessment by a certain date or his license would be suspended.

There are other concerning behaviors cropping up both at home and at work. Over the past one to two years, he has become increasingly forgetful about the details of the various deals in which his company has been involved and tends to blame the other partners or his secretary if there are negative consequences to his forgetfulness.

At home, his wife has noticed increased forgetfulness for names and places. As well, there have been episodes of unexplained anger, usually directed against her; these outbursts are usually short-lived.

When he finally showed up for his medical assessment, he appeared well-groomed, with a suit and tie. He kept repeating the question, "You are not going to take away my driver's license, are you?"

Objectives

- To learn the basics of the information system architecture of the human CNS
- To learn the principal anatomic components of the central nervous system related to higher cognitive function
- To learn the effects of damage on these components and the associated clinical manifestations
- To learn the major dementia types and their clinical manifestations
- To become sensitized to the medical, physical and psychosocial implications of dementia

9.1 Clinical Data Extraction

What other information would you want to gather by history?

Before proceeding with this chapter, you should first carefully review Armand's story in order to extract key information concerning the history of the illness and the physical examination. On the DVD accompanying this text you will find a sequence of worksheets to assist you in this process.

With this information, the neurological examination section of the Extended Localization Matrix Worksheet can be completed.

9.2 The Main Clinical Points

The principal clinical features of Armand's presentation are

- Difficulties with memory, attention and emotional control, all gradually developing over one to two years
- Moderate systemic controlled hypertension
- Bilateral carotid bruits
- Decreased score on the mental status examination
- Difficulty with upward gaze
- Unsteady gait
- Impaired vibration sense in the toes
- Hypoactive ankle jerks
- Brisk bilateral palmomental response

9.3 Relevant Neuroanatomy, Physiology, and Pathophysiology

Since Armand's principal deficits are in memory and attention, we must consider some anatomical and pathophysiological aspects of cognition.

A review of his past medical history revealed that he had had hypertension for many years and was currently on three different antihypertensive medications. Otherwise his neurological and systemic systems review was negative, in particular for stroke and myocardial infarction. His mother died in her 80s of "old age"; his father died of TB in the 1920s when Armand was still an infant.

His general physical examination showed a blood pressure of 160/90 in both arms; his heart rate was 84 and regular. Auscultation of the chest and neck revealed a slight aortic murmur and bilateral carotid bruits, left louder than right. The rest of the exam was normal for age.

Armand's neurological examination showed a Folstein Mini-Mental score of 22 (refer to Chapter 2). He lost points for delayed recall, serial sevens, and naming. He became quite frustrated and angry during the mental status testing so it was stopped and then resumed later. The neurological examination showed some decreased ability to look upward on testing the extra-ocular movements; otherwise the cranial nerves were normal.

The motor exam showed normal power with a slight increase in his reflexes throughout except for his ankle jerks which were decreased. His plantar responses were flexor. He had a brisk palmomental response on both sides (a reflex contraction of the ipsilateral mentalis muscle when the palm of the hand is vigorously stroked—a frontal lobe release sign). The muscle tone was quite variable; he seemed to have trouble with understanding the instructions to relax.

There was slight decrease in vibration sensation (at 256 Hz) in his toes. His upper and lower limb coordination for finger–nose and heel–shin was slow.

His gait was slightly unsteady for tandem walking but otherwise normal.

9.3.1 *Cognitive Function*

Human cognitive function represents the most sophisticated information system of which we are aware among living creatures on earth. Other animals have interesting and diverse adaptations that are necessary for their survival and reproduction, but none rivals human cerebral function in terms of complexity, adaptability and use of language.

The basic elements of human cognitive function include:

- Ability to sense visual, audio, olfactory, position, acceleration and somatosensory information
- Ability to perform motor tasks of locomotion, speech and object manipulation
- Ability to comprehend and codify sensory information
- Ability to comprehend language
- Ability to express language
- Ability to remember information
- Ability to perform mathematical calculations
- Ability to multiprocess
- Ability to codify basic information objects
- Ability to sense, codify, and learn new situations and responses to those situations
- Ability to detect and defend against threats
- Ability to perform executive functions of command and control in order to maintain a system of priorities that best ensures the protection, growth and evolution of the organism

The performance of these tasks 24 hours a day, seven days per week, for an average 85-year life span requires an information system architecture that must have significant redundancy and survivability, to be able to regenerate itself and work within the metabolic framework of nutrients and waste disposal systems provided by the circulation. These functions are accomplished by a massive parallel system of neurons and axons, which communicate among themselves and to muscles and nerves.

The neurons are the microprocessors of the brain. As with personal PCs or large-scale systems, the neuron is essentially a central processing unit (CPU). The basic unit of information processing in any computing system is a central processing unit having the basic functional elements of input, memory, computation, logic processing and output. The sophistication of the processing depends on whether the processor is designed for generalized use or for specific applications.

The processing capability of a given CPU chip depends on the instruction set that is designed into the processor.

The instruction set for a given chip is called the microcode; it constitutes the smallest unit of computational work performed by the chip. For example, to do the most basic operation of adding two double precision numbers together using a high language such as C or Visual Basic®, literally hundreds of operations are required at the microcode level on the chip; nevertheless, the chip does all this very quickly. In human neurons, the microcode is different from one neuron population to another, thus allowing for different functions.

Thus, the human brain consists of a vast network of interconnected microprocessors, with differences in function depending on their location and connections.

The cerebral cortex contains billions of these neuronal microprocessors; they perform complex computations on the input from all of the sensory systems and from other areas of the cortex. The output of these computations is transmitted via white matter tracts either to other areas of the brain, to brainstem or to spinal cord, resulting in complex behaviors such as limb movements or language production.

The human nervous system is therefore a massive parallel processing system that uses the basic neuron as a template and allows different functionalities based on anatomical location and interconnections.

In the process of embryogenesis, the cerebral cortex forms from neurons that migrate outward from a matrix in the periventricular region; under the control of genetic inducers, neuronal stem cells migrate and differentiate into cortical neurons. Instead of designing multiple types of neurons with different functionalities, a basic template is used, which then is chemically and electrically induced to perform different functionalities.

For example, the neurons in the occipital cortex, receiving visual information from the retina via the lateral geniculate bodies, perform a very limited and specific function of creating a matrix of the visual image, which represents the reconstitution of either the left or the right visual field. No other neurons in the CNS can perform this function but the occipital cortical neurons cannot do anything else.

These neurons are in contrast to those in the angular gyrus; if the latter are damaged, there is impairment in the ability to calculate, know left from right and identify by touch a given finger on the hand. These integrative functions are very different from the image processing functions of the occipital cortex but the hardware is drawn from the same basic design template. One of the great challenges for neuroscientists is to discover what makes the functional capabilities of specific neuron populations different. How are the inner workings of each neuron—its instruction set—made different and how can we measure these differences?

There is a term used to describe this difference in function of different neurons in different areas of the brain: *neuronal cytoarchitectonics.* Under the microscope, the layers of cerebral cortical cells (described in Chapter 10) appear slightly differently arranged in each area, thus giving rise to the analogy of continental plate tectonics in the form of layers of cortex. Once laid down after embryonic life and programmed during the early years of development, most areas of cortex assume their given function which, if destroyed, will not recover; the function in that part of the brain is lost. There are some regions of what are called hypervariable function neurons; these seem to be able to take over the function of adjacent neurons that have been damaged.

The axons are the output cables of the neurons; they direct output information either to other neurons for more processing or to motor neurons for motor output. Motor output can be in the form of locomotion, use of hands and tools, or vocal output via the muscles of the upper airway, chest and diaphragm.

The human cortex has been organized functionally such that certain neurons perform certain specialized functions. Table 3.9 is a summary of the localization of cortical functions that are important clinically. The dominance column assumes right-handedness.

9.3.2 Pathophysiology of Dementia

The definition of *dementia* is an irreversible loss of cognitive function due to either structural or metabolic damage to the neurons or their axonal connections. Dementia does not refer to transient states of cognitive dysfunction or to confusion. However, the presence of an underlying dementia may amplify the intensity of an acute confusional state.

Loss of cognitive function therefore depends on which area of the cerebral cortex or white matter connections have degenerated. Pure cortical dementia occurs when large populations of cortical cells in one area lose function and affect the person's cognitive function in that domain. Damage to a specific area from ischemic or hemorrhagic stroke will cause abrupt focal deficits in motor or sensory function, usually on one side of the body, as well as cognitive dysfunction depending on location of damage. The presence of areas of ischemic cortical damage tends to amplify the effects of a developing cortical dementia.

More subtle forms of dementia occur when there are multifocal areas of damage from small vessel disease. This is usually due to hypertension but there are also inherited forms of ischemic white matter disease such as CADASIL (see Chapter 8).

There are some very rare forms of hereditary disease that may produce early-onset dementia. Examples include disorders of myelin synthesis and maintenance (known as leukodystrophies) and of neuronal mitochondrial function (leading to "power failure") such as Leigh's disease.

The pathophysiology of primary dementias has taken on a new language and terminology. The newer terminology refers to various protein entities that tend to accumulate in these various conditions, ultimately ending in neuronal cell death and dementia. For Alzheimer's disease, the term *amyloidopathy* is now often used; for dementia with Lewy bodies, *synucleinopathy*; for some forms of frontotemporal dementia (FTD), *tauopathy*. Life is not that simple, however, in that subdivisions of the more traditional disease states such as Alzheimer's disease do not fit into these neat protein-type categories. Table 9.1 shows the different forms of cortical dementias with their principal neuropathogical findings.

All neurons have metabolic cycles through which the various structural and functional proteins must pass and ultimately be exported or recycled by the neurons themselves or by adjacent glia. As a general rule, most proteins are exported in soluble forms for transport either into astrocytes or to the circulation for eventual reuse. If the isoform of a given protein produced by a neural cell is in an insoluble form, then the protein is not disposed of in the usual fashion and the proteins accumulate either in the neuronal cytoplasm or in the adjacent environment. This accumulation of insoluble protein often forms the various structures that are seen microscopically. There is significant controversy over the significance of these accumulations of insoluble protein, some of which are known as inclusions: Do they represent a pathological abnormality or are they simply a defensive response? For instance, a Lewy body containing synuclein may be an attempt to protect the cell by sequestering the potentially toxic protein. On the other hand, it may just represent the result of how synuclein clumps together based on the physicochemical environment.

The end point is that these insoluble protein by-products eventually accumulate to the point where neuronal cell death occurs; if large populations of neurons die, dementia occurs.

Dementia frequently does not manifest as a single pure entity. It is increasingly appreciated that there is a continuum between the various primary dementias, including Alzheimer's disease, vascular dementia due to small vessel disease, vascular dementia due to large vessel disease causing stroke and other less common forms of dementia, such as dementia with Lewy bodies (DLB) and frontotemporal dementia (FTD).

9.4 Localization Process

The localization matrix indicates findings suggestive of localization in the peripheral nerves (sensory loss in the toes) but the majority of the deficits direct us to structures above the brainstem. Given that Armand's predominant deficits are in cognition, the localization possibilities

Table 9.1 Neuropathological Features of Cortical Dementias

Condition	Location	Protein	Neuronal Features	Astrocyte Features
Alzheimer's disease	Hippocampus, temporoparietal	Amyloid and tau	Amyloid plaques, tangles	
Frontal variant of FTD	Frontal	Tau	Pick bodies	
Semantic or temporal variant of FTD	Temporal			
Non-fluent aphasia of FTD	Temporal			
Lewy body dementia	Diffuse, parieto-occipital	Synuclein	Lewy bodies	
Corticobasal degeneration	Superior frontal parietal, basal ganglia	Tau	Coiled tangles	
Progressive supranuclear palsy	Midbrain	Tau	Globose tangle	Tufted
Creutzfeld–Jacob	Cortex, thalamus	Prion protein	Spongioform changes	

include cerebral cortex, white matter, thalamus, basal ganglia and cerebellum. Within the cortex, there are specific functions based on location; these enable one to localize cognitive domains into frontal, temporal, parietal and occipital areas, with some overlap.

Based on the history and physical examination, the most likely localization in this case would be in the cerebral cortex in those areas serving short-term memory, word finding, executive function and emotional control. Anatomically these would be located in the hippocampus of the temporal lobe for short-term memory, the superior temporal gyrus for word finding and the frontal lobes for executive and emotional function.

Lesions that affect subcortical structures (such as the deep white matter, thalamus and basal ganglia) tend to cause motor and ideational slowing, some elements of which Armand does have.

Lesions in the cerebellum can lead to difficulties with control of the rate and rhythm of motor output but it has been increasingly recognized that the cerebellum has similar modulating functions on cognitive and emotional output.

9.5 Etiology: Armand's Disease Process

In the ongoing assessment of Armand, it would be important to focus on the possibility that his decreased mental status might be due to a reversible cause.

The medication history may be extremely important, especially in the face of a new-onset rapidly progressive picture. This would include a detailed history of exactly what medications have been started and stopped during the period in which the symptoms have developed, including even medications as simple as eye drops. Steroid eye drops used after cataract surgery, for example, have been implicated in the development of a clinical picture of a rapid-onset, progressive dementia; when the ophthalmic steroid is discontinued there is complete reversal of the dementia.

In terms of etiologic diagnosis, it is important to rule out other reversible causes as outlined below, utilizing simple blood tests and performing appropriate imaging studies to rule out any structural causes. Slow-growing tumors such as meningiomas can masquerade as dementia, as can chronic subdural hematomas or normal pressure hydrocephalus (NPH; see Chapter 11, Section 11.6.7).

Specific questioning is required to exclude the presence of systemic diseases such as hypothyroidism, B12 deficiency or alcoholism. Did Armand have exposure to toxins or heavy metals during his working life or when he was in the army? Did he ever suffer from venereal disease such as syphilis?

He has some features of peripheral sensory loss. Could this be related to his cognitive decline?

9.5.1 *Paroxysmal Disorders*

The time course of this disorder is not in keeping with a paroxysmal disorder, although a clinical picture like this may coexist with frequent, uncontrolled seizures. A short-term encephalopathy may occur after a prolonged seizure but this possibility obviously does not fit with Armand's clinical picture.

9.5.2 *Traumatic Brain Injury*

Head trauma (either open or closed) may cause damage to gray or white matter, leading to dementia. The most common form is severe closed head trauma causing diffuse axonal injury (DAI). This is essentially a shear injury to the large axonal bundles due to the effect of acceleration/deceleration forces on the brain. Other mechanisms of brain injury include hemorrhagic and ischemic damage.

Recurrent brain trauma related to sports such as boxing or ultimate fighting has been implicated in dementia pugilistica.

Recurrent falls leading to chronic subdural hematomas can result in cumulative damage to the underlying cortex and a pattern of cortical dementia. This etiology does not fit with the case under consideration.

9.5.3 *Vascular Disorders*

Damage to cortex and white matter secondary to large vessel hemorrhagic or ischemic strokes may result in dementia. As we saw in Chapter 8, strokes are usually associated with risk factors such as advanced age, blood pressure, cholesterol, smoking, diabetes and family history (many of which are present in Armand's case).

Vascular dementia is now recognized to have many different forms. The most common form is the so-called multiple infarct dementia (MID); it results from the accumulation of multiple discrete lobar or subcortical infarctions due to either cerebrovascular disease or recurrent emboli from the heart. The effect of this cumulative damage is dementia due to cumulative loss of neurons from these multiple strokes. The effective treatment of this form of vascular dementia is prevention of the recurrence of stroke by risk factor management.

The second most common form is that of a slowly progressive form of deterioration of the small blood vessels which supply primarily the white matter but also, to a lesser extent, the gray matter. Uncontrolled systemic hypertension is by far the most common risk factor for this form. The genetic disorder CADASIL and others also belong in this category.

Diffuse ischemic white matter injury can also occur after cardiac arrest or prolonged hypotension in septicemic shock. Such injury may also occur in the context of cardiac surgery with prolonged periods on cardiopulmonary bypass or with perioperative hypotension.

Armand's slowness of action might suggest a subcortical vascular dementia due to microvascular degenerative changes considering his risk factor of long-standing inadequately treated hypertension. The CT scan or MRI will aid in determining the extent of this process and its relationship to his dementia.

9.5.4 *Toxic Injury*

The commonest toxic agent causing dementia in North America continues to be alcohol or ethanol. Ethanol destroys cortical neurons and Purkinje cells in the cerebellum as well as depleting vitamins such as thiamine and folate; the latter problem can have both acute and chronic effects on cognitive function. It is indeed a difficult situation in the office when you have to tell people in their late 60s that they have developed dementia and ataxia because of their chronic daily alcohol use. Their response is often: "Why didn't somebody tell me about this earlier?"

The role in the development of dementia of chronic exposure to other nonmedical drugs such as marijuana, cocaine and crystal methamphetamine will be determined in time.

Certain medical agents clearly lead to chronic cognitive dysfunction; these include chemotherapeutic agents such as high dose cyclophosphamide, adriamycin and cisplatin. Radiotherapy to the brain is known to cause cognitive decline and its deleterious effects on the pediatric population have been well documented.

Armand's history does not suggest that there are probable risk factors to support this etiology group.

9.5.5 *Infection*

Infectious diseases caused by bacterial, viral, protozoan and fungal agents and by prions may lead to dementia by virtue of the extent of the damage. These include syphilis, Lyme disease, herpes simplex encephalitis, the arborvirus encephalitides, human immunodeficiency virus (HIV), cytomegalovirus (CMV), and West Nile virus.

There are a few specific infections of white matter that may cause dementia: these include progressive multifocal leukoencephalopathy (PML, due to JC virus, a member of the papova group of viruses), HIV and CMV.

Prion diseases are caused by transmissible insoluble proteins and typically produce dementia: these include Creutzfeld–Jacob disease and its variant form related to bovine spongiform encephalopathy (BSE), fatal familial insomnia (FFI), kuru, and Gerstmann–Straussler–Scheinker disease. Typically these disorders cause a rapidly progressive fatal dementia with myoclonic jerks.

A lumbar puncture would be indicated if there were a high enough suspicion of an infectious disorder.

The time course of Armand's dementia might suggest a slowly progressive infection or a reactivation of a latent infection such as syphilis.

9.5.6 *Metabolic Injury*

Lack of any of the major substrates for brain function such as oxygen, glucose, electrolytes and hormones may cause acute cortical and subcortical dysfunction. These disorders tend to be very acute. Chronic deficiencies of vital substrates can lead to chronic brain deterioration and dementia. Specific metabolic injuries include deficiencies of thyroid hormone and of vitamins B1, B12 and folic acid.

The loss of sensation to vibration in the toes bilaterally might raise the possibility that there could be a metabolic injury or deficiency such as B12, which could affect both the peripheral nerves and cortical functions. A complete blood count with a smear would help to indicate if there were large hypersegmented polymorphs; blood assays for vitamin B12 and folate would show whether Armand has a vitamin deficiency.

9.5.7 *Inflammatory/Autoimmune*

Systemic inflammatory disorders such as systemic lupus erythematosus (SLE), Sjögren disease and isolated cerebral angiitis may lead to dementia.

The most common noninfectious inflammatory white matter disease seen in Caucasian populations is multiple sclerosis. The dementia of MS occurs when the disease burden of MS plaques becomes so severe that interneuronal communication becomes severely degraded.

The time course of the history, Armand's age and the lack of focal white matter findings makes MS unlikely. For the vasculitis group, there were no specific findings of systemic involvement.

9.5.8 Neoplastic Disorders

Malignancies, whether primary CNS or metastatic, usually present with focal defects or seizures but may later produce cognitive decline and eventually dementia. Occasionally, however, dementia is the presenting symptom, by which time there is either a large tumor or multifocal primary or metastatic tumors. The most common primary tumors presenting with acute cognitive decline are malignant astrocytomas.

Progressive dementia can sometimes present as a manifestation of a paraneoplatic disorder.

9.5.9 Degenerative Disorders

Specific criteria for the diagnosis of the various forms of dementia have been developed and are based on some common criteria, modified by specifics for the various subtypes; all of these are outlined in the *Diagnostic and Statistical Manual of Mental Disorders*, Fourth Edition (*DSM IV*). The *DSM IV* criteria for Alzheimer's disease are shown in Table 9.2.

The original NINCDS-ADRDA criteria for Alzheimer's Disease (McKhann et al., 1984) have been refined to include clinical, neuroradiological and other biomarkers (Dubois et al., 2007).

Alzheimer's disease tends to systematically destroy structures in the forebrain that use acetylcholine as the primary neurotransmitter. These structures include the nucleus basalis of Meynert and the diffuse cholinergic projections from this nucleus (see Figure 12.6D). The degenerative process in the neurons leads to the accumulation of several materials within the neurons, extracellular space and the blood vessels of the brain. Neurofibrillary tangles in the neuron cell bodies represent tau protein. Insoluble amyloid is exported into the extracellular space and deposited as amyloid plaques and also as congophilic material in blood vessels (Figure 9.1A).

Cerebral blood vessel walls infiltrated by amyloid are more friable, leading to spontaneous cortical hemorrhages. This disorder is called congophilic angiopathy.

Alzheimer's disease initially tends to target the temporal and parietal lobes more than the frontal lobes. This leads to the initial clinical picture of memory loss and difficulty with the reception of speech. Eventually, as it progresses, the disease affects the cortex throughout, including the frontal lobes, with relative sparing of the occipital lobes. The individual loses the ability to perform basic activities of daily living. When Alzheimer's disease patients become bedridden and unable to look after their basic needs, they become at high risk for infections, especially pneumonia, the usual cause of death. The time from diagnosis is variable but is usually in the order of five to seven years.

The introduction of acetylcholinesterase enzyme inhibitors specific to the brain has led to some success in stabilizing cognitive performance, delaying the need for institutional care.

Armand's clinical picture is consistent with Alzheimer's disease with a component of vascular dementia.

Frontotemporal dementia (FTD), previously called Pick's disease, is now known to have three specific subtypes, each having different clinical, radiological and pathological findings. These are the frontal variant of FTD, semantic dementia and non-fluent aphasia.

The frontal variant of FTD is characterized by behavioral changes that include disinhibition and/or apathy, dietary changes, social withdrawal and depression. This form has pathological changes mainly in the left premotor and right dorsolateral frontal cortex. Neuroimaging shows mainly bilateral frontal lobe atrophy. The pathology for this type is shown in Figure 9.1B.

The semantic or temporal variant manifests with compulsions, mental rigidity, emotional withdrawal, breakdown of semantic knowledge, long-term memory loss and agnosia for faces and objects. The imaging for this type shows asymmetric atrophy in the temporal lobes.

The third type, non-fluent aphasia, presents with significant deficits in verbal fluency, grammar and word repetition, and with social withdrawal and depression; it occurs later than the other two subtypes. The imaging for this type shows asymmetric atrophy in the frontal lobes.

Armand's history did indicate some emotional lability but, since most of his deficits are in the area of memory and comprehension, this makes FTD less likely.

Table 9.2 *DSM IV* Criteria for Alzheimer's Disease

Alzheimer's Disease
Multiple Cognitive Deficits
Criterion A
A1 Memory Impairment
A2 One or more of the following
Aphasia
Apraxia
Agnosia
Disturbed executive function
Criterion B
Cognitive deficits in criteria A1 and A2 each:
Cause impairment in social or occupational functioning
Are not due to CNS disease
Do not occur solely during the course of delirium
Criterion C
Gradual and continued cognitive decline
Criterion D
Other systemic or neurological and psychiatric illnesses should be eliminated
Criterion E
Alzheimer's disease should not be diagnosed in the presence of delirium
Reprinted with permission from the *Diagnostic and Statistical Manual of Mental Disorders*, Text Revision, Fourth Edition, (Copyright 2000). American Psychiatric Association.

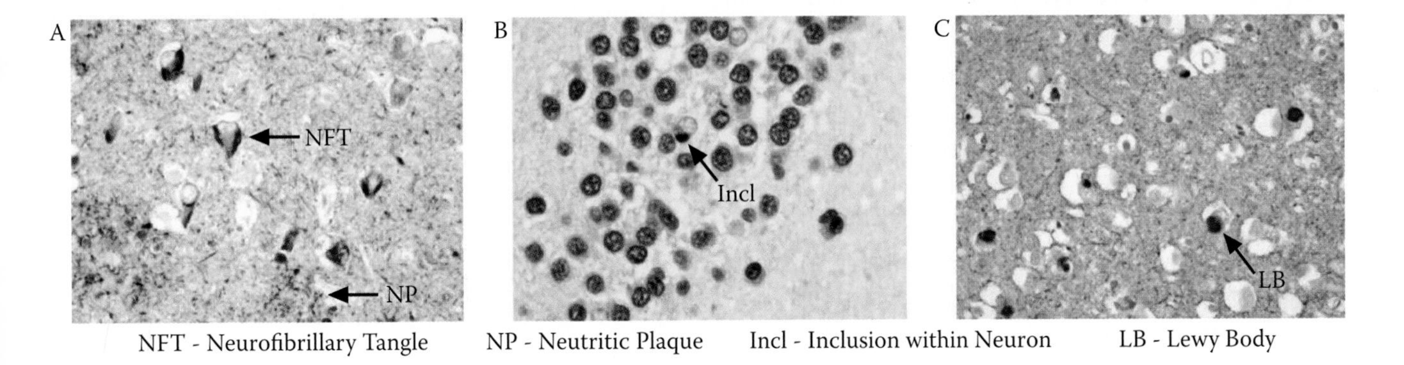

FIGURE 9.1 Pathology of dementias. (A) Alzheimer's dementia: amyloid plaques and neurofibrillary tangles; tau immunostaining (hippocampus). (B) Frontotemporal dementia: FTLD-U; TDP-43 staining (dentate gyrus). Notice the loss of normal nuclear staining in the cell with inclusion (arrow). (C). Dementia with Lewy bodies: cortical Lewy bodies; synuclein staining (cingulate gyrus). (Courtesy of Dr. J. Woulfe.)

Dementia with Lewy bodies (DLB) is characterized by the accumulation of Lewy bodies in cortical and limbic neurons as well as in their neurites (axons and dendrites). The protein associated with this disorder is alpha-synuclein, which accumulates in the neurons mentioned above, as well as in specific populations of monoaminergic neurons in the brainstem (Figure 9.1C).

The core features that distinguish DLB from the other dementias include fluctuation of cognitive function, recurrent visual hallucinations and parkinsonism.

Other clinical features of DLB include the REM behavior disorder (RBD). In this disorder, symptoms occur while the patients are in the rapid eye movement (REM) stage of sleep. They appear to be acting out their dreams. Normal individuals during REM sleep are effectively paralyzed except for the muscles of respiration. There is functional paralysis of the spinal cord produced by a descending pathway from the subcereulean nucleus in the pons to the spinal cord lower motor neurons. This pathway is thought to be glycinergic.

In DLB, the subcereulean nucleus becomes dysfunctional because of alpha-synuclein deposition that destroys most of its neurons. This leads to a loss of the normal state of paralysis that occurs in REM sleep, leading to the behavior of appearing to act out dreams.

Fortunately, there is a very effective medical treatment for this. A small dose of clonazepam, a benzodiazepine with some glycinergic properties, is very effective in stopping the distressing (especially for the bed partner) symptoms of RBD.

Patients with DLB tend to be very sensitive to neuroleptics and have repeated falls, transient disturbances of consciousness and autonomic dysfunction. There is also occipital lobe hypometabolism seen on PET scan and posterior slowing on EEG.

There is no specific treatment for DLB except supportive care. Given the parkinsonian features, special attention is required in order to ensure the safety of the patient from falls. The parkinsonian features are treated in a manner similar to that used in someone with idiopathic Parkinson's disease; it is important to avoid over-treatment.

Multisystem atrophy (MSA) represents an extreme form of alpha-synucleinopathy in which the cortex, basal ganglia, brainstem, cerebellum and autonomic nervous system are all progressively affected. Clinically, the deficits start in one system and then spread to the others at varying rates. Similar to DLB, these patients often develop RBD many years before the other manifestations appear. The presence of RBD is therefore a potential predictor of a future disorder such as Parkinson's disease, DLB or MSA.

The treatment of the depression, hallucinations and delusional behavioral disturbances can be very difficult in that MSA patients are also very sensitive to classical antidepressants and neuroleptics. Agents such as selective serotonin reuptake inhibitors (SSRIs), risperidone, clozapine and olanzapine are useful but must be started in small doses and titrated slowly while watching for side effects.

Armand does not have any of the cardinal features of DLB.

The term *subcortical dementia* is used to represent disorders of progressive cognitive decline secondary to primary disease of white matter. This is really a group of disorders that can be categorized using the etiology matrix.

White matter damage may be caused acutely by drugs or toxins (e.g., organic solvents) causing white matter damage and subacutely or chronically by drugs such as cyclosporine and amphotericin-B.

In summary, the clinical manifestations of a given dementia depend on which area of the brain is involved, its given function and the type of protein that is accumulating. Collateral factors such as the presence of large or small vessel ischemic disease and the presence of other medical or metabolic problems common in the elderly all tend to amplify the effect of the neuronal cell death on the behavioral performance of the individual.

9.5.10 Genetic Disorders

Pure genetic forms of dementia are rare; the best known is Huntington's disease (see Chapter 14 and its Appendix on the DVD). In this autosomal dominant disorder, there is degeneration of medium-sized cells in the caudate nucleus and more diffusely in the cortex. It presents in an individual in the mid-20s with involuntary movements called chorea, characterized by small amplitude sudden jerking of distal muscles. At first it can be very subtle and the individual tries to conceal the movement by incorporating the involuntary movement into a voluntary action. Later there may be writhing movements of the proximal upper and lower limbs referred to as athetosis. The dementia can be difficult to quantify at first but usually by age 35 there is a measurable decrement in the Mini Mental Status Exam; by this age significant deficits can be found on neuropsychological testing.

Huntington's disease patients tend to become progressively disabled by their dementia, more so than by the movement disorder, and are usually institutionalized by their forties or fifties. Since the advent of reliable genetic testing, it is now possible to determine if a given fetus is affected.

Other genetic forms of early-onset dementia include abnormalities of lysosomal metabolism, including Tay-Sachs disease and the so-called ceroid lipofuscinoses. Mitochondrial genetic disorders cause dementia due to cumulative damage of either gray or white matter; examples include Leigh's disease, MELAS, or MERRF (see Table 3.15).

Genetic disorders of white matter also include the various leukodystrophies (such as adrenoleukodystrophy and metachromatic leukodystophy). These tend to declare themselves during childhood but a few forms can have adult onset and present as a dementia.

Armand does not have a family history of premature dementia or clinical features of any of the most common genetic disorders causing dementia.

9.6 Case Summary

In summary, Armand is an 82-year-old man with a one- to two-year history of cognitive decline, with features of memory loss for recent events and difficulty with word finding, executive function and emotional control. There is a long history of treated hypertension although his blood pressure is quite high in the office on his first visit.

The clinical exam confirms impairment in the cognitive domains of memory and executive function, and also suggests some slowness in his motor coordination and gait. There were no discrete focal signs suggestive of previous stroke. He has mild sensory loss in his feet.

9.7 Case Evolution

A metabolic workup showed no obvious reversible cause of dementia such as hypothyroidism, vitamin B12 deficiency or syphilis.

Armand's MRI (Figure 9.2) demonstrated diffuse cortical atrophy, more so in the temporal and parietal lobes and especially the hippocampus bilaterally. Other areas on his MRI scan (not shown) demonstrated multiple small areas of increased signal in the white matter bilaterally, consistent with small vessel microangiopathic disease. There was no sign of tumor or subdural hematoma.

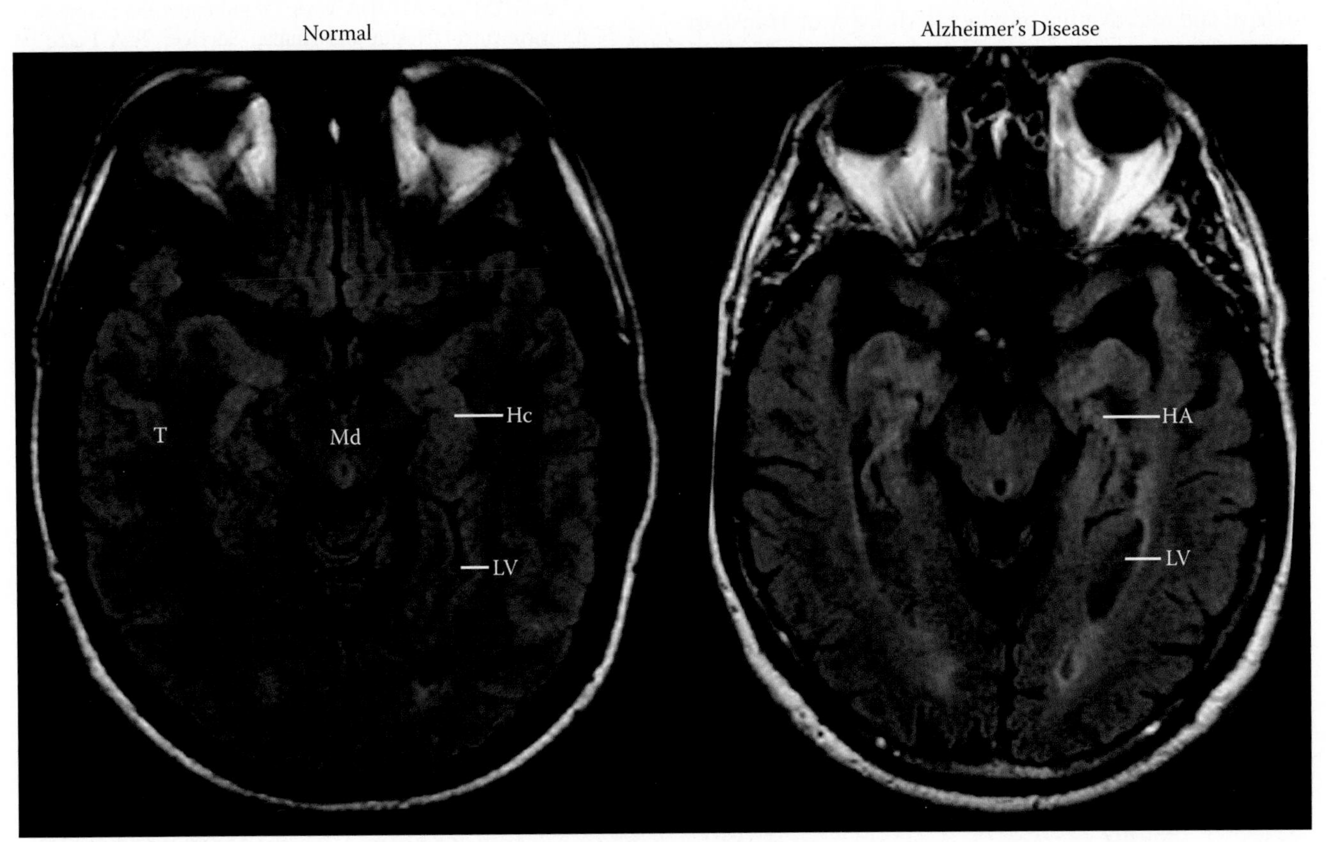

FIGURE 9.2 MRI scan of the brain. An axial view through the temporal lobes of a normal brain (on the left) compared to a brain with Alzheimer's dementia. Note the atrophy of the cerebral cortex, with enlargement of the ventricles and selective severe atrophy of the hippocampus on both sides of the Alzheimer brain.

He also had some findings of decreased upgaze and mild sensory loss in his extremities. These latter two findings were probably age-related and, by lab investigations and imaging, were excluded as being related to the major clinical problem, his memory deficits.

The conclusion was that Armand was probably suffering from a mixed form of dementia: primarily Alzheimer's disease but with a component of vascular dementia resulting from the progressive occlusion of multiple small cerebral vessels.

He was referred to a specialized memory disorder clinic in his community; this provided multidisciplinary diagnostic and treatment support. He had a formal neuropsychological assessment as well as an evaluation by social services and by the local geriatric outreach program. He was started on a cholinesterase inhibitor medication, which seemed to help his memory for a couple of years, after which it seemed to have no benefit. He was told not to drive and told that the provincial Ministry of Transport would be contacted (as required by law) to assess his ability to continue driving. After reviewing the specialist's report, the ministry subsequently suspended his license on medical grounds.

Armand remained at home with support of family and social services for five years after his diagnosis, during which time there was progressive deterioration in his activities of daily living and the appearance of incontinence. He eventually became bedridden and was then transferred to a long-term care facility, where he died 18 months later of pneumonia.

Related Cases

There are e-cases related to this chapter for review on the DVD and Web site.

Suggested Reading

American Psychiatric Association. *Diagnostic and Statistical Manual of Mental Disorders, 4th ed. Rev text*. Washington, DC: American Psychiatric Association, 2000.

Blacker, D. et al. Reliability and validity of NINCDS-ADRDA criteria for Alzheimer's disease. The National Institute of Mental Health Genetics Initiative. *Arch. Neurol.* 51, no. 12 (1994): 1198–1204.

Bogaerts, V. et al. A novel locus for dementia with Lewy bodies: A clinically and genetically heterogeneous disorder. *Brain* 130, Pt. 9 (2007): 2277–2291.

Chabriat, H. et al. Clinical spectrum of CADASIL: A study of 7 families. *Lancet* 346 (1995): 934–939.

Dubois, B. et al. Research criteria for the diagnosis of Alzheimer's disease: Revising the NINCDS-ADRDA criteria. *Lancet Neurol.* 6, no. 8 (2007): 734–746.

Hodges, J.R. et al. Clinicopathological correlates in frontotemporal dementia. *Ann. Neurol.* 56, no. 3 (2004): 399–406.

Heidebrink, J.L. Is dementia with Lewy bodies the second most common cause of dementia? *J. Geriatr. Psychiatry Neurol.* 15, no. 4 (2002): 182–187.

McKhann, G. et al. Clinical diagnosis of Alzheimer's disease: Report of the NINCDS-ADRDA Work Group under the auspices of Department of Health and Human Services Task Force on Alzheimer's Disease. *Neurology* 34, no. 7 (1984): 939–944.

Mendez, M.F. et al. Psychotic symptoms in frontotemporal dementia: Prevalence and review. *Dement. Geriatr. Cogn. Disord.* 25, no. 3 (2008): 206–211.

Mioshi, E. et al. Activities of daily living in frontotemporal dementia. *Neurology* 68 (2007): 2077–2084.

Neary, D. et al. Frontotemporal lobar degeneration: A consensus on clinical diagnostic criteria. *Neurology* 51, no. 6 (1998): 1546–1554.

Ropper, A.H., and R.H. Brown, eds. Cerebrovascular diseases. In *Adams and Victor's Principles of Neurology*. New York: McGraw-Hill, 2005.

Seeley, W.W. et al. Early frontotemporal dementia targets neurons unique to apes and humans. *Ann. Neurol.* 60, no. 6 (2006): 660–667.

Tournier-Lasserve, E. et al. Cerebral autosomal dominant arteriopathy with subcortical infarcts and leukoencephalopathy maps on chromosome 19q12. *Nature Genet.* 3 (1993): 256–259.

Zaccai, J., C. McCracken, and C. Brayne. A systematic review of prevalence and incidence studies of dementia with Lewy bodies. *Age Ageing* 34, no. 6 (2005): 561–566.

Chapter 10

Didi

DIDI

The time has come to tell you something about Crash and Fifi's children. Dimitri (affectionately known as Didi) is older, currently age six. His sister Antoinette (Tonton) is a year younger. Like nearly everyone else in this remarkable but unfortunate family, Didi and Tonton have medical problems.

Didi's particular problem was first noted three months ago by his teacher, at the beginning of first grade. Whereas Didi had performed well during his senior kindergarten year, being above the class average, this year he has been having difficulty paying attention in class and is having trouble keeping up with the work assigned. In addition, the teacher has noticed that, several times a day, Didi will abruptly stop what he is doing and stare straight ahead for a few seconds. During that time he does not respond to his name being called or appear to be aware of activities going on around him. Just as abruptly, he then returns to his previous activity as if nothing had happened. Sometimes, however, he appears to be transiently perplexed following an episode, as if he does not know where he is or what he is doing. After weeks of observing this behavior, and watching Didi's work evaluations continue to decline, the teacher called Didi's mother to share her concerns.

As soon as Fifi heard the teacher's comments, she realized with a start that she had witnessed similar episodes in the evening over the same time period. Perhaps because, at some level, she did not want to recognize what was going on, Fifi ascribed the episodes she had seen as "day-dreaming," or due to fatigue. She now realized that Didi's "spells" were similar in nature to episodes Tonton had first developed at age four. When Tonton's staring episodes began to interfere with normal conversation, Fifi realized there was a significant

Objectives

- To learn the principal anatomical components of the central nervous system that are involved in the maintenance of consciousness
- To review the neurophysiology of neuronal membranes
- To review the mechanisms of synaptic transmission and the types and functions of neurotransmitter molecules
- To learn definitions for *coma*, *stupor* and *confusion*
- To develop the ability to distinguish the various causes of transient loss of consciousness
- To learn the main mechanisms involved in the production of epileptic seizures
- To review the classification of epileptic seizure types
- To learn the basis of electroencephalographic (EEG) "brain-wave" activity and the main EEG abnormalities associated with epileptic disorders
- To review the principal mechanisms of action of the various anti-epileptic drugs

10.1 Clinical Data Extraction

Didi's neurological disorder is clearly quite different from what you have encountered in previous chapters. His symptoms are intermittent rather than relentlessly progressive; he appears to be completely normal when his symptoms are not present. Nevertheless, it is important that you begin the problem-solving process by reviewing the clinical data provided and working through the history, examination, localization and etiology worksheets in the accompanying DVD. Once you have completed this process, you are ready to proceed with the remainder of the chapter.

10.2 The Main Clinical Points

In comparison with many of the previous problems, Didi's history is relatively short, and summarizing the clinical features relatively straightforward:

- Three-month history of recurrent brief episodes of unresponsiveness (seconds only)
- Episodes begin and end abruptly, without any incoordination or loss of balance, and with either

problem and took her daughter to the family physician. There followed a referral to a local neurologist, who witnessed an episode and arranged for an electroencephalogram (EEG). The EEG revealed the presence of an epileptic disorder. Tonton was then started on an antiseizure medication, following which the staring spells gradually disappeared.

Concerned that Didi has developed the same disorder as his sister, Fifi took Didi to see the family physician, who turns out to be *guess who*?

Upon reviewing your file on Dimitri, you remind yourself that he was a term baby following an uncomplicated pregnancy, weighing 3.1 kg at birth. He was walking by age 12 months, speaking in short sentences by age two, and toilet-trained at age three years, three months. He learned to ride a two-wheel bicycle 18 months ago and to tie his own shoe laces last summer. He now knows all the letters of the alphabet, reads four- to five-word sentences, writes his first and last names, can draw stick figures, and can name all the players on the Ottawa Senators. In the past, Didi has been in good health except for several ear infections.

Your age-appropriate general physical and neurological examinations are perfectly normal. Having learned from your experience with Tonton, you have Didi carry out one more test maneuver. You ask him to pretend that his right index finger is one of those trick birthday candles that resists being blown out. Didi is to try to blow out the candle as hard as he can and to keep puffing on it until you tell him to stop. After about 45 seconds of vigorous huffing and puffing, Didi suddenly stops blowing and stares vacantly into space. When you call his name there is no response. You ask him to look over at you but, again, he does nothing. You notice a very fine tremor of his head. After about 10 seconds of this behavior, he suddenly changes his sitting position, blinks once, and then resumes blowing out the imaginary candle.

extremely transient or no disorientation following the episode

- Difficulty with paying attention in the classroom
- Unexpected problems in keeping up with class work
- Normal pregnancy and birth history
- Normal developmental milestones
- History of a seizure disorder in the younger sister
- Normal neurological examination
- Witnessed episode triggered by voluntary hyperventilation: blank stare, unresponsiveness to questions or commands; slight rhythmic head tremor; no confusion afterward

In other words, Didi appears to be having intermittent disruptions in conscious awareness and, except for academic difficulties, seems to be neurologically intact.

This being the case, we can quickly slash our way through much of the localization matrix by concluding, on empiric grounds, that the problem in Didi's nervous system is unlikely to be located below the neck; that is, a primary dysfunction of the spinal cord, peripheral nerves and muscles is highly unlikely.

Having made this decision, our first task is to consider what parts of the central nervous system are concerned with the maintenance of consciousness.

10.3 Relevant Neuroanatomy

Our understanding of how the brain maintains a state of vigilant awareness has evolved from a consideration of what parts of the brain, when damaged, are accompanied by sustained loss of consciousness, or *coma*.

Coma may be defined as a persistent sleep-like state from which a person cannot be aroused. In fact, a state of coma is the far end of a gradation of impaired consciousness from full awareness to complete absence of awareness. Between these extremes are

1. *Confusion*, in which an individual appears awake but is slow-thinking and incoherent, and has impaired or absent memory for events occurring during the confused state.
2. *Stupor*, in which an individual appears asleep, can be partially roused, but is unable to respond appropriately, or to follow instructions.

From a clinicopathological analysis of comatose patients, it has been recognized that coma may result from damage or loss of function in the upper brainstem (pons, midbrain), both thalami, or both cerebral hemispheres (cerebral cortex and subcortical white matter, diffusely). Damage to or dysfunction of one cerebral hemisphere or thalamus is not normally accompanied by loss of consciousness; the only exception, on occasion, is an extensive injury to most of the dominant hemisphere.

What specific components of the brainstem, thalami and cerebral hemispheres comprise the system, or network, that maintains the conscious state? Briefly put, the components are

1. Pons and midbrain—the so-called reticular activating system, or reticular formation
2. The "nonspecific" or intralaminar and lateral reticular thalamic nuclei
3. The diffuse thalamocortical projection system, connecting the nonspecific thalamic nuclei and the cerebral cortex
4. A corticothalamic feedback system connecting the cerebral cortex back to the nonspecific thalamic nuclei

Let us consider some of the components in more detail.

10.3.1 Brainstem Reticular Formation

Located in the central portions of the pontine and midbrain tegmentum, the reticular formation, as the name implies, is (in histological preparations) a lacy network of neurons of varying size, for the most part without obvious neuronal aggregations ("nuclei"). The reticular formation is highly conserved through vertebrate evolution and is therefore the most "primitive" part of the brainstem (see also Section 1.2.5; Figure 1.6C).

The location of the reticular formation is shown in Figure 10.1. As can be seen, the reticular formation also extends into the central medulla. At this level, however, the reticular formation is not concerned with the mechanism of arousal, but with projections to cerebellum and spinal cord involved with motor, sensory and autonomic functions.

As Figure 10.1 illustrates, there are some aggregates of larger neurons within the reticular formation that are recognizable at the microscopic level, and that have specific functions. Examples include the nucleus raphe magnus and the nucleus of the locus ceruleus. These two nuclei are involved with the regulation of consciousness insofar as they play a role in sleep–wake regulation, among other functions. They form important components of the so-called monoaminergic projection systems and will be reviewed in some detail in Chapter 12 when we consider disorders of "higher function."

The reticular formation neurons participating in the consciousness system are usually small, and have a characteristic morphology. They have an extensive local dendritic network, allowing them to receive simultaneous input from a variety of sources. It is important to note, for example, that second-order neurons in the somatic and visceral pain pathways send collaterals to the reticular formation (shown in Figure 2.6B). These have an obvious function, as signals of noxious stimuli, in increasing levels of arousal so as to put in motion behaviors designed to avoid or remove the source of pain.

Reticular formation axons may be short, with local connections, or long, projecting to the thalamus or even higher. In either case there is an elaborate network of terminal branches. This allows a single reticular formation neuron to potentially modify the function of many neurons at a higher level, whether within the consciousness system (e.g. thalamic intralaminar nuclei) or without (e.g. specific thalamic nuclei, basal ganglia, hippocampi, and cerebral cortex). Considering this arrangement, it is easy to see how a small, localized painful stimulus, such as a bee-sting on one foot, could result in a massive, diffuse effect on cerebral function.

10.3.2 Nonspecific Thalamic Nuclei

We have already considered some of the components of the thalamus, the largest gray matter complex in the diencephalon, in previous chapters (see, for example, Figure 1.8). The ventral posterolateral (VPL) and ventral posteromedial (VPM) nuclei of the thalamus relay sensory information from the body and face, respectively, to the somatosensory areas of the cerebral cortex. Chapter 7 included a discussion of the ventral anterior (VA) and ventral lateral (VL) nuclei, both involved in the initiation and coordination of motor function. All of these thalamic nuclei, as well as others to be considered later, have "specific" functions in that they are connected to discrete regions of the cerebral cortex.

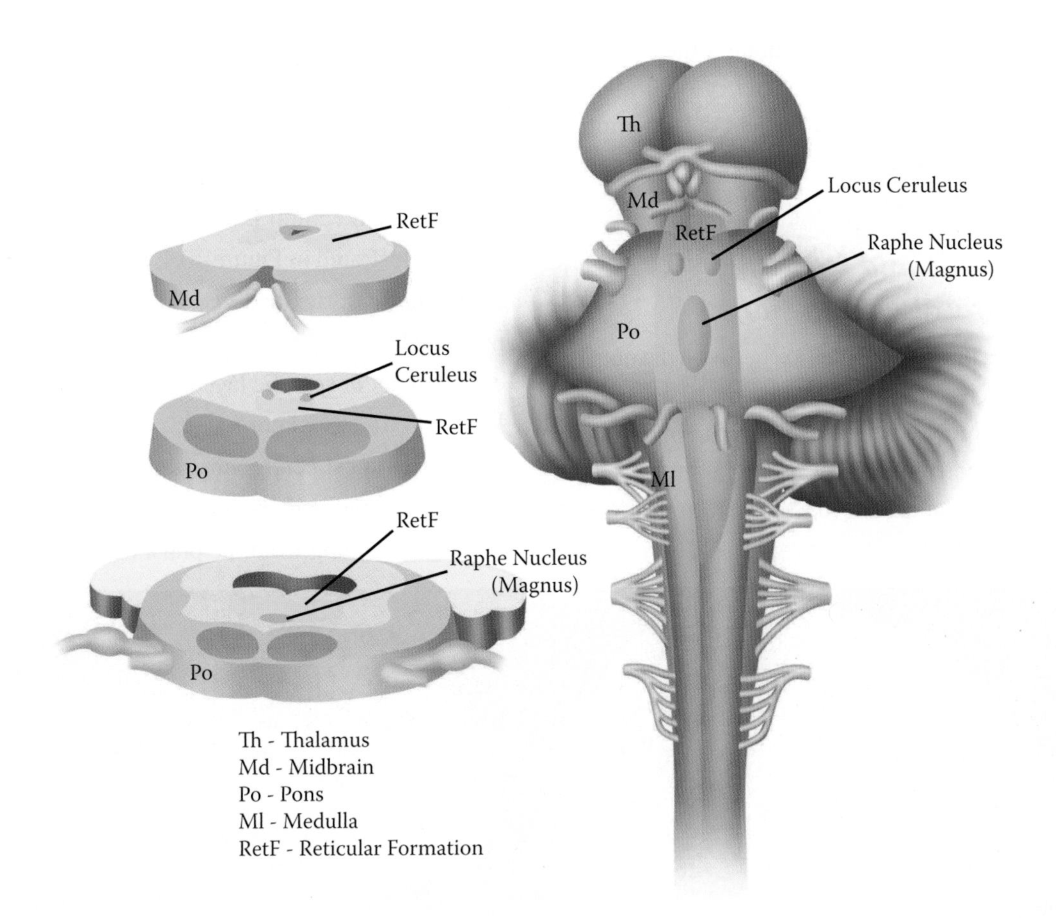

FIGURE 10.1 Illustration of the location of the locus ceruleus (upper pons) and raphe nucleus (magnus, mid-pons) within the brainstem. The reticular formation is shown in pale yellow, the two named reticular nuclei in darker yellow.

Nonspecific thalamic nuclei, on the other hand, project in a diffuse fashion to the entire cerebral cortex, as might be predicted for structures involved in the regulation of consciousness. As is demonstrated in Figure 10.2, the nonspecific thalamic nuclei consist of three main components:

1. A medial thalamic area adjacent to the third ventricle medially and the dorsomedial thalamic nucleus laterally
2. The intralaminar nuclei, located in the Y-shaped intermediate band of nervous tissue separating the medial and lateral thalamic nuclear clusters and the anterior nucleus (in the fork of the Y) from all the others
3. A lateral "reticular" nucleus covering the lateral part of the thalamus.

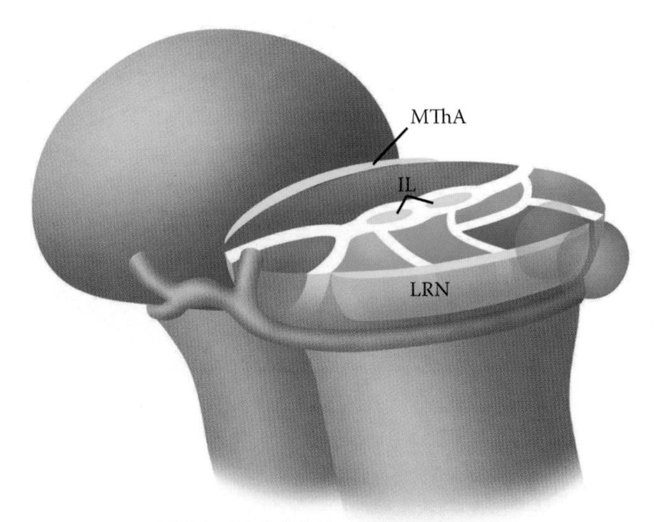

FIGURE 10.2 Axial section of the thalamus showing the location of the nonspecific "reticular" nuclei.

The nonspecific thalamic nuclei receive input from both the pontine and mesencephalic reticular formation, and from the cerebral cortex, among other sources. Neurons in the various reticular nuclei of the thalamus resemble those of the brainstem reticular formation in having elaborate dendritic trees and widespread axonal projections to higher levels.

10.3.3 Thalamic and Cortical Reciprocal Connections

Figure 10.3 shows the interconnection between the nonspecific thalamic nuclei and the cerebral cortex. Essentially there is a feedback loop involving the two brain regions, a diffuse reciprocal interaction that is crucial in the regulation of states of vigilance varying from sleep to focused, task-oriented awareness.

Axons derived from nonspecific thalamic nuclei terminate in the most superficial layer of the cerebral cortex, establishing contact with dendrites of cortical neurons whose cell bodies are located in deeper layers of the cortex. The architectural layers of the "typical" cerebral cortex are illustrated in Figure 10.4. Some cortical areas (such as the motor cortex and visual cortex) are highly specialized and have atypical layering arrangements, with marked enlargement of specific layers (layer 5 in the motor cortex; layer 4 in the visual cortex).

In brief, most cerebrocortical areas have six fairly distinct layers defined by a preponderance of certain cell types. Proceeding from the pial surface inward we have:

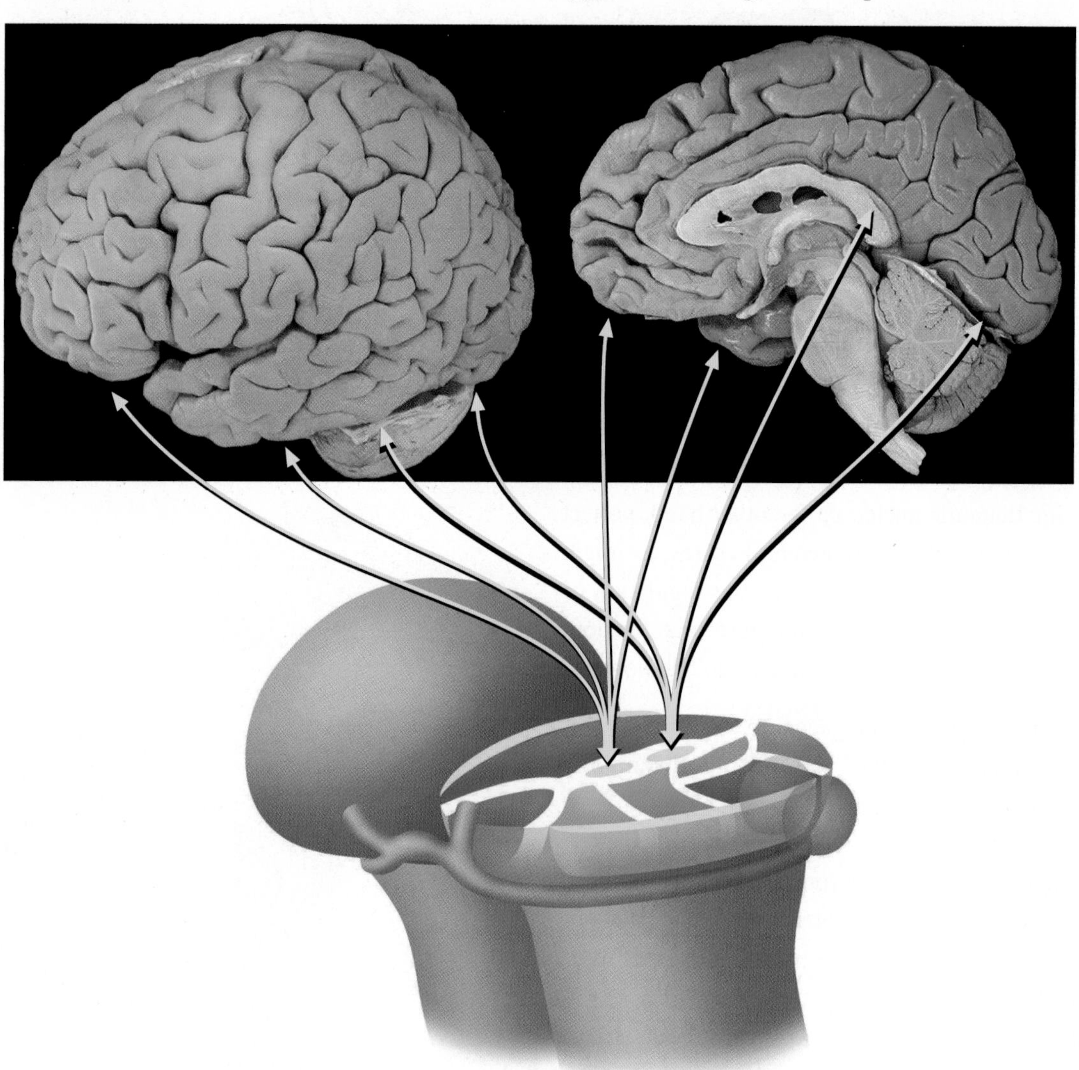

FIGURE 10.3 Reciprocal connections of the nonspecific thalamic nuclei with the cerebral cortex.

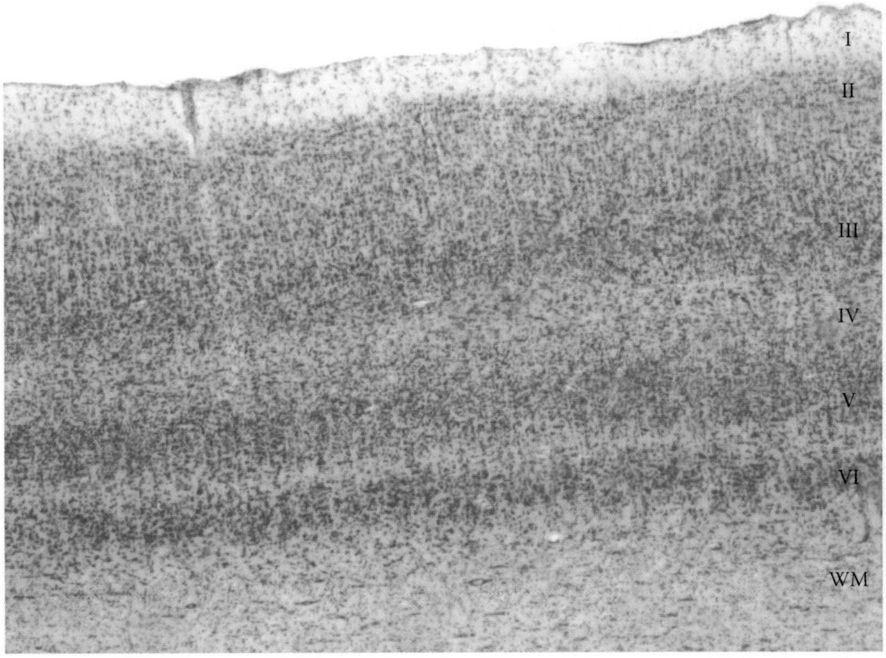

WM - White Matter

FIGURE 10.4 **Photomicrograph of human cerebral cortex, cresyl violet stain, showing the six architectural cell layers (indicated by roman numerals), with the white matter (WM) below. (Courtesy of Dr. J. Michaud.)**

Layer 1—relatively free of cell bodies, primarily consisting of nonspecific thalamic axon terminations and cortical neuronal dendrites that have ascended from neurons in deeper layers

Layer 2—small neurons primarily involved in local cortical circuitry

Layer 3—medium to large neurons whose axons project to adjacent cortical areas, homologous areas of the opposite cerebral hemisphere (via the interhemispheric callosal connections) and thalamic nuclei

Layer 4—small neurons involved in the receipt of input from *specific* thalamic nuclei, for example, VPL nucleus for somatosensory information from the neck down; lateral geniculate body for visual information

Layer 5—large neurons whose axons project primarily to brainstem and spinal cord, for example, upper motor neurons

Layer 6—variably sized neurons also involved in intracortical circuitry

Cortical neurons are organized in columns extending the full width of the six-layered cortex; these columns represent the basic functional units of the cortex.

For our purposes, it is the relationship between the axon terminals from nonspecific thalamic neurons and dendrites from layer 3 (thalamic projection) neurons that requires our attention. Electrical activity in the thalamic axon terminals induces, through synaptic connections with cortical layer 3 dendrites, fluctuating degrees of electrical charge in the dendrites, referred to as dendritic potentials. Dendritic potentials are unlike axon potentials in that they are not all-or-none phenomena: they fluctuate locally without necessarily committing their cell body (and axon) to a complete membrane depolarization, or action potential. (The phenomenon of the resting membrane potential, and the way neurons communicate with one another, has been considered briefly in Chapter 1 and will be revisited in more detail later in this chapter.)

Located as they are in the most superficial layer of the cerebral cortex and, therefore, in many areas, quite close to the scalp skin surface, dendritic potentials in the cortex can be recorded and amplified from electrodes placed on the scalp. Fluctuations in cortical dendritic potentials form the basis of background rhythms in electroencephalography (EEG).

In a relatively low vigilance state (such as lying awake with eyes closed), there is a relatively large rhythmic

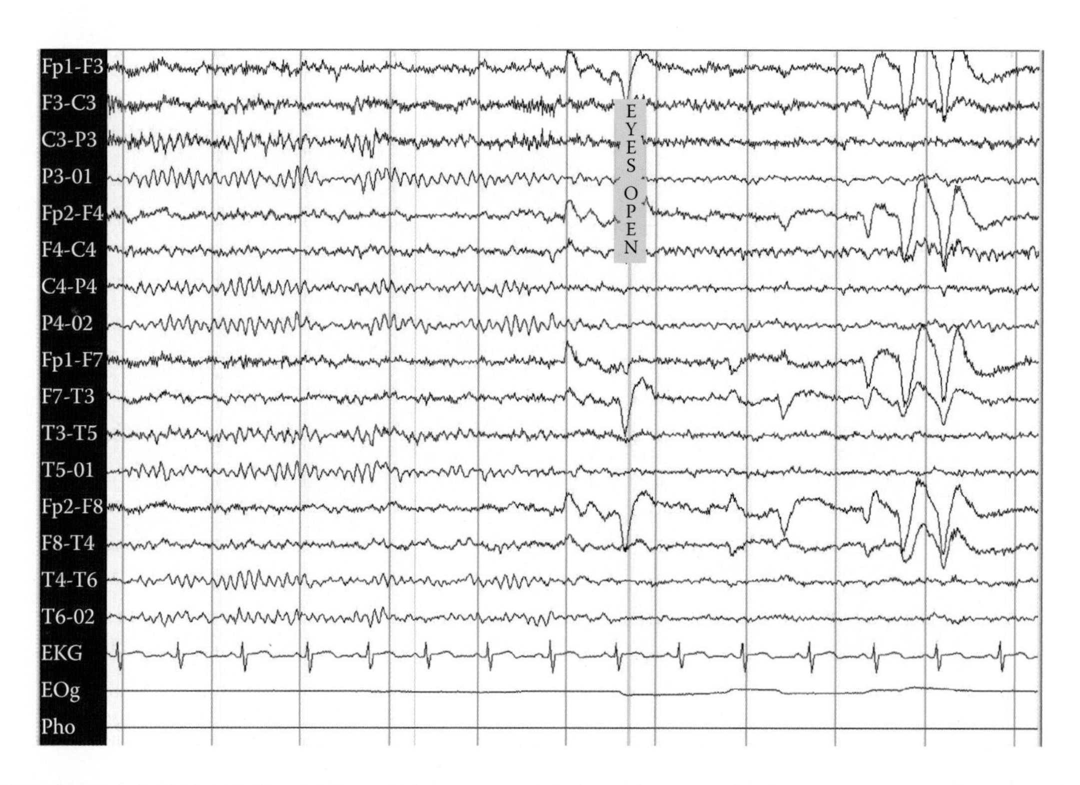

FIGURE 10.5 Normal EEG tracing in the awake state, showing the pattern change in the posterior head regions accompanying eye opening. The letters and numbers on the left of the image refer to electrode positions on the scalp. F = frontal, C =central, T = temporal, P = parietal, O = occipital; odd numbers apply to the left hemisphere, even numbers to the right. (Courtesy of S. Bulusu and Dr. S. Whiting.)

oscillation in cortical dendritic potentials. This reflects a rhythmic fluctuation in the thalamo-cortico-thalamic feedback loop we have just described (Figure 10.5). With the eyes open, and a state of enhanced, focused attention, the rate of oscillation in dendritic potentials increases while the amplitude decreases, that is, the scalp electrode recording becomes desynchronized. In contrast, in a non-rapid eye movement sleeping state, the rhythmic oscillation in dendritic potentials slows while the amplitude increases, a scalp electrode pattern of hypersynchronization.

These varying degrees of hypo- and hypersynchronization of cortical dendritic activity reflect changes in activity of the thalamo-cortico-thalamic feedback loop and are the pedestrian equivalents of changes in the level of consciousness.

Thus, if the proverbial bee-sting to the foot occurs while one is asleep, the sudden, massive volley of pain data surging up the somatosensory pain pathway will spread to the pontomesencephalic reticular formation and the nonspecific thalamic nuclei. This enhanced level of reticular formation activity will disrupt the slow, rhythmic oscillation in the thalamo-cortico-thalamic consciousness system feedback loop, leading to a sudden desynchronization of cortical dendritic activity accompanied by an abrupt state of focused, panic-stricken awareness!

10.4 Localization Process

Now that we have defined the location of the consciousness system, we can proceed to try to localize the pathological process that has been causing Didi to have recurrent, transient disturbances in conscious awareness. Based on what we have learned, we must consider processes located at the various levels of the consciousness system, either at one level or any combination of levels:

1. Upper brainstem
2. Thalamus and/or thalamocortical white matter connections
3. Cerebral cortex

10.4.1 Brainstem

The pons and midbrain are relatively small, complex, densely structured parts of the brain. The pontine reticular formation is centered amidst a large number of important structures such as the Vth to VIIIth cranial nerve nuclei and the medial lemnisci (Figure 10.1, Figure 6.4, and Figure 1.6C).

Similarly the mesencephalic reticular formation is adjacent to the IIIrd and IVth cranial nerve nuclei, the ascending somatosensory pathways and both the superior cerebellar and the cerebral peduncles. This proximity means that a pathological process located in the pontomesencephalic reticular formation would almost certainly also produce one or more of cranial nerve pareses, somatic sensory loss and limb ataxia or spasticity. It is therefore unlikely that Didi has a lesion, or at least a destructive lesion, in this region.

It is conceivable, of course, that our patient has a functional disorder, rather than a destructive process, confined to brainstem reticular neurons. This possibility will be considered once we have completed our anatomical survey.

10.4.2 Thalamus and/or Thalamocortical White Matter Connections

When we move to the reticular nuclear components of the thalamus, we encounter the same problem of multiple adjacent functionally important structures that we saw in the brainstem. The intralaminar nuclei alone are cheek-by-jowl with nearly all of the specific thalamic nuclei, not the least of which are the anterior nucleus (AN), the dorsomedial nucleus (DM), VA, VL and VPL. Damage to the last three structures would result in motor or sensory disturbances in the opposite side of the body. As we shall see later (Chapter 12), the anterior nucleus projects to the cingulate gyrus, a part of the limbic system, while DM projects to the frontal lobe. Pathology in these regions would be accompanied by higher function disturbances such as emotional disorders, memory dysfunction or impaired judgment. Didi has no such problems: between his periods of apparent unconsciousness, he is completely intact.

Furthermore, since both thalamic nonspecific nuclear projection systems would have to be compromised for loss of consciousness to occur, a putative destructive process would have to involve specific thalamic nuclei bilaterally, with profound sensory, motor and behavioral consequences. As with the brainstem, therefore, we must conclude that Didi's neurological disorder is either not located in the thalamic area, or is somehow confined to the reticular neurons of the consciousness system.

10.4.3 Cerebral Cortex

Pathology at the level of the cerebral cortex, in order to produce loss of consciousness, would typically involve most of the cortex in both hemispheres. On rare occasions, widespread damage to the cortex of the dominant (left) hemisphere alone may produce a profound reduction in conscious awareness. As was the case with the thalamic region, unilateral or bilateral cerebrocortical pathology in this boy seems improbable in the absence of sensorimotor dysfunction and profound disturbances in mentation.

Accustomed as you have become to localizing disease processes by analysis of motor and sensory deficits, it may appear strange to you that Didi clearly must have a disorder involving the consciousness system, yet has no deficits of any kind, when in an awake state, to help us in our localization exercise. At the very least, as we have already suggested, his "disease process" must be strictly confined to the consciousness system, either in the brainstem, thalamus or cortex, or in some combination of all three. Furthermore, his "disease process" must be of a type that can result in intermittent disturbances of consciousness without affecting the function of the consciousness system, to even a small degree, on a continuous basis.

Having localized Didi's neurological disorder to the best of our ability, even if imperfectly, we must seek the answer to our apparent riddle by proceeding to consider what his disease process might be.

10.5 Etiology: Didi's Disease Process

Returning to Table 3.13, we quickly recognize that Didi's history presents an ambiguity. On the one hand, the duration of each period of unresponsiveness lasts only seconds; on the other hand, he has been having these episodes for at least three months. Based on the former criterion, Didi has an acute process; based on the latter, a chronic process. While both conclusions, in effect, are correct, it is the former conclusion—that Didi has an acute, or even hyperacute process—that is most helpful in working out a differential diagnosis. As we discussed in Chapter 3, longstanding disorders that produce brief symptomatic bursts, separated by long asymptomatic intervals, are collectively termed *paroxysmal disorders*.

Since Didi's disorder is clearly paroxysmal in nature, we will abbreviate the systematic review of all 10 etiologic categories that has formed a component of the previous problem-based chapters in this text. As will be seen shortly, the toxic, metabolic and genetic categories may contribute to paroxysmal disorders and will be considered in varying amounts of detail.

The remaining etiologic categories may, on occasion, disrupt the consciousness system but not in a paroxysmal fashion. Trauma to the head routinely interrupts consciousness, most often due to contusion of the midbrain. Vascular pathologies, whether ischemic or hemorrhagic, also produce coma if they extensively damage the brainstem or both thalami. Coma in viral encephalitis or bacterial meningitis is secondary to diffuse cerebral dysfunction. Postinfectious autoimmune disorders such as acute disseminated encephalomyelitis (ADEM) often present with a rapid-onset comatose state that may last several days. Other inflammatory/autoimmune disorders such as multiple sclerosis and CNS lupus disturb consciousness only when they are far advanced, in the terminal phase of the illness. The same is true for neoplasia and for degenerative diseases.

Returning to the paroxysmal category, there are a variety of such disorders, each having quite distinct mechanisms. Our task will be to consider the most likely items on the list. In working through this list we must bear in mind that, whatever Didi's problem is, it appears to be anatomically confined to the consciousness system.

In general, central nervous system paroxysmal disorders may be defined according to the mechanism of transient disruption of neurologic function:

1. Lack of substrates required for maintenance of neuronal function. For the most part there are only two such substrates, *glucose* and *oxygen.*
2. Pharmacological suppression of neuronal function, for example, drugs and toxins.
3. Disturbances in neuronal membrane function, either hyperexcitability with inappropriate depolarization of neurons or hypoexcitability with inappropriate hyperpolarization.

Let us consider these possibilities in turn.

10.5.1 Lack of Substrate

10.5.1.1 Glucose All neurons require a continuous supply of glucose in order to maintain a resting membrane potential. If the supply of glucose is inadequate, neuronal synaptic function (about which more will be said later in this chapter) ceases within seconds, followed by loss of resting membrane potentials and persistent neuronal depolarization. The severely hypoglycemic patient becomes rapidly unconscious and unable to move. If glucose is then replenished within a satisfactory period of time (minutes), normal neuronal function will recommence and the patient will resume normal activities unscathed. Severe, prolonged hypoglycemia (more than 8 to 10 minutes duration) may result in the death of neurons and permanent deficits.

Thus, as may occur in patients with diabetes mellitus whose insulin dose is too large for the amount of glucose consumed, hypoglycemia may produce recurrent episodes of unconsciousness. Typically, however, such episodes are preceded by warning symptoms of impending neuronal failure: brief confusion, pallor, sweating, and anxiety. In addition, the period of unconsciousness with hypoglycemia lasts for minutes, not seconds, and the patient, if sitting or standing up, will collapse, unable to maintain postural stability.

Since Didi's episodes of unresponsiveness only last about 10 seconds, during which time he maintains his sitting or standing posture, a hypoglycemic etiology seems unlikely.

10.5.1.2 Oxygen Neuronal function is disturbed just as rapidly when the brain is deprived of oxygen as when it is deprived of glucose. Lack of oxygen may occur as a consequence of reduced atmospheric oxygen (as in sudden airplane decompression at high altitude), impaired pulmonary function (as in drowning or severe pneumonia), impaired red blood cell oxygen carrying capacity (as in severe anemia or carbon monoxide poisoning), or impaired perfusion of blood to the brain (as in infection-induced shock or cardiac arrest).

None of these putative etiologies, however, is likely to produce brief, recurrent episodes of diffuse central nervous system hypoxia. On the other hand, localized blood vessel disease involving arteries supplying the brainstem reticular formation or the thalami might result in transient periods of impaired perfusion to these structures or, in other words, TIAs (see Chapter 8). TIAs associated with episodic unconsciousness can be due to cerebrovascular disease in the vertebrobasilar system (see Chapter 8). In general, however, patients with TIAs due to vertebrobasilar insufficiency also have symptoms and signs of other brainstem deficits, for example, diplopia, vertigo and ataxia.

Episodic unconsciousness may also occur in the context of transient hypotension, such as may occur with intermittent cardiac arrhythmias (cardiac conduction block or *Stokes–Adams attack*), or with reflex hypotension in response to a strong emotionally laden stimulus in predisposed individuals: the common "fainting spell," otherwise known as *vasovagal* or *neurovascular syncope*.

With the exception of the last-mentioned cause of transient cerebral anoxia or ischemia, however, none of these possible etiologies is very likely in a previously well six-year-old boy! In addition, all of the objections applied to episodic unconsciousness secondary to hypoglycemia also apply to anoxia.

10.5.2 Drugs and Toxins

Any drug capable of inducing neuronal suppression (e.g. barbiturates, benzodiazepines), as well as any chemical compound inducing hypoglycemia or anoxia (e.g., oral hypoglycemic agents, carbon monoxide, cyanide) may produce transient unconsciousness. In practice, however, such pharmacologically induced comatose states last for hours to days rather than seconds.

10.5.3 Disturbances in Neuronal Membrane Function

In this instance we are considering a group of disorders in which there is an intrinsic disturbance in neuronal function in which cell membranes are either (1) unable to consistently stabilize resting membrane potentials when normal functioning neuronal networks require this, and thus produce action potentials inappropriately, or (2) are excessively hyperpolarized and unable to generate appropriate action potentials to allow participation in normal neuronal network function.

An episodic, inappropriate, excessive electrical discharge of cortical neurons is an accepted definition for an *epileptic seizure*. Epileptic seizures, depending on the area or areas of the brain participating, are accompanied by a variety of transient disturbances in neurological function, including twitching movements, sensory phenomena (e.g. tingling sensations, visual or auditory hallucinations), stereotyped semi-purposeful movements (automatisms) and *loss of conscious awareness*. In the latter case, the epileptic seizures either originate in or spread to involve the consciousness system as we have defined it.

Epileptic seizures typically last seconds to minutes, occasionally longer. Thus, as a pathophysiological phenomenon, they are in keeping with the transient disturbances described in Didi's history.

On the other hand, transient suppression of normal neurological function usually implies a distinctly different form of neurological disorder. While it is true that some forms of epileptic seizure, in some areas of the brain, may produce a hyperpolarization (rather than depolarization) of cells in a specific neuronal network leading, for example, to transient paralysis of an arm or leg, this is distinctly rare. A far more common scenario involving transient neuronal hyperpolarization is the phenomenon of cerebrocortical spreading depression, originally described by Leão and bearing his name.

Transient spreading depression of cortical neurons, by which is meant the slow sequential suppression of function across the surface of the cortex, is the hallmark of *classical migraine* (or migraine with aura). Migraine is a common paroxysmal disorder of genetic origin characterized by recurrent attacks of pounding headache, photophobia, nausea and vomiting. In classical migraine, prior to the onset of headache, there may be a premonitory period lasting several minutes during which there is a transient sensory disturbance (hallucination of a flickering, saw-toothed line; numbness of the face and arm on one side) known as an aura; it is this phenomenon that results from spreading cortical depression. Migraine without aura is experienced much more frequently and is called *common migraine*.

Some forms of migraine headache do involve disturbances in consciousness, either with confusion and loss of memory for events during the attack (confusional migraine), or with transient unconsciousness. This disorder is called *basilar artery migraine* because the symptoms suggest dysfunction in midbrain reticular formation, thalami and temporal lobes, all supplied by the basilar artery and its superior branches (see Chapter 8). Symptoms of neuronal depression in migraine, however, last for minutes or hours, not seconds.

In conclusion, our review of possible etiologies of transient, recurrent neurological dysfunction (or paroxysmal disorders) leaves us in the "electrical" column of the etiology matrix, and the most likely diagnosis of some form of *recurrent epileptic seizures*.

Before considering what form of epileptic seizure might fit best with Didi's history, however, we will need to back up a little and consider how and why neurons develop electrical charges that "run amok" when seizures occur. In this section, we will be elaborating on basic concepts introduced in Chapter 1.

10.5.3.1 The Neurophysiology of Neuronal Membranes What enables a neuron to develop and maintain an electrical charge is its semipermeable and highly specialized cell membrane. Neuronal membranes are completely impermeable to large organic molecules but selectively permeable, under certain circumstances, to small inorganic molecules such as sodium, potassium, calcium and chloride. In living systems, all of these molecules, large and small, are present in their ionic forms: sodium, potassium and calcium have positive charges, chloride and large inorganic molecules negative charges.

In fact, neuronal membranes, consisting of lipid bilayers enclosing a central protein component (Figure 10.6), are, as such, completely impermeable to Na^+, K^+, Ca^{2+}, and Cl^-. That they are capable of intermittent permeability to these ions is due to the presence of specialized ionic pores, normally closed but capable of opening briefly to permit passage of the ion across the membrane (Figure 10.6).

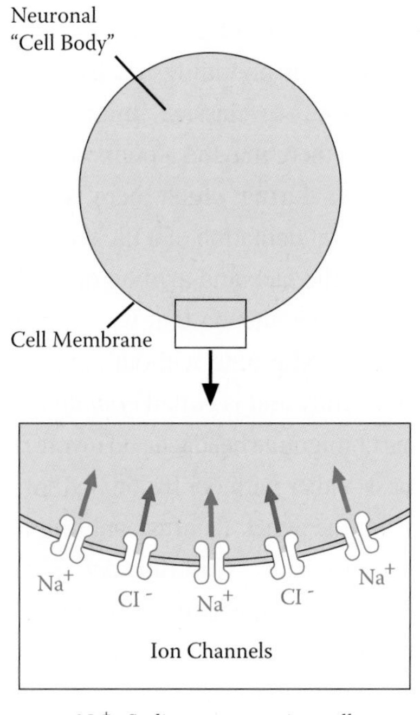

FIGURE 10.6 Schematic of a simplified neuron shown as a circle with blue "cytoplasm," without axon or dendrites. The lower panel shows a magnified view of the "cell" membrane revealing sodium (Na) and chloride (Cl) ion channels.

There are specific pores, known as ion channels, for each of the four small charged molecules, that is, specific sodium, potassium, calcium and chloride channels. Each type of ion channel is a highly individualized structure, designed specifically to permit passage, for example, of sodium ions, but not the other three ions. All four types of ion channels are constructed of a number of protein components in a stereochemical relationship that determines pore size and ion specificity. A typical sodium channel is constructed of α ($n = 2$), β ($n = 2$) and γ ($n = 1$) subunits, the three subunit types being encoded by separate genes in the nuclear genome.

In a resting (not actively discharging) neuron, several ionic gradients are maintained by the semipermeable cell membrane (see Figure 10.7):

Na^+—high concentration outside the cell, low inside
K^+—high concentration *inside* the cell, low outside
Ca^{2+}—high concentration outside the cell, low inside
Cl^-—high concentration outside the cell, low inside
$Protein^-$—high concentration *inside* the cell, low outside

The net effect of these various ionic gradients is to create a negative charge inside the neuron with respect to the

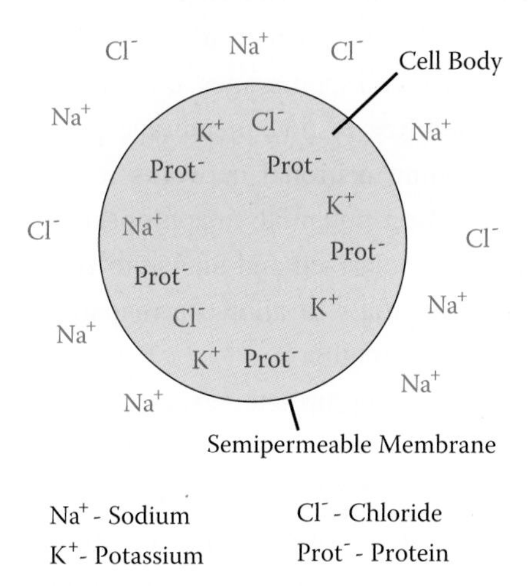

FIGURE 10.7 "Neuron" cartoon showing the relative concentrations of different ions within and outside the cell; these ionic gradients generate the resting membrane potential.

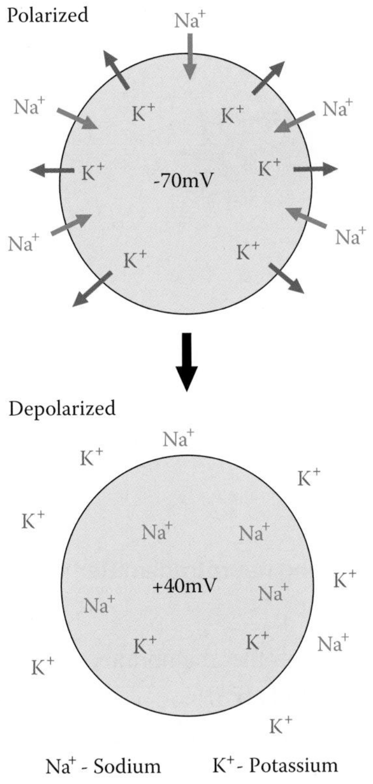

FIGURE 10.8 **"Neuron" schematic illustrating a polarized neuron and a depolarized state.**

outside, typically about –70 mV. It is this negative charge that is referred to as the resting membrane potential.

When the neuron receives a message inciting it to discharge its accumulated voltage, a threshold is eventually crossed and two almost simultaneous events occur. The sodium channels open briefly, permitting Na+ ions to enter the cell along the established concentration gradient. This abrupt ionic shift results in the loss of the negative intracellular charge, replaced by a weakly positive charge (about +40 mV). The sudden depolarization of the cell creates an action potential that then propagates over the adjacent cell body membrane and down the axonal membrane centrifugally, with Na+ channels opening in succession in a rapidly spreading wave (Figure 10.8). In this manner our sample neuron sends an electrical message to other neurons in its network.

Microseconds after the local action potential is produced, the potassium channels open, resulting in K+ ions quitting the cell for the extracellular fluid space and reestablishing the resting membrane potential. The calcium and chloride channels usually remain closed throughout this process.

The ionic shifts and associated axon potential have no sooner occurred than the neuron sets about reestablishing the status quo ante. Sodium ions are rapidly pumped outside the cell and potassium ions inward, to recreate the baseline state and ready the cell to repeat the process. Until this occurs, the cell is incapable of generating another action potential; that is, it is in a refractory period.

The active reestablishment of the original concentration gradients requires a great deal of energy, in the form of ATP, acting on a kind of ion-exchange pump, the Na+/K+ ATPase system. The ionic shifts and exchanges occur incredibly quickly, permitting a typical neuron to produce action potentials, or "fire," repeatedly in the space of one second.

You will notice that, despite their existence having been mentioned, the chloride and calcium channels did not play a role in the production of action potentials. Their participation in this story will come later, after we have considered how one neuron is able to affect the excitation status of another. This "communication" mechanism is synaptic transmission.

10.5.3.2 Synaptic Transmission Action potentials propagated along our neuron's axon may influence the activity of a large number of other neurons. Toward the end of its course, the axon typically splits into a number of branches, each of which may abut against dendrites or cell bodies of downstream neurons. At the point of contact, or synapse, the axon branch terminates in a highly specialized bulb, the synaptic bouton (Figures 1.2D and 10.9).

Synaptic boutons contain many mitochondria and a collection of vesicles in which are stored one or more neurotransmitter chemicals designed to influence the excitability status of the adjacent neuron (see Chapter 1). Most neurotransmitters are synthesized in the neuron's cell body and transported down the axon to the synaptic boutons where they are packaged in the synaptic vesicles for subsequent use.

The outer membrane of the synaptic bouton is studded with calcium channels. When an action potential reaches the bouton, the calcium channels open, thus permitting calcium to enter the cell. This, in turn, triggers a complex sequence of events that terminates in the synaptic vesicles' moving to the portion of the cell membrane adjacent to the contact point with the next cell, the synaptic cleft (see Figure 10.9). The vesicles then fuse with

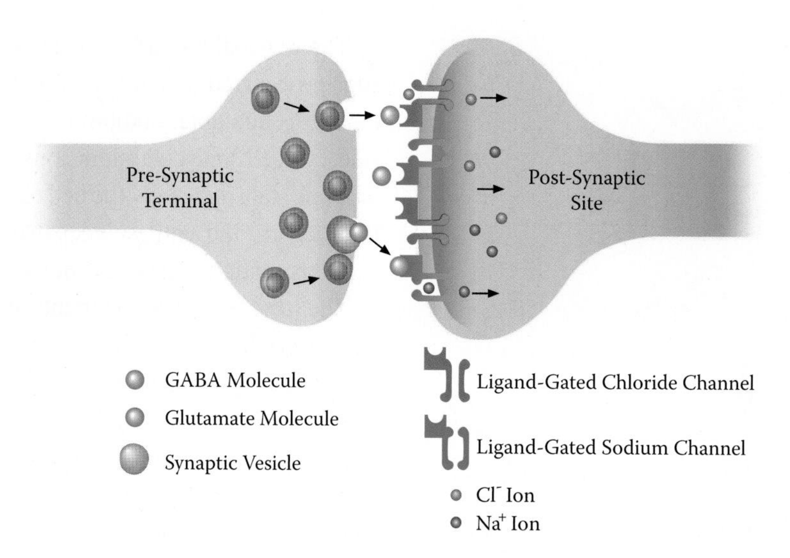

FIGURE 10.9 Schematic of a stylized "multipurpose" CNS synapse illustrating neurotransmitter molecule release and the generation of excitatory and inhibitory postsynaptic potentials.

the membrane, releasing their neurotransmitter packages into the synaptic cleft. This process is referred to as *calcium-modulated exocytosis.*

In the membrane of the dendrite or neuronal cell body opposite the synaptic bouton are complex structures known as *neurotransmitter receptors.* Typically there are several different types of receptors in any given postsynaptic site. Receptor molecules are membrane proteins frequently linked with adjacent ion channels. The nature of the ion channel will determine what effect the transmitter molecule has on the downstream neuron. Ion channels linked to a neurotransmitter receptor structure are known as *ligand-gated* channels, whereas the remaining (majority) ion channels in a given neuronal cell membrane—responding only to the passage of an action potential along the membrane—are known as *voltage-gated* channels.

Receptor molecules linked with sodium channels typically, when activated by the neurotransmitter, cause the channels to open, admitting Na^+ to the inside of the cell. This sodium influx is strictly a local process, leading to a small drop in the resting membrane potential, but not below the threshold point required to trigger an action potential. Such localized transient potential changes are called *excitatory postsynaptic potentials* (EPSPs).

On the other hand, a receptor molecule linked to a chloride channel will have the opposite effect. If, in response to the presence of a transmitter molecule in the synaptic cleft, chloride channels selectively open, Cl^- ions will be admitted to the postsynaptic cell, resulting in localized hyperpolarization of the membrane, and an inhibitory postsynaptic potential (IPSP).

Receptor molecules can also be linked to potassium channels in which case the local effect of receptor activation will be inhibitory. Some receptor molecules are not linked to ion channels but to a different class of membrane molecule known as G-proteins. Activation of G-proteins leads, in turn, to modifications in cell metabolism through adenylate cyclase (cyclic AMP) that will make the cell less sensitive to excitation, thus effectively inhibiting the cell.

Since it is the neurotransmitter receptor and its intramembranous linkages that determine whether the response to the transmitter will be excitatory or inhibitory, it follows that a given transmitter molecule may induce either an EPSP or an IPSP, depending on which type of receptor is present at a given synapse. In general, axodendritic synapses tend to give rise to EPSPs (through Na^+ channel activation) while axosomatic (cell body) synapses tend to be associated with IPSPs (through Cl^- channel activation). We have already seen this duality in the mechanism of action of dopamine on the accelerator and brake circuits in the motor system (Chapter 7).

If we return briefly to our representative neuron, we are now in a position to demonstrate how our neuron is persuaded to generate an action potential or, as the case may be, is dissuaded from doing the same. Our neuron, with its elaborate dendritic tree, has thousands of synaptic boutons from other neurons that impinge variously upon the dendrites and cell body, even its axon. Sufficient excitatory input from other neurons may generate enough EPSPs in the dendritic tree to cause our neuron's resting membrane potential to cross the threshold and trigger an action potential. This sequence could still be vetoed, as it were, if there were sufficient contravening IPSPs generated through chloride channel–linked receptors on the cell body. Such a veto would come about through the influence of inhibitory neurons (through inhibitory transmitters) acting to counterbalance the influence of excitatory neurons within the network to which our neuron belongs. Fluctuating countervailing synaptic influences are the basic machinery that permits normal operation of the nervous system.

10.5.3.3 Neurotransmitters Although transmitter molecules, depending on the nature of the receptor on which they act, may function in an excitatory or inhibitory capacity, most transmitters tend to be primarily either excitatory or inhibitory. There are four main categories of transmitter molecules: acetylcholine, amino acids, amines and peptides. The most important transmitters, their modes of action, and their principal locations in the nervous system are given in Table 10.1.

As can be seen in Table 10.1, acetylcholine is primarily excitatory and functions chiefly in the peripheral motor system and in the autonomic nervous system. There is also a very important role for acetylcholine in the central nervous system in association with memory, a role that was introduced in Chapter 9 and will be further explored in Chapter 12.

The amino acid transmitters, particularly glutamic acid (glutamate) and gamma-amino butyric acid (GABA), are highly relevant to Didi's case and will be considered shortly.

The biogenic amines dopamine, norepinephrine and 5-hydroxytryptamine (serotonin) play important roles in many central nervous system functions, including basal ganglia motor circuitry (dopamine), sleep–wake cycle regulation (norepinephrine, serotonin) and regulation of attention, emotion and other higher functions (all three compounds). Their roles will also be considered in Chapter 12.

Peptide transmitters are numerous, tend to influence neuronal function through G-proteins, and are present throughout the nervous system, whether peripheral somatosensory, autonomic/visceral or central. They play key roles, for example, in intestinal motility, pain transmission and

Table 10.1 Neurotransmitters

Transmitter Type	Location of Cell Body and Projection
Acetylcholine	Basal forebrain (e.g., nucleus of Meynert) → cerebral cortex
	Striatum interneurons (muscarinic)
	Brainstem and spinal cord lower motor neurons (nicotinic)
	Preganglionic autonomic (muscarinic)
	Postganglionic parasympathetic (muscarinic)
Amino acids	
Glutamic acid	Throughout CNS, excitatory
Gamma-amino butyric acid	Throughout CNS, inhibitory
Glycine	Spinal cord and medulla (inhibitory)
Amines	
Dopamine	Midbrain (SN) → corpus striatum, limbic cortex, frontal cortex (neuromodulatory)
Norepinephrine	Pons (locus ceruleus, lateral tegmental area) → cerebral cortex, central gray, limbic system, cerebellum, spinal cord (neuromodulatory)
Serotonin	Pons, medulla, cord, hippocampus → cortex, central gray, limbic system, cerebellum (neuromodulatory)
Histamine	Posterior hypothalamus, midbrain → forebrain cortex, thalamus (excitatory)
Peptides	Throughout CNS, neurenteric NS (neuromodulatory) (includes substance P, enkephalins, somatostatins, neuropeptide Y, vasoactive intestinal polypeptide, dynorphin, vasopressin, cholecystokinin)

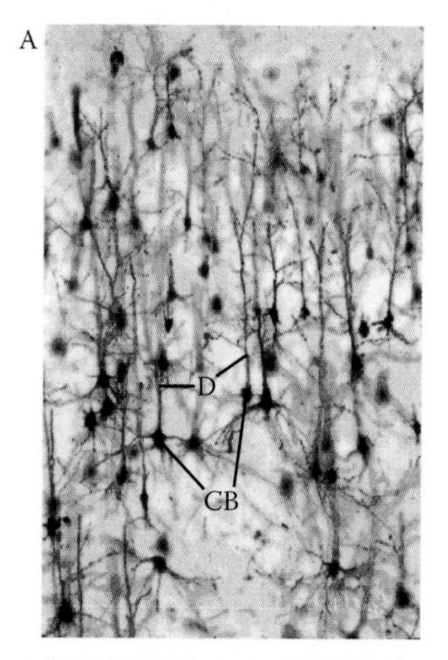

CB - Cell Body D - Dendrite

FIGURE 10.10 Photomicrographs showing typical pyramidal (glutamatergic) neurons. (A) Pyramidal neurons in the motor cortex (Golgi stain); note the prominent apical dendrites (D) and round to triangular (pyramidal) shape of the cell bodies (CB). (B) High power view of a single cerebrocortical pyramidal neuron (Bielschowsky stain); note the broad apical dendrite (D). (Courtesy of Dr. J. Michaud.)

modulation, basal ganglia circuitry, hypothalamic-pituitary axis function and cognition.

Returning, then, to the amino acid neurotransmitters, glutamate is the most ubiquitous excitatory transmitter in the central nervous system. We have already considered its role in the motor system, in upper motor neuron function and in basal ganglia circuitry (Chapter 7). Glutamate is also the main transmitter employed by nonmotor cortical neurons projecting to adjacent or distant cortical gyri within a hemisphere, or to homologous areas in the contralateral hemisphere (via the corpus callosum). As such, it plays an important excitatory role in association cortices and is implicated in learning and behavior.

Glutamate has varying effects on neuronal networks depending on the type of glutamate receptor present: NMDA, kainate, AMPA or metabotropic receptors. The first three receptors are linked to ligand-gated sodium channels while the fourth (metabotropic) is linked to a G-protein and thus functions as a neuromodulator rather than in an excitatory fashion.

GABA, on the other hand, is the principal inhibitory transmitter in the cerebrum. Neurons producing GABA as their synaptic messenger are typically small and have relatively short axons, while glutamatergic neurons are large and have long axons (Figure 10.10). GABAergic neurons are widely distributed in the nervous system, as befits their crucial role, and are found in cerebral cortex, basal ganglia, brainstem, cerebellum, and spinal cord. GABA receptors, as might be predicted from our previous discussion, are linked to chloride channels; their activation tends to elevate resting membrane potentials, thus inhibiting the production of action potentials.

Glycine's activity is primarily inhibitory at the spinal cord level (in modulating motor function) and excitatory at the cerebral level. Its roles do not directly impinge on Didi's problem.

There is a close and interesting chemical relationship between glutamate and GABA, the paramount excitatory and inhibitory central nervous system transmitters. As demonstrated in Figure 10.11, GABA is synthesized directly from glutamate by the removal of a CO_2 moiety. This biochemical conversion is catalyzed by the enzyme glutamate decarboxylase, using pyridoxine (vitamin B6) as a cofactor. Thus, through the action of a single enzyme, a delicate balance is maintained between excitation and inhibition. On the chessboard of synaptic transmission, glutamate and GABA are like opposing queens on adjacent squares, each preventing the one side from overwhelming the other.

10.5.3.4 Mechanisms of Epileptic Seizures Now that we have considered how neurons develop electrical charges, and how, through action potentials, they are able to excite or inhibit one another, we must return to

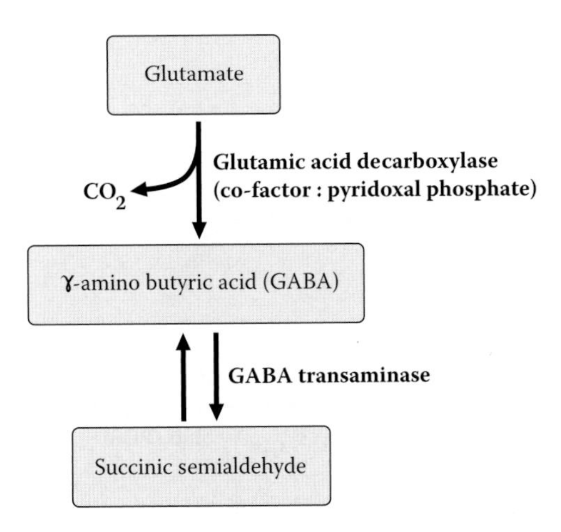

FIGURE 10.11 GABA synthesis biochemical pathway.

our definition of an epileptic seizure. Given that epileptic seizures consist of abnormal, excessive electrical discharges by cortical neurons, we must ask the next obvious question: Why do cortical neurons discharge excessively?

The answer to this question cannot be given in its entirety as some aspects remain unclear. For simplicity's sake, it is helpful to review the main circumstances in which humans develop epileptic seizures. These essentially break down into two main categories:

1. Focal cerebral cortical pathology in which a cluster of cortical neurons becomes unstable and produces synchronous, massed action potentials that interrupt normal *regional* brain functions.
2. A lowered threshold in both the cerebral cortex and the thalamic regions with which it has reciprocal connections, such that the cortex as a whole has a tendency to develop synchronous neuronal discharges that *diffusely* interrupt brain functions, including consciousness.

Both of these epileptogenic factors may produce epileptic seizures in isolation, as well as by acting in collaboration. In the case of the former, almost any type of focal cortical pathology (whether ischemic injury, hemorrhage, trauma, infection, or dysplasia) may produce focal epileptic discharge. In the case of the latter, any systemic metabolic dysfunction (e.g., anoxia, hypoglycemia, neuroexcitatory pharmacologic agent) may trigger generalized epileptic discharge, as may the existence of a genetically determined low threshold for such discharges.

Focal Pathology Focal epileptic discharges, regardless of the type of pathology, originate in large cortical glutamatergic neurons. These neurons have a unique capacity to fire in bursts, once goaded to discharge. Micropipette recordings from single neurons of this type reveal an initial decline in resting membrane potential followed by a burst of action potentials, then a return to a normal resting membrane potential. This phenomenon is known as a *paroxysmal depolarization shift* (see Figure 10.12). The mass production of synchronous paroxysmal depolarization in a large cohort of epileptogenic cortical neurons produces a summative discharge that appears via regional scalp electrodes as a focal *spike discharge*.

Isolated sharp or spike discharges may be recorded from patients with focal-origin epilepsy in the asymptomatic state. During clinical seizures, however, long trains of repeated spike discharges occur, reflecting the uncontrolled sustained synchronous discharge of thousands of large cortical neurons, the trains lasting seconds to minutes (see Figure 10.13).

Low Epileptic Threshold Generalized, synchronous epileptic discharges involving all cortical regions appear to require the collaboration of the nonspecific thalamic nuclei and cortical neurons, acting through a feedback loop: cortico-reticulo-cortical. In essence, this means that generalized epileptic discharges are based, at least to some extent, in the upper components of the consciousness system, as described earlier in this chapter.

The participation of the nonspecific cortico-thalamo-cortical feedback loop in the genesis of generalized epileptic seizures was elucidated through an elegant series of experiments in cats. Penicillin, a potent epileptogenic compound if applied topically to the brain surface, was instilled on the pial surface of awake cats; this resulted in the development of recurrent episodes of unresponsiveness, lasting seconds, during which there was flickering of the eyelids or twitching of the whiskers. During these unresponsive episodes, surface electrodes on the cerebral cortex recorded synchronous discharges from all cortical areas consisting of repetitive spike bursts, each individual spike followed by a high voltage slow wave. Rhythmic spike-and-wave discharges are the hallmark of generalized epilepsy in the penicillin cat model, and in EEG recordings of generalized seizures in humans (see Figure 10.14).

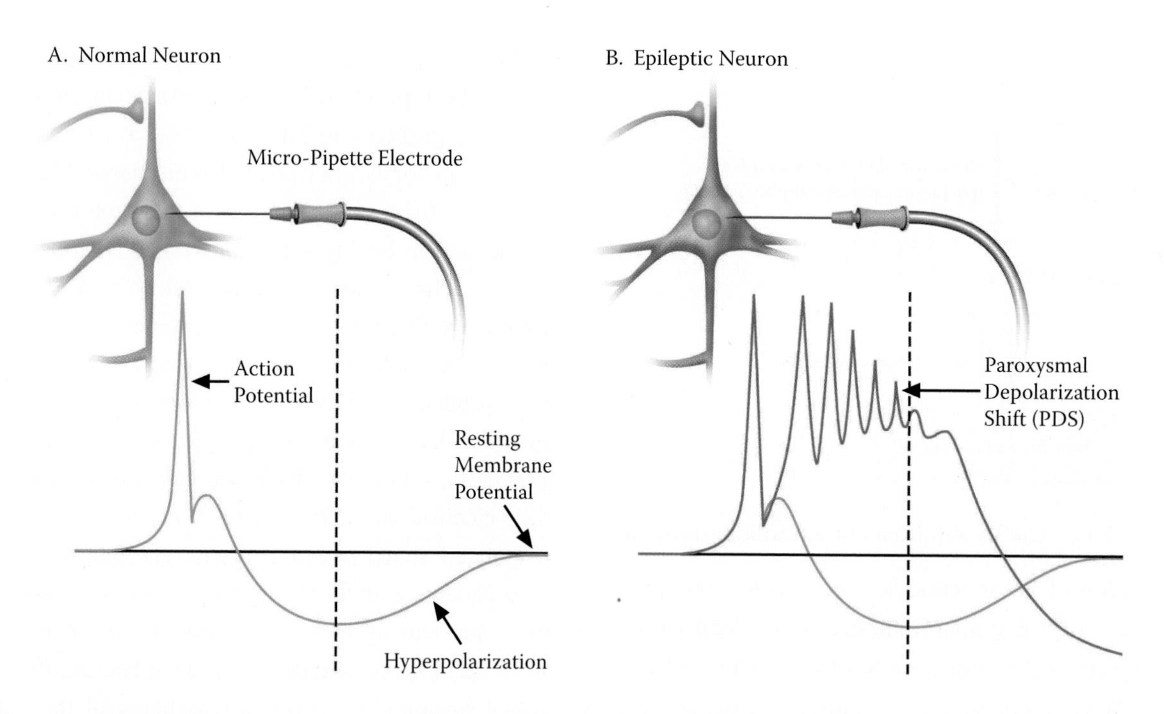

FIGURE 10.12 Intracellular recording from a pyramidal cell neuron (in the cerebral cortex). (A) Normal depolarization action potential; (B) "epileptogenic" neuron shows a paroxysmal depolarization shift (PDS).

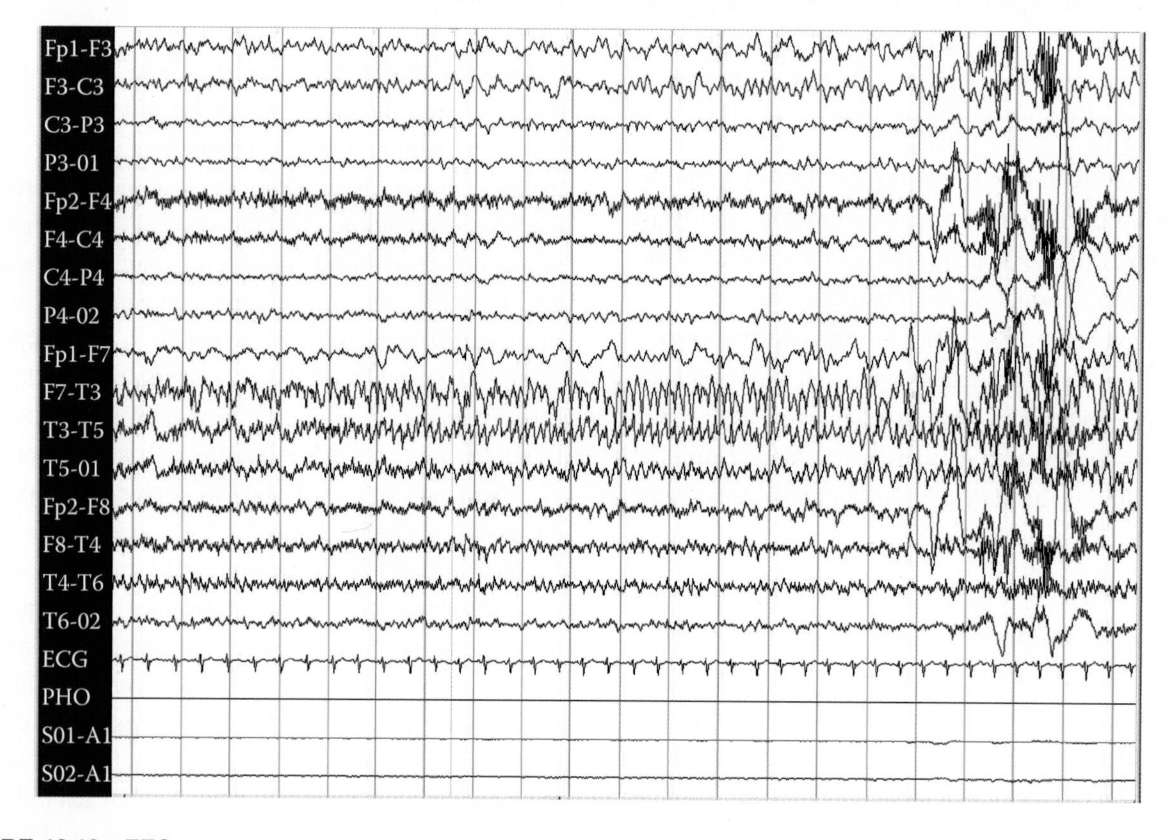

FIGURE 10.13 EEG tracing illustrating a focal-onset epileptic seizure: the beginning of the left-sided (electrode positions F7, T3, T5), rapid, gradually augmenting focal discharge was accompanied, in the patient, by a sudden loss of responsiveness, a blank stare and by intense facial flushing. (Courtesy of S. Bulusu and Dr. S. Whiting.)

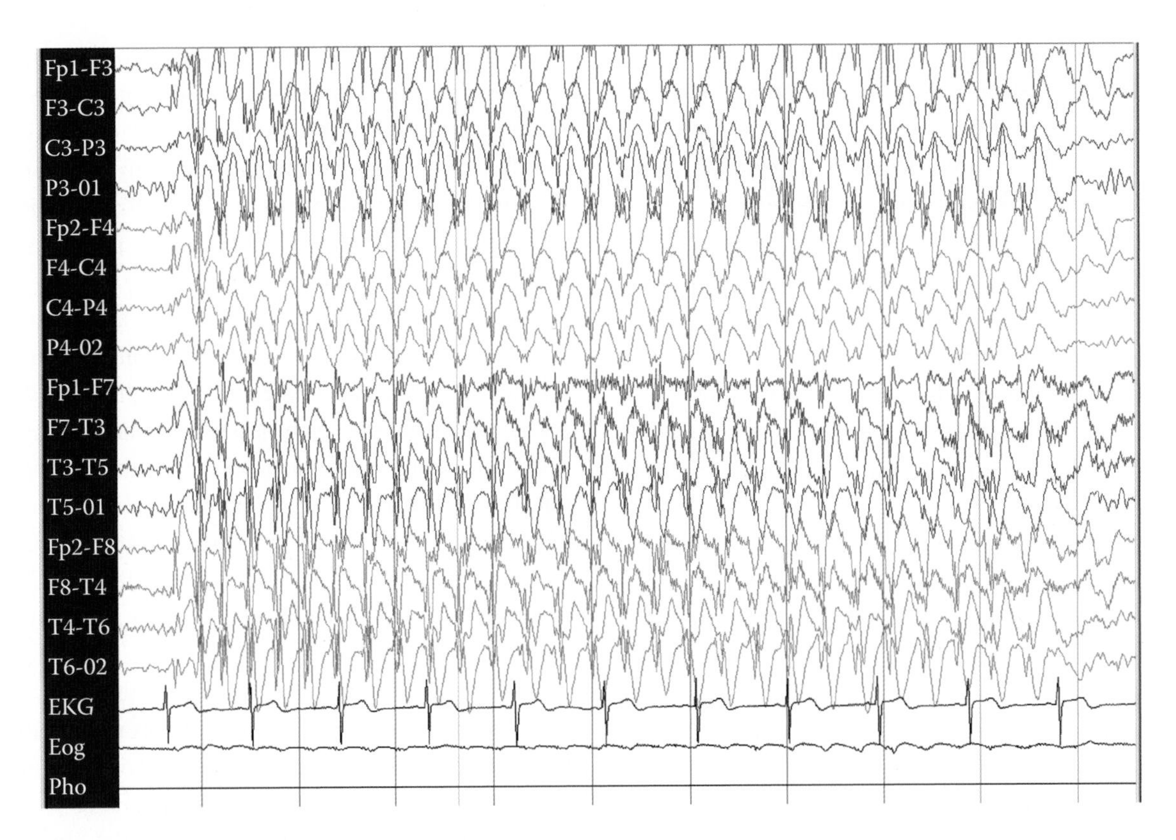

FIGURE 10.14 EEG tracing of a typical absence seizure in childhood absence epilepsy syndrome: the background activity is abruptly replaced by a generalized three per second spike-and-wave discharge that terminates equally abruptly. (Courtesy of S. Bulusu and Dr. S. Whiting.)

Sequential recordings following local instillation of penicillin to cat cortex demonstrated an interesting evolution. Initially there was a generalized high voltage rhythmic slow-wave discharge of steadily increasing amplitude, reminiscent in its appearance of what is seen in humans (and cats!) as they become drowsy and fall asleep. This increasingly hypersynchronous discharge gradually transformed into the spike-and-wave discharge, suggesting that the reticulo-cortical mechanism of generalized epileptic discharge was functionally related to the normal mechanism of deep sleep.

If you have remained alert, and not narcotized by the above dissertation, you will have recognized that the staring-blinking episodes seen in the penicillin model of generalized feline epilepsy are remarkably reminiscent of the episodes that have so perturbed Didi's teacher! In fact, Didi's episodes are secondary to an abnormally low, presumably genetic, threshold in the same cortico-reticulo-cortical epileptogenic mechanism uncovered by cortical penicillin application in the cat.

Having finally returned from a basic consideration of neuronal membrane function and mechanisms of epileptogenesis to the intermittent behavioral abnormalities in Didi ascribed as epileptic seizures, it is now necessary to briefly look at the sundry patterns of clinical epileptic seizures that the two fundamental epileptogenic mechanisms may evoke.

10.5.3.5 Types of Epileptic Seizure

Since epileptic seizures originate in populations of excessively discharging neurons, it would be predictable that the symptoms produced during the seizure would derive from loss of normal function of the neurons involved. In general, this is what occurs. One has to distinguish, however, between *positive* symptoms, in which behaviors are seen that are abnormal or inappropriate, and *negative* symptoms, in which there is a temporary loss of normal function.

In the case of positive symptoms, there may be uncontrollable movements, sensory hallucinations or sudden, unexpected emotional experiences such as groundless fear.

Negative symptoms may include temporary paresis of an arm or leg, or transient loss of consciousness, as occurs with Didi.

When epileptic seizures occur in a region of focal cortical pathology, the symptoms produced will depend entirely on the normal function of the area involved. Thus, a focal seizure beginning in a cortical area devoted to hand movement may consist of uncontrolled rhythmic twitching of the thumb and fingers. In contrast, a seizure involving visual cortex may produce a hallucination of lights, geometric forms or complex images such as a face, while seizures involving the amygdala may produce a transient, abrupt state of panic.

On the other hand, seizures originating in a hypersynchronous cortico-reticulo-cortical loop produce generalized rather than focal symptoms: tonic stiffening or rhythmic twitching of all four limbs, sudden loss of control of limb and trunk muscles with an abrupt fall to the ground, or transient loss of awareness without loss of postural control.

At the present time, seizure patterns are classified according to a generally accepted system developed by the International League Against Epilepsy (see Table 10.2). In essence, focal-onset seizures are classified according to whether or not they also interfere with consciousness: simple partial seizures if consciousness is maintained, complex partial seizures if it is not. Simple partial seizures are also categorized according to the principal type of symptom present, whether motor, sensory, autonomic or psychic.

Generalized seizures are classified according to the type of behavior seen by observers. Thus, a typical generalized seizure may consist of an abrupt loss of consciousness, a simultaneous uncontrolled contraction of all body musculature manifested as generalized tonic stiffening of the limbs, then a period of rapid violent twitching of all four limbs in synchrony. This sequence is usually summarized as a generalized tonic-clonic seizure and often referred to by the lay public as "grand mal seizures," an archaic terminology.

Table 10.2 Main Seizure Patterns

Focal onset	Simple partial—motor, sensory, vegetative, psychic
	Complex partial—impaired consciousness
	Simple or complex partial with secondary generalization
Generalized	Tonic-clonic
	Tonic
	Clonic
	Absence
	Atonic
	Myoclonic

A generalized seizure consisting of a sudden loss of muscle control and a fall is called an atonic seizure, while a sudden, violent, shock-like jerk of the limbs (sometimes without loss of consciousness) is termed a myoclonic seizure. Finally, a transient loss of consciousness without loss of trunk support is called an absence seizure—*absence* being the French term for absent awareness. The archaic, but commonly used term for the latter is "petit mal seizure," to contrast it with the flagrant, dramatic manifestations of the grand mal seizure. Absence type seizures best fit the description given for Didi's attacks.

At this point in our review, it is important to distinguish between epileptic seizures and the generic term *epilepsy*. By definition, *epilepsy* means the appearance in an individual of two or more unprovoked epileptic seizures. This means that a patient who has a single epileptic seizure, or more than one seizure in the context of a drug overdose, does not have epilepsy and is not "epileptic."

We have already alluded to the fact that some individuals with epilepsy may have a combination of focal cortical pathology and a genetically determined low threshold for generalized or corticoreticular seizures. In such individuals a seizure may begin in a cortical area, with focal motor or sensory symptoms, and then spread preferentially to the nonspecific thalamic nuclei; this generates a second phase in which there is a generalized tonic-clonic seizure. This phenomenon is known as focal onset with secondary generalization. If the initial, simple partial seizure produces a transient state of fear, followed by a generalized seizure, the simple partial component or premonitory feeling is termed an aura.

Alternatively, a focal-onset seizure may simply spread to adjacent cortical regions without projecting to the thalamus. If the initial symptom of the seizure consists of rhythmic twitching of the thumb, sequential localized spread will lead to twitching of the hand, arm, shoulder and ipsilateral face, a phenomenon called a Jacksonian march, named after Hughlings Jackson, who originally described it. Likewise, a seizure beginning focally in the amygdala (with a sensation of fear) may then spread to the hippocampus (with associated confusion and loss of memory registration) and to the insular cortex (with retching and drooling behavior). In this case, the seizure commences as simple partial, then transforms to complex partial, consciousness having become partially impaired.

10.6 Case Summary

Didi has been experiencing frequent brief (10 seconds) episodes of staring, inattention and unresponsiveness for at least three months; they appear to be interfering with his ability to function appropriately in the classroom. His developmental status and neurological examination are entirely normal. Given that the episodes can be triggered by deliberate hyperventilation and that Didi's sister has absence seizures, the probable diagnosis is *childhood absence epilepsy.*

10.7 Additional Information

Childhood absence epilepsy is a specific epilepsy syndrome with diagnostic criteria outlined in the International League Against Epilepsy classification system of epileptic disorders (see Table 10.3).

Faced with a child suspected of having absence epilepsy, the physician's first task, before undertaking a treatment regime, is to confirm the diagnosis by documenting an absence episode during an EEG recording. This step is essential because, after all, why couldn't Didi simply be having day-dreaming episodes? Anyone with any experience of six-year-olds will recognize that children this age can be so self-absorbed, all environmental "filters" fully activated, that they fail to respond to having their names called repeatedly. In addition, children with chronic airway obstruction, due either to allergic rhinitis or tonsillar hypertrophy, may have interrupted sleep with obstructive sleep apnea; they are thus chronically sleep deprived, and inattentive during the daytime. All of these alternative possibilities need to be considered.

The EEG findings in absence epilepsy are quite dramatic, even between seizures (the interictal state). While the patient is lying quietly, awake but with closed eyes, the normal background EEG rhythms are frequently interrupted by one- to two-second generalized spike-wave bursts, in other words, shortened versions of the prolonged discharge shown in Figure 10.14. Recorded synchronously over all scalp regions and occurring at a rate of three spike-wave complexes per second, these discharges are not accompanied by any obvious change in behavior or awareness. On occasion, as well as in response to hyperventilation, longer periods (8 to 10 seconds) of three per second (or 3 Herz) spike-and-wave discharge will occur (Figure 10.14), this time accompanied by unresponsiveness to name-calling and often by rapid flickering of the eyelids.

Table 10.3 Selected Epilepsy Syndromes

Localization-related	
Idiopathic:	Benign epilepsy of childhood with centro-temporal (rolandic) spikes
	Benign epilepsy of childhood with occipital spikes
Symptomatic:	Temporo-limbic epilepsy with mesial temporal sclerosis
Non-localization-related	
Idiopathic:	Benign neonatal familial convulsions
	Benign sporadic neonatal convulsions
	Benign myoclonic epilepsy of infancy
	Severe myoclonic epilepsy of infancy (Dravet syndrome)
	Childhood absence epilepsy
	Juvenile myoclonic epilepsy (Janz syndrome)
	Juvenile-onset absence epilepsy
Symptomatic:	Infantile spasms (West syndrome)
	Lennox–Gastaut syndrome

Adapted from the epilepsy syndrome classification system of the International League Against Epilepsy.

The mechanism of the alternating spike-and-slow-wave discharge in absence epilepsy is thought to be a rhythmic oscillation in cortical neurons, alternating between an excitatory spike potential and an inhibitory slow wave. This rapid cycling between excitation and inhibition appears to be driven by the nonspecific thalamic nuclei. It is not correct to assume, however, that generalized epileptic discharge of this type originates in the thalamus; both the cortex and the thalamus, acting in tandem, are required for this phenomenon.

A generalized 3 Hz spike-and-wave EEG pattern is typical of childhood absence epilepsy and is entirely different from what we would observe if we recorded complex partial seizure. Typically originating focally in one temporal lobe, such a seizure may resemble superficially a childhood absence attack in that the child will be inattentive and stare blankly for the duration of the episode. In general, however, the loss of consciousness in a complex partial seizure of this type will last one or two minutes, not 10 seconds, and will be followed by a period of confused behavior. Automatic behaviors such as lip-smacking and purposeless fidgeting with clothes may also occur during the seizure itself. An EEG during a complex partial seizure will reveal a sustained focal rhythmic sharp wave discharge in the temporal lobe of origin, with some spread to surrounding areas as the seizure proceeds (as shown in Figure 10.13).

If the diagnosis of childhood absence epilepsy is confirmed electroencephalographically, you may well ask whether other investigations are required, for example, an imaging study of the brain. Decades of accumulated experience, both with CT and MR imaging, however, have demonstrated that children with any form of primary generalized epilepsy nearly always have normal-appearing brains. Hence, for well-documented absence epilepsy, an imaging study is unnecessary. Imaging studies are definitely indicated, however, for most patients whose seizure patterns or EEG findings suggest a focal origin. The only exception to this dictum is a group of benign, familial localization-related epilepsies, for example, benign epilepsy of childhood with rolandic spikes; benign epilepsy with occipital spikes—see Table 10.3.

Once the diagnosis of childhood absence epilepsy has been made, the next step is to determine whether Didi should be started on an antiepileptic medication. For childhood absence epilepsy, starting such medication is not an automatic step, given that patients do not fall during attacks and are at low risk of injury. In most instances, it is the frequency of the absence attacks and their possible negative effect on attention and learning that determine whether antiseizure medication is to be used. As you will recall, Didi's presenting problem is a striking decline in school performance; this and the numerous witnessed absence seizures strongly suggest that his epilepsy should be treated.

At the present time there is a large number and variety of antiepileptic medications; the principal drugs, and their indications for use, are listed in Table 10.4. Which drug or drugs should we consider using? To some extent, our decision may be guided by a consideration of mechanisms of action of antiepileptic medications.

There are three principal mechanisms of action of antiepileptic drugs thus far identified:

1. Prolongation of inactivation of sodium channels in circumstances of high frequency neuronal discharge. In essence, this mechanism involves the stabilization of neuronal membranes and is the main mode of action of many of the older antiepileptic medications such as phenytoin and carbamazepine.
2. Augmentation of GABA-ergic inhibition. Drugs acting through this mechanism attach to GABA receptors, thus augmenting chloride conduction, and hyperpolarizing neuronal memebranes. The benzodiazepine class of medication (e.g. clonazepam, nitrazepam, clobazam), as well as phenobarbital, fall into this category.
3. Inhibition of T-type calcium channels. T-type Ca^{2+} channels are particularly prominent in neurons located in the nonspecific (reticular) thalamic nuclei. Inhibition of these channels appears to suppress generalized corticoreticular epileptogenic mechanisms, particularly in the case of absence seizures, rather than generalized tonic-clonic seizures. Ethosuximide is an antiepileptic drug that has its main effect through this mechanism.

A fourth antiepileptic mechanism, in principal, is the inhibition of excitatory glutamate receptors; drugs designed to address this mechanism are currently being developed.

A consideration of the above mechanisms of action, and a perusal of Table 10.4, would lead us, for a child with absence epilepsy, to settle on ethosuximide. Alternative

Table 10.4 Antiepileptic Medications

Seizure Type	Main Therapeutic Options
Generalized	
Tonic-clonic	**valproate, carbamazepine, phenytoin**, clobazam*, phenobarbital, lamotrigine
Absence	**ethosuximide, valproate**, lamotrigine
Myoclonic	**valproate**, topiramate, clonazepam, nitrazepam, acetazolamide, levetiracetam, ketogenic diet
Atonic	**valproate**, lamotrigine, topiramate, zonisamide
Partial	**carbamazepine, phenytoin, clobazam***, valproate, lamotrigine, topiramate, gabapentin, vigabatrin, phenobarbital
Syndromic	
Benign rolandic	**carbamazepine**, valproate, phenytoin
Juvenile myoclonic	**valproate**, lamotrigine, topiramate
Lennox–Gastaut	valproate, lamotrigine, clonazepam, topiramate
West syndrome	**vigabatrin**, ACTH, steroids, valproate

Medications shown in bold are those used most frequently.
* Not available in the U.S.

agents, in the event that ethosuximide were to be either ineffective or poorly tolerated, would be valproate and lamotrigine. Most patients with absence epilepsy will respond to one or more of these three medications.

10.8 Investigations

As expected, Didi's baseline EEG in the awake state showed frequent brief bursts of 3 Hz spike-and-wave discharges; no clinical absence seizures occurred.

When Didi was directed by the neurophysiology technologist to hyperventilate, however, he had a 10-second episode identical to the one we saw in your office. During that 10-second epoque, there was sustained 3 Hz generalized spike-wave discharge with a slowing in frequency just before the offset. The instant the spike-wave discharge ceased, the EEG resumed a normal background and Didi was immediately alert.

No other investigations were performed.

10.9 Treatment/Outcome

Didi was started on a small amount of ethosuximide, given in two divided doses per day. The dose was gradually increased over a period of two weeks during which time the absence seizures declined in frequency and eventually became undetectable. At the same time, there was a dramatic improvement in Didi's school performance and he was soon functioning at his usual superlative level. Initially he complained of feeling nauseated shortly after taking each dose of medication; this side effect disappeared after about a month.

After Didi had been seizure-free for a year, an attempt was made to taper his medication. The absence seizures almost immediately returned so the medication was restarted. A year later, the ethosuximide was again tapered, this time with no recurrence of seizures.

Related Cases

For further localization and etiological exercises on material related to this chapter, please see the e-cases on the text DVD and Web site.

Suggested Readings

Browne, T.R., and G.L. Holmes. Epilepsy. *N. Engl. J. Med.* 344 (2001): 1145–1151.

Chang, B.S., and D.H. Lowenstein. Epilepsy. *N. Engl. J. Med.* 349 (2003): 1257–1266.

Gardiner, M. Genetics of idiopathic generalized epilepsies. *Epilepsia* 46, Suppl. 9 (2005): 15–20.

Sillanpaa, M., et al. Long-term prognosis of seizures with onset in childhood. *N. Engl. J. Med.* 338 (1998): 1715–1722.

Wirrell, E.C. Natural history of absence epilepsy in children. *Can. J. Neurol. Sci.* 30 (2003): 184–188.

Web Sites

National Institute for Neurological Diseases and Stroke: www.ninds.nih.gov/disorders/epilepsy/detail_epilepsy.htm

Epilepsy.com: http://www.epilepsy.com

Chapter 11

Chantal

CHANTAL

Chantal is a joyful, almost boisterous person who loves to talk, smoke, and eat. She is always invited to family get-togethers in the countryside around Quebec City, particularly at the holiday season, and although only distantly related to the McCool family (via a long-lost cousin), she is now welcomed as one of the family.

Those who knew Chantal as a young girl remember her as a slim, somewhat shy figure. Over the years, she has become more and more outgoing but at the same time quite overweight. There were a few opportunities for her to get involved with a member of the opposite sex but somehow things never worked out; there were obligations to her dogs, or to a project she was working on, and so Chantal remains unattached. Now that she is almost 40 years old, everyone has stopped pestering her to get married or trying to fix her up with some male of dubious character.

About three years ago Chantal began having occasional headaches, sometimes associated with her menses. Her family physician had noted the occurrence of these and the fact that her blood pressure readings were averaging about 160/95. She instructed Chantal to watch her diet, to avoid foods with a high salt content, and to cut down on her smoking—all to no avail.

The physician felt that the headaches were migrainous in nature, as there is a positive family history of migraine, and recommended that, when the headache strikes, she take two extra-strength acetaminophens and stay away from noisy and brightly lit rooms; no migraine-specific medication has been prescribed. About six months ago hydrochlorthiazide was prescribed to control her blood pressure but Chantal admits that she has only taken the medication "from time to time."

Over the past few months the headache frequency has gradually increased and Chantal has noted a few episodes where the headaches have not gone away for over a day, despite taking acetaminophen every four to six hours. She says that the headaches are generally in the forehead area, although the most recent ones have been located mostly at the back of her neck; there has been no radiation of the neck pain into her arms. Flashes of light or other visual phenomena have not preceded the onset of the headaches, nor are they accompanied by nausea or vomiting.

Her current visit has been precipitated by a recent worrisome event: she was driving home from a family gathering (it was snowing and the country road was totally dark) and her car slipped off the road and into a ditch. Luckily, there was enough snow to cushion her car from serious damage but Chantal was not wearing her seat belt (it had become too uncomfortable over her heavy coat). Fortunately she was found by a passing motorist driving a four-wheel-drive panel truck.

Initially Chantal did not respond to the shouting of her rescuers but quickly regained consciousness. The driver of the truck and his friend easily managed to pull the car out of the ditch; afterward they all went back to his nearby farm house where there was a friendly gathering. Chantal was soon back on the road and arrived home safely. Her mother, hearing the story, insisted that she seek immediate medical attention and Chantal is now in the emergency room at the local community hospital, the morning following the incident.

Physical Examination

On examination, Chantal presents as a pleasant if somewhat subdued individual who is clearly worried about what could have happened to her. On recounting her story it seems that she had failed to see a sign indicating that there was a curve in the road. She had had only one beer at the party (with lots of finger food) and avows that she never has more than this to drink if she knows that she is going to drive home afterward. Her retelling of the previous night's incident is quite clear and consistent, and she does remember her car sliding toward the ditch. Her next recollection is some person yelling at her and reaching over her to shut off the car engine. She has no idea how much time elapsed between the moment the car slid off the road and the time when she was found.

Chantal is clearly very overweight (BMI 38 kg/m^2; see also Etienne, Chapter 8) and her blood pressure is 170/105 sitting and 165/95 reclining. Chantal has a slight bruise on the top of her forehead, on the right side, just at the hairline. Her mental status examination is completely within normal limits—she is fully oriented to person, time and place. She is fully cooperative with the examiner.

Cranial Nerve Examination

CN I: Smell testing is normal

CN II: Using the confrontation technique, no visual field defect is noted. Visual acuity with her glasses is within normal limits. The pupillary reaction to light is normal and equal in both eyes.

CN III, IV, VI: Movements of the eyes in the cardinal positions are within normal limits

CN V: Sensory—touch and pain are normal in the three divisions of the trigeminal nerve; motor—teeth clenching is symmetrical and the jaw reflex is normal.

CN VII: No facial asymmetry is noted and facial movements are normal.

CN VIII: No defect in hearing between the two sides is noted with tuning fork testing.

CN IX, X: The gag reflex is normal and no hoarseness of voice is noted.

CN XI: Head turning and shoulder shrugging are normal and symmetrical.

CN XII: Tongue movements are normal.

Motor Examination

Muscle strength is appropriate and symmetrical in all four limbs. Reflexes are somewhat "sluggish" but symmetrical; the plantar reflex is downgoing bilaterally. There is no ankle clonus.

Coordination and gait are within normal limits. The Romberg sign is not present.

Sensory Examination

No deficits in sensory testing are noted over the limbs and trunk; in particular, there is normal vibration sense and position sense in all four limbs.

Fundoscopy

With the lights in the room dimmed, the optic disc margins are found to be definitely blurred bilaterally. (See Figure 11.1, right panel.)

The assessment is that Chantal has bilateral *papilledema*, which is considered an emergency and requires an immediate consultation. An arrangement is made for her to see the senior ophthalmologist at the regional tertiary

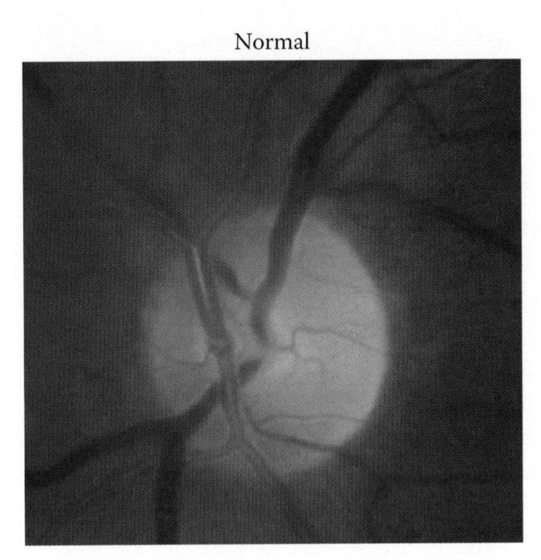

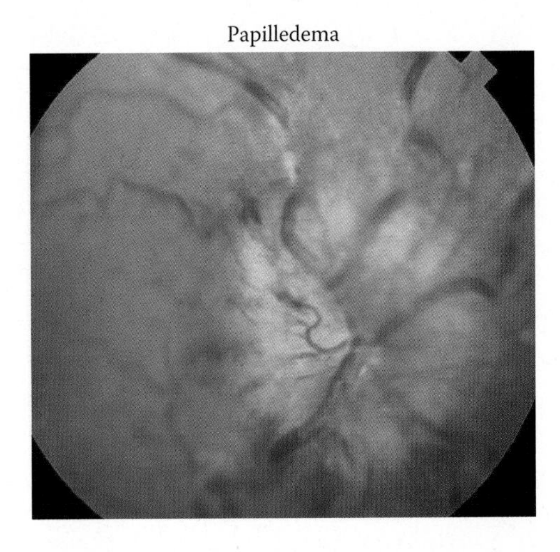

FIGURE 11.1 **Optic disc: Photographic images of the retina as seen with the ophthalmoscope. The normal optic disc (shown on the left) has a whitish appearance with blood vessels, both arteries and veins, emerging from its center; note the clearly defined disc margin. An abnormal disc with papilledema is shown on the right; note the swollen appearance of the disc and the absence of any clear margins. (Courtesy of Dr. M. O'Connor.)**

referral hospital early that afternoon. The bilateral papilledema is confirmed along with the presence of venous engorgement. The patient is sent immediately for testing of the visual fields and for neuroimaging (CT scan), and with a strong likelihood of needing a lumbar puncture (LP, see below).

Visual fields can be thoroughly and accurately mapped using a special instrument called a Goldmann perimeter. With this apparatus, the patient focuses on the center of a rounded screen and a small light stimulus is projected at random on the screen. Each eye is tested separately and the visual field is mapped and charted in a standardized manner. The testing can be done by an automated machine, available at most major hospitals, and takes 10 to 15 minutes; testing can also be done by a technician if the automated testing is not satisfactory or in order to confirm any abnormal findings. Chantal's visual field testing reveals a surprise! A significant enlargement of the physiological blind spot is found, in both eyes (compare a normal in Figure 11.2A with an abnormal, Figure 11.2B). In effect, there is a large region just lateral to the fovea where Chantal can see nothing whatsoever.

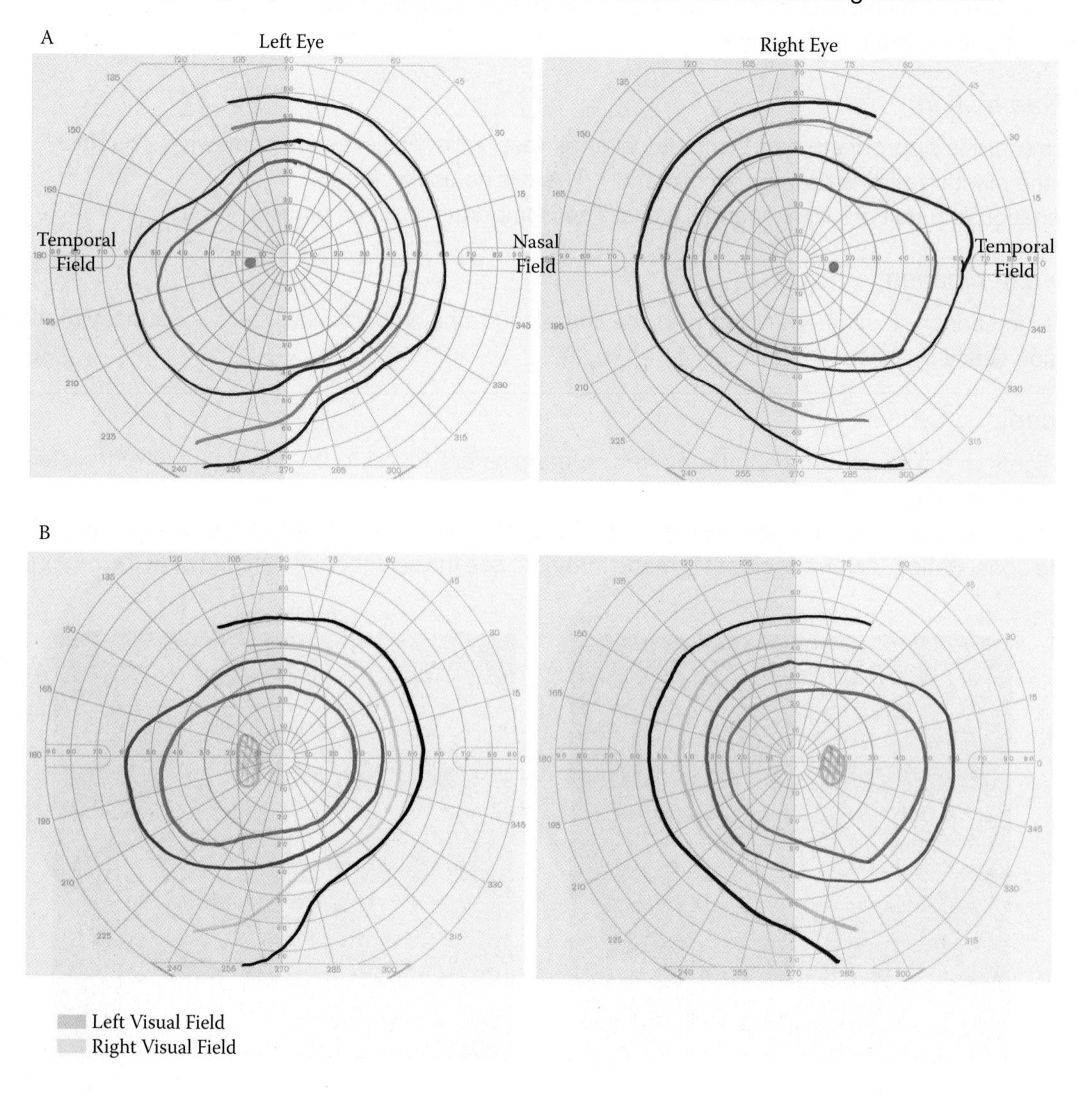

FIGURE 11.2 Visual field (Goldmann) perimetry. In presenting the visual fields, the convention is to display them in the same way that the patient is viewing (right eye on the right side and left eye on the left side). (A) The normal visual fields, with the blind spot indicated (in red) in the temporal retina. (B) The visual fields of a patient with papilledema; note the enlarged blind spot bilaterally (shown in yellow). (Courtesy of Dr. M. O'Connor.)

Objectives

- To understand the anatomy of the cranial cavity and its contents other than the brain, including the meninges, the cerebrospinal fluid and its circulation, and venous drainage, as well as the involvement of these components in clinical disease states
- To comprehend the visual pathway and the usefulness of the clinical assessment of vision in the diagnosis of intracranial lesions

11.1 Clinical Data Extraction

At this point, the student should proceed to the DVD and complete the history, physical examination, localization and etiology worksheets, utilizing the information just provided concerning Chantal's problem.

11.2 Review of the Main Clinical Points

- This is a 37-year-old obese female with a history of hypertension and a long-standing neurological complaint of headache.
- There has been a recent motor vehicle incident, associated with a period of unconsciousness of unknown duration.
- There is a small bruise over the right forehead.
- The significant finding is bilateral papilledema (more prominent in the right eye), while the rest of the neurological examination is within normal limits.
- Formal visual field testing has revealed enlarged blind spots bilaterally.

11.3 Relevant Neuroanatomy

In order to understand Chantal's complaint of headache, we need to know which structures of the head and in the interior of the skull are pain sensitive. Pain can originate from any of the structures of the head, including the skin, teeth, mucous membranes (including the gums), the sinuses, and from the eye and orbit. The brain itself has no pain fibers, but the brain coverings, the meninges, are pain sensitive, as are the walls of the major cerebral blood vessels. Any irritation (infection, blood) or stretching (pulling) on the meninges or cerebral vessels will give rise to pain.

In the light of this information, it will be necessary first for us to review the anatomy of the cranial and spinal meninges. Second, as part of the meningeal compartment, the cerebrospinal fluid and its circulation need to be considered. Third, in order to appreciate the significance of bilateral papilledema and enlarged blind spots bilaterally, the visual system and its pathway should be understood. Finally, we must also consider some further aspects of the cerebrovascular supply to the skull and brain, both arterial and venous.

11.3.1 The Meninges and Cerebrospinal Fluid

The meninges are the connective tissue coverings of the brain and offer a certain measure of protection to the underlying brain tissue. The meninges consist of three layers, the dura, the arachnoid and the pia, with spaces or potential spaces between the layers. The cerebrospinal fluid, the CSF, lies within the meninges such that the brain is in fact "floating" in fluid.

11.3.1.1 Cranial Meninges The outermost layer of the meninges is the dura, a thick strong sheet of connective tissue. Within the cranial cavity, the dura and periosteum of the skull bones adhere to one another (Figure 11.3A). Bleeding from the large artery that supplies the meninges, the middle meningeal artery, occurs between the periosteum and the dura and is appropriately called an epidural hemorrhage; usually this is caused by trauma to the side of the head in the temporal region of the skull.

Two dural sheaths separate the parts of the brain from each other and "hold" the cerebral hemispheres in place. The first is the falx cerebri, situated in the midline between the two cerebral hemispheres, above the corpus callosum. The other is the tentorium cerebelli, a horizontal sheath of dura that lies between the occipital lobe of the hemispheres above and the cerebellum below (see Figure 11.4). The large venous sinuses are located at the attachments of the falx cerebri and tentorium cerebelli (see Figure 11.3A).

The next layer is the arachnoid. There is a potential space between the dura and arachnoid, and the cerebral veins pass through this space. These bridging veins, as they are called, course from the surface of the brain into

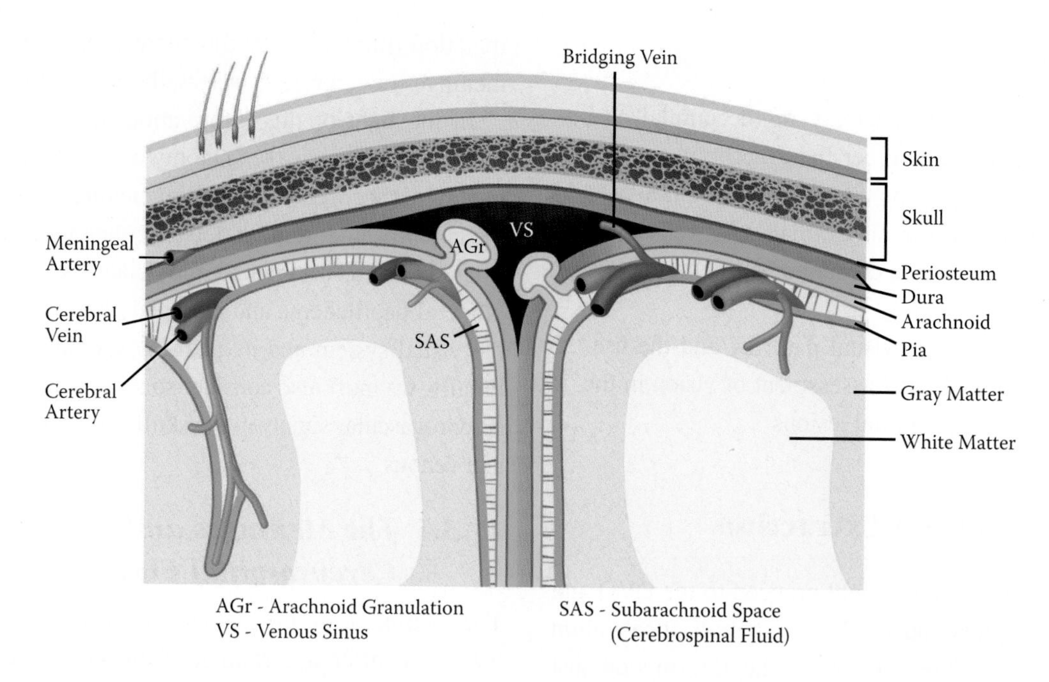

FIGURE 11.3A The meninges of the cranial cavity are shown in a coronal view, including the venous sinus (VS) and the CSF compartment. Note the bridging vein and the arachnoid granulation (AGr), as well as the location of the meningeal artery.

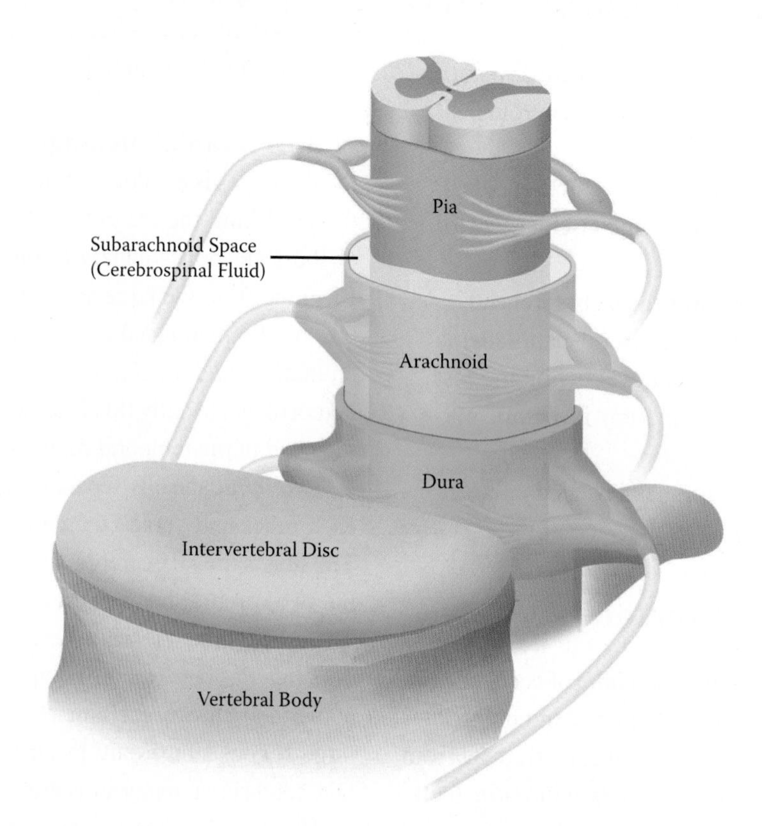

FIGURE 11.3B The spinal meninges are shown, noting that there is a space between the dura and the vertebra. The meninges, and CSF, continue just beyond the dorsal root ganglion, where the dorsal (sensory) and ventral (motor) roots unite.

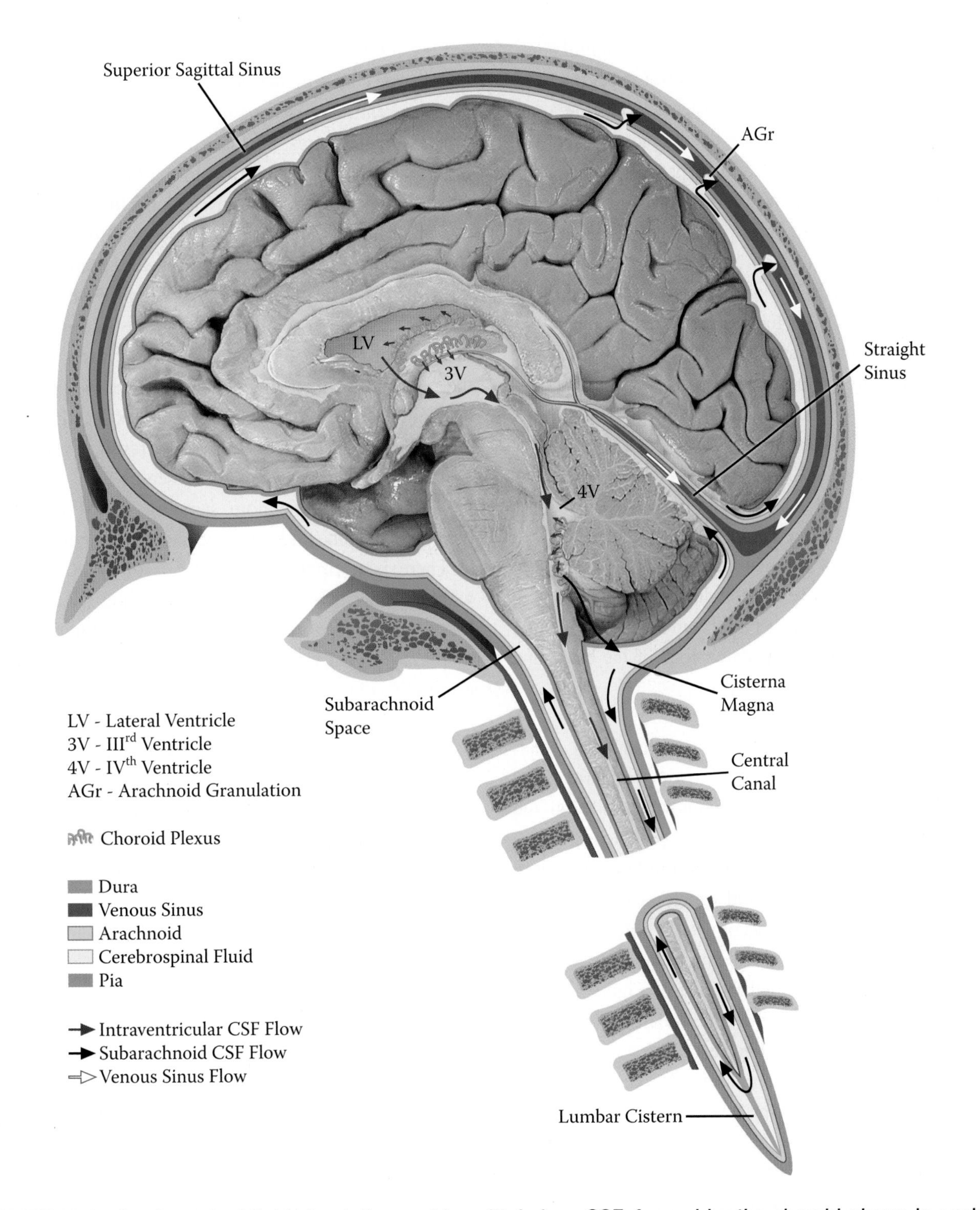

FIGURE 11.4 Cerebrospinal fluid circulation—mid-sagittal view: CSF, formed by the choroid plexus in each of the ventricles, flows though the ventricular system (as indicated by the gray-black arrows). At the lower end of the IVth ventricle, CSF enters the subarachnoid space into the cisterna magna, and flows around the brain and down and around the spinal cord (dark black arrows), and into the lumbar cistern. CSF is returned to the venous circulation via the arachnoid granulations (AGr) in the superior sagittal sinus (with white arrows).

the venous sinuses, particularly the superior sagittal sinus. It is here that they may be disrupted by trauma to the head, in which case blood leaks into this space, producing a subdural hemorrhage. This usually occurs slowly over time (subacute or chronic) but may also present acutely; this type of bleed is more common in the elderly. (The superior sagittal sinus is shown in Figure 11.3A and in mid-sagittal view in Figure 11.4.)

The innermost layer, the pia, lies on the surface of the brain and follows all its folds. The subarachnoid space, between the arachnoid and pia, contains the CSF; large arteries and veins are also found in this space. Cisterns are enlargements of the subarachnoid space, and several exist around the brainstem; the largest of these is the cisterna magna, located behind the brainstem and below the cerebellum, in the posterior cranial fossa, just above the foramen magnum (see Figure 11.4).

All three layers of the meninges, with the CSF, continue onto the optic nerve and the dural layer merges with the outermost layer of the eyeball. From the developmental perspective, the retina is an "extension" of the brain and the optic nerve is in fact a CNS tract.

11.3.1.2 Spinal Meninges The meninges continue around the spinal cord within the vertebral canal (Figure 11.3B). The spinal cord dura is separated from the periosteum of the vertebra (and the intervertebral discs) by a space, which is filled with fat in the lower vertebral region (shown in Figure 5.2) and which contains a plexus of veins. A sleeve of dura accompanies the ventral and dorsal roots of the spinal nerves, until they come together in the intervertebral foraminal region (see Figure 11.3B).

The CSF continues within the subarachnoid space around the spinal cord (see Figure 11.4). The spinal cord with its pia ends at the vertebral level of L2, whereas the dura and arachnoid end at the level of S2. The large cistern in the vertebral canal below the level of the spinal cord is called the lumbar cistern (see Figures 1.4A and 5.1, and also Figure 5.2) and is the site for sampling of CSF (discussed below).

11.3.2 The Ventricular System and CSF Circulation

The cerebral ventricles of the brain and the central canal of the spinal cord are found within the nervous tissue and are the remnants of the original neural tube from which the nervous system developed (see Figure 11.4). There are four ventricles: one (ventricles I and II) in each of the cerebral hemispheres, also called the lateral ventricles (see Figure 1.11), the IIIrd ventricle in the thalamic (diencephalic) region, and the IVth ventricle in the brainstem region.

Cerebrospinal fluid, CSF, is formed within the ventricles from specialized tissue, the choroid plexus, located in each of the ventricles. The composition of CSF differs from plasma because of a blood-CSF barrier; the site of this barrier is the cells lining the choroid plexus. Notwithstanding this, the cells lining the remainder of the ventricular system do allow a free exchange between the extracellular space of the brain and the CSF fluid. As a result, the CSF will accumulate certain proteins with diseases, such as in multiple sclerosis (MS); therefore, sampling of CSF may aid in determining some disease processes of the CNS.

There is a circulation of CSF within the ventricles, with the flow being from ventricles I and II to the IIIrd ventricle, and then via a narrow aqueduct in the midbrain to the IVth ventricle in the pontine region (gray-black arrows in Figure 11.4). At the lower end of this ventricle, CSF "escapes" into the subarachnoid space, the cerebello-pontine cistern, or cisterna magna (see above).

CSF flows from the cisterna magna within the subarachnoid space around the brain and also downward around the spinal cord where it accumulates in the lumbar cistern, below the termination of the spinal cord. It is almost always the lumbar cistern that is used to sample CSF and to take the CSF pressure, a procedure called a lumbar puncture, or LP. CSF pressure is measured when the needle initially penetrates into the lumbar cistern and is normally 7 to 19 cm H_2O or 5 to 14 mmHg.

The normal amount of CSF in the ventricles and craniospinal subarachnoid spaces is estimated to be around 150 mL and is replaced roughly every six to eight hours, which indicates that there is a continuous process of production and absorption of CSF, in effect a (slow) CSF circulation. CSF is returned to the venous circulation via the arachnoid granulations (see Figure 11.3A) which protrude into the venous sinuses, particularly into the superior sagittal sinus (white arrows in Figure 11.4). A small pressure differential is thought to account for the transport of CSF across these villi and into the venous sinuses, thus completing the circulation of CSF.

Blockage of CSF flow, for example, at the level of the midbrain where the aqueduct connecting the IIIrd with the IVth ventricle is very narrow, would cause an increase in size, and of pressure, in the lateral and IIIrd ventricles. This blockage, called obstructive or noncommunicating hydrocephalus, would be visualized with CT or MRI as an enlargement of the lateral ventricles of the hemispheres. In an adult, because the sutures of the skull are fused, this process would be accompanied by raised intracranial pressure, whereas in a young child (e.g., the first two years) with nonfused cranial sutures, the head itself will enlarge and there would be a separation of the bones of the skull; at a very early age, the anterior fontanelle would bulge.

Blockage of the CSF flow can also occur at the level of the arachnoid granulations, and in fact this does occur following meningitis (further discussed in the etiology section of this chapter). If the villi are nonfunctional or blocked, or if there is a blockage of the venous sinuses (e.g., a venous sinus thrombosis) or a stenosis of the sinuses, then CSF can no longer return to the venous circulation, resulting in an increase in CSF pressure. Hydrocephalus developing from CSF flow obstruction at a point outside the brain is called communicating hydrocephalus.

11.3.3 The Visual Pathway

Vision starts in the retina with the photoreceptors, the rods and cones. The fovea, a small area of the retina in the central axis of visual input, is the visual area required for fine vision, including reading and color vision (with the cone photoreceptors). The vast peripheral region of the retina captures peripheral vision, using the rod photoreceptors, and is used in conditions with poor illumination.

The cones and rods are specialized receptor cells located in the deepest part of the retina and are activated by light. They connect with first-order sensory neurons, also in the retina; after processing by other neurons within the retina, the messages are passed on to the ganglion cells, the second-order neurons in the visual system, which are also located in the retina. It is these neurons whose axons form the optic nerve, CN II.

The axons of the ganglion neurons exit at the optic disc, an area of the retina with no photoreceptors and hence responsible for the physiological blind spot (see Figure 11.1 and Figure 11.2, panel A). It marks the beginning of the optic nerve; CN II is in fact a pathway of the CNS, with the myelin of the nerve formed by oligodendrocytes. In its path through the orbit it is ensheathed by the meninges of the brain, with a typical subarachnoid space containing CSF. Any increase of intracranial pressure may be reflected via this space onto the optic nerve and cause its compression, as well as compromising the blood vessels supplying the retina (the arteries and veins) running within the nerve.

Pathway (Figure 11.5): After exiting the eyeball, the optic nerves cross the orbit, pass through the optic foramen, and enter the interior of skull. In the area above the pituitary gland, the nerves undergo a partial crossing (decussation) of fibers in a structure called the optic chiasm. (There is no synapse in the optic chiasm.) The fibers from the nasal retina on one side cross the midline and join with those from the temporal retina from the other eye (which do not cross) to form the optic tract. Thus, the image of one-half of the visual world which started in different parts of the retina of the two eyes, is now brought back together in the optic tract (described in detail below).

The visual world, usually called the visual field, is divided into quadrants; temporal and nasal, upper and lower, for each eye. Because of the lens, the upper visual field projects to the lower retina (and the converse for the lower visual field), while the temporal visual field projects to the nasal retina, and the nasal visual field to the temporal retina.

Using a specific example, the visual world on the left side consists of the temporal field for the left eye and the nasal field for the right eye. The temporal half of the visual field of the left eye projects to the nasal retina of the left eye, while the nasal (medial) half of the visual field of the right eye projects to the temporal (lateral) retina of the right eye. (Suggestion: making a sketch diagram of the verbal description is a simple and effective way of understanding the visual pathway.)

Within the cranial cavity, because of the optic chiasm, the information from the nasal portions of each retina—representing the temporal halves of both visual fields—crosses to the opposite side. The optic tract is now formed, bringing together information from the contralateral visual world, consisting of the temporal visual field of the contralateral eye and the nasal (medial) visual field of the ipsilateral eye.

Using our example, the right optic tract will carry information from the ipsilateral temporal retina (representing the medial or nasal visual field of the right eye projecting onto the temporal retina) and the crossed fibers

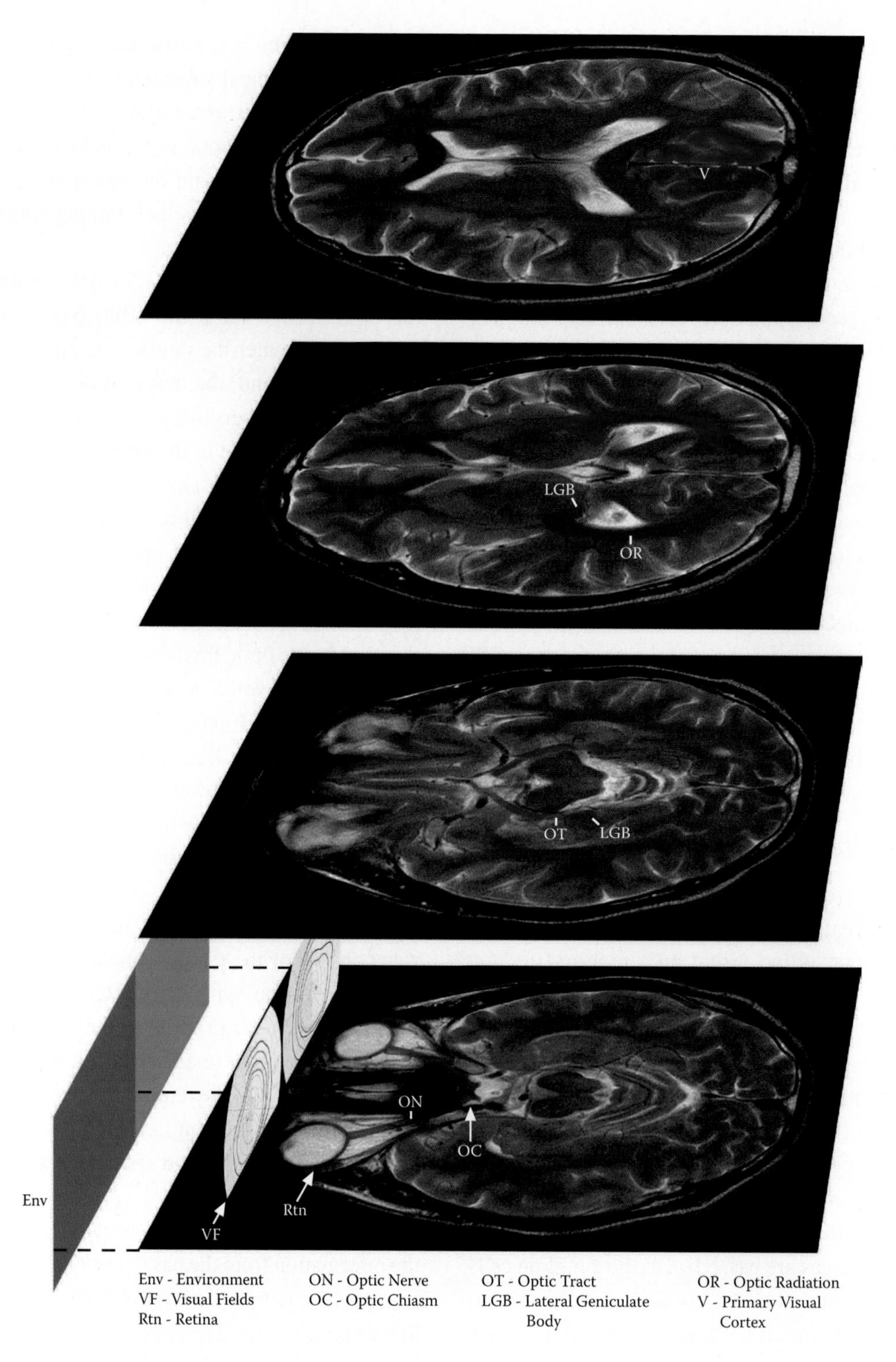

FIGURE 11.5 The visual pathway: overlay on T2-weighted MRIs of the brain; the images are in order from inferior (lowest) to superior. An image in the left visual field—of each eye—is conveyed to the retinal fields, the nasal portion in the left eye and the temporal portion of the right eye. Because of the crossing of the nasal retinal fibers in the optic chiasm, both images are united in the right optic tract. From here, the image relays in the LGB of the thalamus and projects as the optic radiation through the white matter of the hemispheres to the visual cortex in the occipital lobe. (Courtesy of Dr. R. Grover.) This pathway is animated on the text DVD and Web site, from visual fields to the primary visual cortex.

from the nasal retina of the left eye (representing the lateral or temporal visual field of the left eye projecting onto the medial retina). This explains how the right optic tract is carrying information from the contralateral left visual fields of each eye, and includes the fibers composing one half of each retina.

The visual fibers in the optic tract have a specific relay nucleus (the lateral geniculate nucleus or body, the LGN or LGB) in the thalamus; this is the site of the third-order neurons whose axons project as the optic radiations. Some of the fibers project directly posteriorly deep in the white matter of the parietal lobe to the visual cortex in the occipital lobe, the calcarine cortex; other fibers from the lower retina (from the upper visual field) project from the thalamus forward in the temporal lobe before proceeding posteriorly to the occipital cortex (a route also known as Meyer's loop). Note that the fibers that carry the information for the pupillary light reflex leave the optic tract before the thalamic relay nucleus and head for the midbrain, where the center for the light reflex is located (as described in Chapter 2, see Figure 2.2).

The final destination for the visual fibers is the cortex along the calcarine fissure of the occipital lobe, located on the medial surface of the brain; this is the primary visual area, V1 or calcarine cortex, also known as area 17. Cortical areas adjacent to the calcarine cortex further process the visual information; additional visual regions in the inferior aspect of the brain deal with specific aspects of vision, such as the recognition of faces.

Note that the visual pathway—from cornea to the calcarine cortex—extends through the whole brain (excluding the frontal lobe); hence its importance in the assessment of nervous system integrity.

11.3.4 Cerebral Vasculature

The arterial supply of the brain has been considered in Chapter 8. The larger cerebral arteries course through the subarachnoid space on the way to their destinations. Pathological localized dilations or bulges of the arterial wall of arteries, called (cerebral) aneurysms, may occur on the larger arteries, particularly of the circle of Willis; these may bleed. Usually this occurs precipitously, as a subarachnoid hemorrhage, resulting in an intense headache and frequently by a loss of consciousness. The CSF is tinged red, as it is filled with arterial blood.

At this point it is relevant to consider the venous system. The venous return from the brain courses via a deep system within the substance of the brain tissue and a superficial system over the surface of the hemispheres. Both systems drain into the venous sinuses of the skull. These sinuses are formed along the attachments of the dura to the skull bones; the dura splits to enclose these large venous spaces. The major one is the superior sagittal sinus, which is found along the upper (attached) border of the falx cerebri, in the midline.

Most of the superficial veins of the hemispheres empty into the superior sagittal sinus. This sinus continues posteriorly, and at the back of the interior of the skull it divides to become the laterally placed transverse venous sinuses, attached to the skull at the lateral edges of the tentorium cerebelli, one on each side. Venous blood exits the skull via the sigmoid sinuses, which continue as the internal jugular veins on each side of the neck.

The system of veins that drains the deep structures of the brain emerge medially as the internal cerebral veins, which join to form the great cerebral vein (of Galen) in the midline. This vein empties into the straight sinus, a midline sinus lying above the cerebellum and within the tenorium. This sinus joins with the superior sagittal sinus posteriorly as it divides, and the blood flows into the transverse sinuses.

11.4 Localization Process

Chantal's headache symptoms, the presence of papilledema, and the associated enlargement of the blind spot all clearly point to something occurring within the skull, affecting the visual system but not the motor or sensory functions, and not affecting cognition or Chantal's personality or behavior. The structures within the skull, other than brain tissue, include the meninges, cerebrospinal fluid (the ventricles and cisterns, and CSF circulation), and the arteries and veins. All of these structures must be considered in the localization exercise.

Let us now consider the localization implications for the three main clinical features of Chantal's case: headache, papilledema and enlarged blind spots.

11.4.1 Headache

An analysis of headache characteristics by itself does not usually contribute much to the localization of a neurological problem. On occasion, a localized head pain will reflect the presence of pathology in the immediate region of the pain, for example, a left frontal lobe tumor in someone with left frontal head pain, the stretching of the regional meninges and major blood vessels stimulating selected CN V sensory fibers whose input can be localized by thalamic VPM and somatosensory cortex. In general, however, a bilateral frontal headache like Chantal's could result from pathology anywhere inside the head.

11.4.2 Papilledema

Optic disc edema may result from a variety of causes, including optic nerve vascular compromise, inflammation, infiltration and infection. Raised intracranial pressure is also transmitted to the optic nerves, resulting in optic disc edema. Optic disc edema caused by raised intracranial pressure is known as papilledema. Because the raised intracranial pressure is usually transmitted to both optic nerves, papilledema is almost always bilateral; in contrast, optic nerve inflammation (papillitis) is often unilateral. The cause of this increased pressure could be located anywhere within the skull.

In its early stages, papilledema is not typically associated with significant impairment in visual acuity. In contrast, other causes of optic disc edema commonly produce early, severe loss of visual acuity. In Chantal's case, the finding of bilateral optic disc edema with normal visual acuity is consistent with a diagnosis of papilledema due to elevated intracranial pressure.

In summary, bilateral papilledema without any obvious impairment in visual acuity suggests the presence of increased intracranial pressure but does not, by itself, help one to localize the cause of the pressure elevation.

11.4.3 Enlarged Blind Spots

Having previously reviewed the anatomy of the visual pathways, we must determine the part of that pathway that, if damaged, would present with selective enlargement of the blind spots. By a rapid process of elimination, this turns out to be quite simple. A unilateral optic nerve lesion would result in problems in one eye (e.g., complete loss of vision or a large central scotoma); bilateral optic nerve lesions would produce, if sufficiently severe, almost total blindness. Lesions of the optic chiasm affect vision in the temporal fields (the crossed fibers) while sparing the nasal fields, a visual defect known as bitemporal hemianopia. Lesions of the optic tract, the lateral geniculate bodies or the optic radiations would produce a homonymous visual field defect on the side opposite to the hemisphere affected (otherwise known as a homonymous hemianopia).

Clearly none of the above patterns of visual deficits correspond to those of our patient. With long-lasting papilledema, while there is no loss of visual acuity, the size of the blind spot is eventually increased and peripheral vision becomes progressively restricted; this is quite difficult to detect during a routine clinical neurological examination. Thus, Chantal's visual field abnormalities are likely related to the presence of long-standing raised intracranial pressure.

In summary, the absence of any focal neurological deficit does not enable one to localize the problem to any specific area of the brain. The location, however, for the chronic state that led to the papilledema must be intracranial, leading to an increase in intracranial pressure (ICP). This includes any expanding intracranial mass, often called a space-occupying lesion (to be discussed below). Clearly, however, if Chantal has such a mass, it cannot be located within or impinging upon CN II, the optic chiasm, optic tracts, thalamus, optic radiations or the visual cortex.

There is no indication, as yet and without any investigation, as to where the problem is localized within the cranial cavity. Raised intracranial pressure may result from a volumetric enlargement of any one of the following intracranial components (or a combination thereof), including:

- The brain tissue
- The meninges
- The vasculature, either arterial or venous
- The ventricles and/or the quantity or pressure of CSF (CSF circulation)

In other words, any pathophysiology or lesion that leads to an increase in the volume of the contents of the intracranial cavity will cause a mass effect and lead to an increase in intracranial pressure (given that the cranial cavity is a closed "box" of fixed size in the adult). As we will see, preoccupation with the possibility of a space-occupying lesion may divert one's thinking away from some intracranial pathology that would not fit under the category of a mass lesion (discussed in Section 11.5). An LP is necessary to measure whether there is an increase in CSF pressure (discussed in Section 11.8).

11.5 Etiology: Chantal's Disease Process

In considering the possible etiology in this patient, we are now certain that the raised intracranial pressure is of long-standing duration, that is, chronic, whereas the trauma is an acute event. Papilledema takes time to develop and one would not associate its presence with a lesion that occurred acutely, such as Chantal hitting her head as a result of the incident the previous night. In other words, as causes of papilledema, paroxysmal and traumatic disorders are ruled out. Of the various chronic disease categories, which ones could be associated with an increase in intracranial pressure (see Table 3.13)? The etiologic possibilities associated with the acute event will be considered subsequently.

Disease processes related to the meninges should also be taken into account when discussing raised intracranial pressure, including any interference with CSF production, circulation or reabsorption.

11.5.1 Vascular

In the description of the meninges, mention was made of the possibility of bleeding between the various layers. Although vascular disease processes usually produce acute-onset symptoms, bleeding from the veins draining the brain may occur over a prolonged period of time in the form of a chronic subdural hematoma; again, this is a space-occupying lesion that might give rise to headache and raised intracranial pressure. This condition usually affects the elderly, possibly because of the fragility of the veins at this age associated with an age-related shrinkage of the brain, and may occur following even a minor head trauma. A chronic subdural hematoma is most unlikely in a 40-year-old woman without a history of a remote head injury or of alcoholism. The possibility of an acute subdural will have to be revisited because of the history of the acute head trauma.

Another vascular-related possibility is that CSF circulation can be obstructed due to a problem with CSF reabsorption; this could be impaired because of a reduction in venous return, such as a venous sinus thrombosis, leading to an increase in intracranial pressure.

Spontaneous venous sinus thrombosis can occur in individuals with a high fever, a hereditary predisposition to clotting (e.g., among many disorders, protein C and protein S deficiencies; factor V Leiden mutations), or with chronic autoimmune vasculitides (see below). Recent studies indicate that narrowing or stenosis of a venous sinus, rather than complete occlusion, may be associated with intracranial hypertension.

11.5.2 Toxic

Under toxic conditions, it is well known that a drug used for the treatment of acne, isotretinoin (a form of vitamin A), has been associated with an increase in intracranial pressure; Chantal is beyond the usual age for this condition.

It should be noted that the following agents have also been associated with otherwise unexplained intracranial hypertension: tetracycline and chlortetracycline (both used to treat acne), nitrofurantoin and nalidixic acid (used to treat urinary tract infections), and oral contraceptive agents. Chantal is not using any of these medications.

11.5.3 Infectious

Meningitis is an inflammation of the meninges accompanied by an inflammatory reaction of neutrophiles (with a bacterial infection) or lymphocytes (with a viral infection) which are found in the CSF. Although most cases are acute, bacterial meningitis can be subacute to chronic in nature. Sometimes these infectious processes can interfere with the CSF circulation, that is, reabsorption, by obstructing the arachnoid granulations in the walls of the venous sinuses (discussed above).

A chronic intracerebral abscess from an infectious source (e.g., from an infection on the heart valves) may cause a cerebral mass anywhere within the brain; there is, however, no history of fever or of an infectious process elsewhere in the body in this case, and no heart murmur (which might be expected, for example, in the context of rheumatic fever with a diseased, secondarily infected mitral valve). Parasitic cysts, sometimes multiple, can also occur within the brain and produce raised intracranial pressure (e.g., cysticercosis). Such diseases are relatively rare in North America but common in Central and South America, and in Africa. In North America parasitic infections of the brain must be considered in recent immigrants from endemic areas who present with raised intracranial pressure. Parasitic infection is highly unlikely in this case as Chantal has never travelled abroad, and these usually present with an acute-onset epileptic seizure (see below).

11.5.4 Metabolic Conditions

Metabolic conditions are not known to lead to raised intracranial pressure, unless associated with brain edema.

11.5.5 Inflammatory/Autoimmune

Inflammatory/autoimmune disease processes that would give rise to chronic increased intracranial pressure without focal neurological signs are very rare, and do so through spontaneous clotting in the cerebral venous sinuses (see above); examples include systemic lupus erythematosus and Behçet's disease. Acute disseminated encephalomyelitis (ADEM) often presents with raised intracranial pressure (see Chapter 10) but, as the name implies, does so acutely (days) and with focal neurological signs or diminished consciousness.

11.5.6 Neoplastic

With the chronic history in this case, the possibility of an intracranial tumor must be considered, either primary (a meningioma or glioma) or secondary (a metastasis from elsewhere). A tumor may produce increased intracranial pressure either because of its size or because it obstructs the circulation of the cerebrospinal fluid within the ventricular system, leading to progressive ventricular distension (discussed previously). In addition, one could explain the accident as having occurred because of a brief seizure caused by a tumor (to be discussed below). Neuroimaging studies will be needed to rule out this possibility.

11.5.7 Degenerative Conditions

Degenerative brain conditions would not likely be associated with an increase in intracranial pressure. In the elderly, where there is significant cerebral atrophy, an obstruction to CSF flow may produce progressive hydrocephalus without symptoms of raised intracranial pressure or papilledema; the atrophic brain simply collapses or distorts to accommodate the enlarged CSF compartment. This disorder is known as normal pressure hydrocephalus and typically presents with urinary incontinence, gait apraxia and dementia (see Chapter 9).

11.5.8 Space-Occupying Lesions

In general, a brain abscess, cyst, or tumor, as space-occupying lesions, would produce a mass effect within the cranial cavity and lead to increased intracranial pressure and bilateral papilledema. It would be expected, however, that these lesions would produce focal signs, some indication of parenchymal (brain) pathology, such as a sensory, motor or language deficit; none of these have been found clinically. Association regions of the brain, such as the frontal lobes, could harbor a lesion without producing any focal or language deficit. A lesion in this area might be linked with a change in personality or behavior and there has not been any hint of this, insofar as one can tell from interviewing Chantal.

11.5.9 Cerebrospinal Fluid Circulation Disturbance

The CSF and its circulation need to be considered in arriving at an etiologic diagnosis since pathology here could occur quite possibly without any focal signs. Any enlargement of the cerebral ventricles due to an obstruction (e.g., a tumor) to the CSF flow within the ventricular system would lead to an increase in intracranial pressure and to papilledema; this would be readily seen with neuroimaging as enlarged ventricles.

An alteration of CSF pressure, caused by an increased production or some change in the flow dynamics of the CSF, could lead to an increase in intracranial pressure, and this would not necessarily be seen with routine CT or MRI. There is a disease entity that fits with this category, known as idiopathic intracranial hypertension, IIH, (*idiopathic* meaning without known cause); this disease was previously called pseudotumor cerebri, as it mimics a tumor in producing an increase in intracranial pressure.

11.5.10 The Acute Incident

With respect to the acute event the previous evening, there are a number of additional possibilities that need to be considered. Under acute (see Table 3.13), one should consider paroxysmal and vascular etiologies, in addition to the obvious traumatic event.

11.5.10.1 Paroxysmal Could there have been an epileptic seizure that caused Chantal to lose control of her automobile? Isolated seizures do occur, often as a prelude to or indicative of other diseases. For example, tumors may cause irritation and/or compression of adjacent normal tissue and do cause seizures. Bleeding into a tumor may lead to its rapid enlargement and cause a seizure. Under such circumstances, however, Chantal would not have retuned to normal behavior by the time she was helped out of her car. In addition, the fact that Chantal remembers the car sliding off the road into the ditch argues against there having been a seizure.

11.5.10.2 Traumatic Injury The presence of a scalp bruise suggests that an additional intracerebral lesion could have occurred due to the motor vehicle incident, since it seems that there was a period of unconsciousness, of unknown duration. One must consider the possibility of some degree of brain bruising, a concussion, however minor. Any brain concussion can lead to headaches and difficulty with concentration, and occasionally mood and behavior problems, lasting from days to months. In Chantal's case, however, the headaches developed long before the head injury occurred.

Not to be forgotten is the possibility of an acute subdural associated with the head trauma, but this would be expected to lead to a rapid deterioration in her level of consciousness over several hours. Neuroimaging would assist in sorting this out.

11.5.10.3 Vascular Disorders Could there have been a vascular event causing a brief loss of consciousness? It would be important to ask Chantal whether she is taking any medication, (including birth control pills), over-the-counter medication, or herbal remedies. There is a known increased risk of cerebral vascular occlusion leading to a transient ischemic attack (a TIA) or a stroke in women on birth control pills, especially if that person is also a smoker. Nothing in the history or physical suggests a TIA or stroke.

Could there have been an acute bleed either from an aneurysm (a subarachnoid hemorrhage) or from an arteriovenous malformation, with a brief loss of consciousness? Both of these would result in an acute intense headache and a patient who is obviously ill. Neither of these possibilities is at all likely in view of the patient's normal mental status and relatively benign neurological examination the following morning. Besides, none of these vascular events would account for the bilateral papilledema.

A more urgent consideration is an intracranial bleed from a tearing of the middle meningeal artery, which usually occurs with a head injury on the side of the head (in the temporal region), resulting in an epidural hemorrhage. The typical case has a brief period of unconsciousness followed by a lucid interval before a rapid deterioration in consciousness, due to the acute arterial bleeding inside the skull. With an epidural hemorrhage, however, Chantal would not have remained conscious for more than one to two hours.

11.5.10.4 Infectious Some acute infectious diseases may produce raised intracranial pressure and papilledema, the classic example being acute hemorrhagic leukoencephalitis. Again, our patient would be seriously ill by this point.

11.6 Case Summary

In summary, the major finding in this case is papilledema, which needs time to develop. Its cause is an increase in intracranial pressure due to either a space-occupying lesion, or to a process interfering with the normal flow of cerebrospinal fluid (including its reabsorption). Out of the various etiological possibilities, the following emerge as the most likely

- Intracranial hypertension caused by a space-occupying lesion
- Venous sinus thrombosis associated with intracranial hypertension
- Intracranial hypertension, cause unknown (idiopathic)

11.7 Evolution of the Case

Since none of the above etiological possibilities can be entirely eliminated simply on the basis of logic or reasoning, further investigations are required to specify the disease causing the papilledema so that one can hopefully find a way to manage the problem before it worsens.

The cause for the vehicular incident and the associated brief loss of consciousness remains speculative at this time, although both Chantal's visual field deficit and the weather seem to be the major contributing factors.

Could Chantal's visual deficit, with her enlarged blind spots, have caused her to miss seeing the road sign? This is an interesting possibility and might be considered should there have been any significant damage to her vehicle

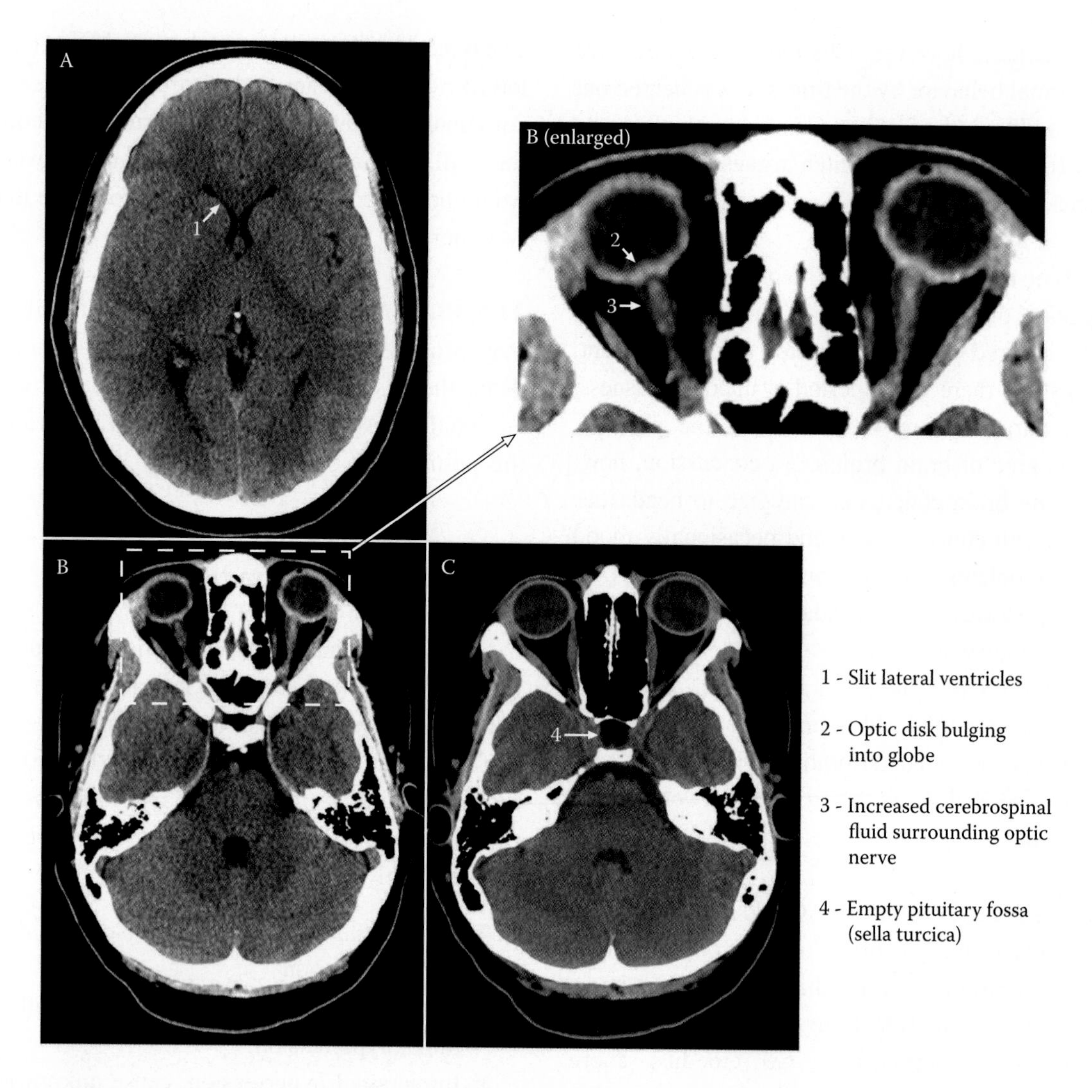

FIGURE 11.6 Radiographs: CT scan of a patient with idiopathic intracranial hypertension. (A) The lateral (cerebral) ventricles appear slit-like (1). (B) The globe of the eye and the optic nerve are shown, within the orbit. (B, enlarged) The head of the optic nerve, the optic disc, is pushed into the eyeball (2), as compared to its normal appearance; the sclera is also flattened posteriorly. The subarachnoid space (with CSF) surrounding the optic nerve is enlarged (3). The pituitary fossa (sella turcica) appears "empty" (4). (Courtesy of Dr. C. Torres.)

resulting in an insurance claim. On a more legal note, if Chantal would have been involved in an accident causing damage to another car and/or injury to a person, the neurologist and ophthalmologist might have been asked to give evidence as to the significance of the visual impairment in the causation of the accident.

It is also possible that the vehicular incident is in fact a true accident caused by the poor driving conditions that night and by the usual reduction of vision with night driving, especially in the countryside.

In any event, Chantal has been ordered off the road, that is, not permitted to drive a car; the Ministry of Transportation of the province was notified concerning her visual field deficit. Although this had a serious impact on Chantal's life in the short term, she was assured by friends that they would take her to places most times and they did; at other times she had to use a taxi or public transportation.

11.8 Investigations

A CT scan was ordered immediately (see Figure 11.6) and the following was reported: no intracranial mass was seen and the ventricles (lateral and IIIrd) appeared reduced in size with a normal configuration. This eliminates several of the possibilities, including a tumor or obstructive hydrocephalus. In addition, the CT revealed swelling of the optic nerve heads and an "empty" sella turcica (the pituitary fossa), consistent with the diagnosis of *idiopathic intracranial hypertension* (IIH).

Based on the normal CT, it was decided that an LP investigation must be done next. Note that in this clinical context the CT study must precede the LP, because should there be an intracranial mass, the LP, by reducing the pressure below, may cause the brain to descend and herniate into the foramen magnum of the skull. Such an event would impair brainstem regulation of respiration and cardiac control, thereby endangering the life of the patient.

An LP was performed with some difficulty (due to Chantal's obesity) but successfully. The opening pressure was measured at 42 cm H_2O (31 mmHg); the CSF was clear and colorless. Laboratory testing of the CSF was entirely within normal limits, with a normal glucose level, no white blood cells, and a normal protein concentration.

Additional neuroimaging was carried out a few days later, including an MRI, an MRA (a magnetic resonance arteriogram), and an MRV (an MR of the cerebral venous sinuses).

- MRI: The results confirmed all of the findings on the CT scan; the lateral ventricles were described as "slit-like."
- MRA: This is an MRI of the major blood vessels of the circle of Willis, often using contrast enhancement. No abnormalities were seen in the vessels of the circle of Willis.
- MRV: The view of the major venous sinuses was obtained later during the same examination. No occlusion or abnormality of the venous sinuses was seen.

11.9 Diagnosis, Management, and Outcome

At this stage, the most likely diagnosis, by a process of exclusion, appears to be idiopathic intracranial hypertension, a disease of still unknown pathophysiology but found to occur predominantly in very obese women between the ages of 20 and 40, often associated with systemic vascular hypertension.

As this disease is being investigated more thoroughly using current imaging techniques, more and more cases previously diagnosed as idiopathic are now being found to be due to a venous sinus thrombosis or some structural abnormality of the venous sinuses. Nevertheless, the association of this disease entity with obesity and the female gender remains unexplained.

Having eliminated all possibilities, we are left with a disease that has no "known" cause. The major danger of progression of the disease process, that is, the papilledema, is irreversible blindness in up to 10 percent of these cases. Treatment must be actively considered in order to reduce the pressure on the optic nerve or to relieve the intracranial hypertension, as well as the associated headache.

Pharmacologic reduction of the production of CSF may be helpful, such as the use of a carbonic anhydrase inhibitor, acetazolamide (Diamox™). In emergency situations one could consider doing repeated therapeutic LPs to relieve the CSF pressure, followed eventually by the shunting of CSF, either from the lateral ventricles or from the lumbar cistern to the peritoneal cavity.

In order to reduce the pressure on the optic nerve a surgical procedure has been devised which makes small holes in the dura and arachnoid of the meningeal sleeve extending along the optic nerve. This operation is called an optic nerve sheath fenestration. Fortunately, the ophthalmologist who saw Chantal is trained in this procedure and has offered to do it almost immediately (within four to six days). Had Chantal's MRV demonstrated a significant transverse venous sinus stenosis, an alternative approach to treatment would have been the placement of a stent in the stenotic region by an interventional neuroradiologist.

After hearing an explanation of what optic nerve sheath fenestration involves and its risks, and weighing her options, which include the possibility of progressive blindness, Chantal gives her consent to the surgical intervention. The operation is done and considered successful, with no ocular complications.

In addition to this emergency surgery, Chantal starts counseling sessions with a nutritionist so that she can lose weight and eat properly to sustain the reduced weight status. Along with this, she is advised by a physiotherapist on an exercise program in order to maintain the weight loss. On the advice of friends, she hires a personal trainer for a six-month period until she develops the lifestyle modifications necessary to change her way of living and cope better with stress. Both measures have been found helpful not only to alleviate the intracranial hypertension, but also to reduce the vascular hypertension. She also enrolls in a smoking cessation program and successfully reduces her smoking habit, hoping to quit entirely by the end of the one-year contract.

All of this has been accomplished with tremendous support from her friends and "family."

Six months following the accident, Chantal visits her family doctor for follow-up. She now has lost over 50 kg, her blood pressure is now 125/85, and she is a much changed person, both in her appearance and in her attitude. The headaches are all but gone and the physician discontinues the antihypertensive medication (tapering over a period of four weeks).

She is also seen at about the same time by the ophthalmologist, who notes that the disc margins are still not sharp but that there is less venous engorgement. The Goldmann perimetry is redone and there has been no further progression in the size of the blind spot.

Her driving license is restored but with the restriction that she is not to drive at night (starting 30 minutes before sunset and extending to 30 minutes after sunrise); Chantal readily complies with this limitation.

Chantal will be followed on a regular basis by both physicians. One wonders what life will have in store for her in the future.

Related Cases

There are e-cases related to this chapter available for review on the text DVD and Web site.

Suggested Readings

Ball, A.K. and C.E. Clarke. Idiopathic intracranial hypertension. *Lancet Neurology* 5 (2006): 433–442.

Friedman, D.I. Idiopathic intracranial hypertension. In *Neurology and Clinical Neuroscience,* A.H.V. Schapira, Ed. 807–816. St Louis, MO: Mosby, 2007.

See also Annotated Bibliography.

Chapter 12

Mickey

MICKEY

A couple of weeks ago Fifi came to see you for an evaluation of her state of recovery from the Guillain-Barré syndrome (she is functioning normally but still demonstrates residual weakness of dorsiflexion of both feet and complete areflexia nine months following the onset of her illness). At the end of the visit Fifi asks if there is any way you could arrange to see her brother-in-law, Michael, with respect to his rage attacks.

Michael McCool ("Mickey") is 21 and still lives with his parents. The McCool's family physician, experiencing burn-out, closed her practice a few months ago and went to work for Health Canada. Thus far, Mickey's parents have not been able to find another physician and are at their wit's end over what to do about their son's behavior.

Fortunately you have a cancellation for this afternoon and are able to spend some time with Mickey and his mother. Mickey's father, Lt. Col. Walt McCool, is unfortunately out of the country evaluating the ongoing logistical requirements for one of the current missions of the Armed Forces.

The McCools have four children, of whom Crash is the oldest; Mickey came along relatively late in his parents' lives as something of an afterthought.

Concerns about Mickey's behavior have been hanging over the family like storm clouds for years. Although he was born at term without difficulty and appeared normal as a toddler, he began to show signs of hyperactive behavior by the time he was four. At that time his parents found him to be fidgety and restless, always on the go. He never seemed to like to play quietly with his toys but was invariably outside in all weathers racing around on his scooter or tricycle from dawn to dusk. When he started kindergarten his teacher found him difficult to deal with; he had trouble staying in his seat and frequently disrupted the rest of the class with vocal outbursts.

During the kindergarten years Mickey's restless behavior was managed, barely, with the help of teacher's aides to help keep him on task. Once he got to grade one, however, his school progress began to suffer because he simply could not concentrate on a specific assignment for more than five minutes. Typically he would get distracted by extraneous events such as his friend Piggy Martin scratching his backside, air bubbling in the classroom fish tank, and city buses passing by the classroom window every few minutes.

Concerns about Mickey's inattentiveness and poor school performance prompted a psychological assessment at the end of second grade. The school board psychologist found that Mickey's overall intellectual abilities were in the high average range but that there was a marked discrepancy in subtest scores: verbal subtest scores were in the superior range while nonverbal scores were in the low average to borderline range. Furthermore, the Connors Rating Scale demonstrated findings consistent with the diagnosis of attention-deficit/hyperactivity disorder.

At the psychologist's suggestion, Mickey was started on the stimulant medication methylphenidate (Ritalin™), with doses given on school days only, at breakfast and lunchtime. The use of this medication was associated with a significant improvement in behavior: Mickey became quite calm, at least during school hours; he ceased disrupting the class and was able to complete assignments successfully without being distracted. The baseline level of activity and distractibility returned in the evenings and, of course, was present on weekends. By the middle of grade three, thanks to the methylphenidate, he had caught up to his classmates in all subjects except mathematics. While he had no difficulty with abstract math concepts he found written math problems a daunting prospect.

At age eight, Mickey developed a new set of symptoms in the form of nervous tics. The problem appeared gradually over several months, and took the form of repetitive blinking, nose wrinkling, mouth opening and brusque turning movements of the head to one side. At the same time, Mickey began to produce a variety of noises, including grunting, throat-clearing, and repetitive sniffing. Eventually all of these stereotyped movements and noises occurred on a daily basis, waxing and waning in frequency, in varying combinations. They occurred more frequently in the evenings, when he was tired, and when he was nervous or stressed. When his exasperated parents told him to "stop making those silly noises," he would comply, but never for more than five minutes, following which they would return. At school, some of his classmates began making fun of him, and he was the object of occasional bullying by older students.

When the tics became a major issue for the parents, the family physician wondered whether the methylphenidate might be the cause of the problem and suggested stopping the medication. Unfortunately, this did not help matters; the tics continued unabated and there was a catastrophic deterioration in school performance accompanied by marked restlessness, verbal aggression toward teachers and other students, and frequent fighting in the schoolyard. Within two months the experiment was terminated and the stimulant medication restarted. School performance returned to the previous level, but the arguments and the fighting continued, leading to visits to the principal's office and occasional day- to week-long expulsions from school.

Pressured by the parents to "do something" about the tics, the family physician tried Mickey on small doses of haloperidol, a medication originally introduced to treat violent behavior in psychotic individuals. This produced a major reduction in tic frequency but made Mickey sleepy and lethargic, "like a zombie." Again, his school work suffered, so the medication was abandoned.

After the age of 12, the frequency of the movement and voice tics began to decline spontaneously. In the classroom they were rarely seen as Mickey had learned to control them, more or less, until he got home, at which point there was a veritable explosion of twitching, blinking, and noise for an hour or so. As the tics decreased, they appeared to be replaced by a number of more complex stereotypic or compulsive behaviors. These included constant twiddling of pencils and pens, thrumming his fingers on his desk, compulsive manipulation of knobs on electronic equipment and cooking appliances, and shooting imaginary baskets above each door he passed through (Mickey had become very accomplished at basketball and was a valued member of the school team, which helped to offset the damage to his self-esteem resulting from his classroom performance).

For a while, in his early teens, Mickey's image at school improved considerably. He had learned, for the most part, to avoid confrontations and to stay out of fights. He had a very bubbly personality, a quick wit, and an ability to make friends easily (although keeping friends was more difficult). He continued to keep up academically by dint of hard work, and with the help of a mathematics tutor.

At 17, however, the behavioral problems grew worse again, particularly at home. As often happens with teenagers, Mickey grew increasingly exasperated with his parents who, he felt, were trying to control him too much. They were always taking him to task in front of his friends and were obliging him to spend long hours doing homework assignments when he wanted to be shooting baskets at the community center. Mickey began having frequent shouting matches with his parents, particularly his mother (his father was often absent on military business). These arguments gradually transformed into sustained outbursts of violent behavior during which Mickey would scream and swear, kicking holes in the wall and smashing furniture. Although the violence was never directed toward his mother, she became increasingly afraid of him. After the explosions of anger were over, Mickey was always very contrite and filled with self-loathing.

At the same time, the parents became concerned about increasing obsessive thinking and behaviors, which reached the point of becoming disabling. Mickey constantly counted and checked things, such as his video game collection, his colored pencils and the ranking of his favorite NBA teams in the newspaper. At bedtime he had an invariable ritual consisting of checking the locks on the front and back doors, then checking under his bed and in his clothes closet, and finally checking to make sure the bedroom window blinds were completely closed. When doing assigned homework he would often read the same sentence over and over, claiming not to understand what it meant, and loudly proclaiming that he is "stupid." Attempts by the parents to interfere with his rituals are one cause of his rage attacks.

At age 18, Mickey dropped out of his final year of high school, stating that he hated it there and always had. On occasion he has worked for a few weeks at a time (in grocery stores or gas stations) but has invariably been fired for insubordination. He has also tried taking courses in adult education programs but always drops out after attending two or three classes. Recently he has been refusing to look for work, stating that he is "too stupid" and "can't do anything useful." Most of his days are spent alone in his room playing video games, sometimes until the wee hours of the morning. He no longer goes out to play basketball and socializes with his friends less and less. It is at this point that Fifi is approached to try to arrange a medical reevaluation.

Before closing her practice, the previous family physician had tried Mickey on two different medications in an attempt to deal with aspects of his behavior. To reduce the violent rage outbursts she tried him on clonidine; this produced a modest reduction in rage attacks but made him so sleepy that he eventually refused to take it. The obsessive behaviors were approached with a trial of the selective serotonin reuptake inhibitor paroxetine (Paxil™). Unfortunately this made him very agitated, even in the presence of methylphenidate, and was withdrawn after a month.

After gleaning all of this historical information from Mrs. McCool, you inquire about other medical problems. It turns out that Mickey has generally been in good health all his life. He has never required hospitalization for any reason and has never had any convulsions, febrile or unprovoked. His only visits to hospital emergency departments have been for scalp lacerations and for a broken finger incurred while playing basketball. He has knocked heads a few times while driving for the hoop but has never had a serious head injury.

The family history turns out to be potentially relevant (beyond what you already know!). Big brother Crash and Walt McCool were both apparently pretty hyperactive as children, although not to the same extent as Mickey. In addition, Mrs. McCool recalls that Crash had a few nervous tics when he was in grade school: a nose-wrinkling tic and a compulsive tendency to clear his throat when nervous. The tics disappeared by the time Crash entered high school. As well, Lt. Col. McCool has always had a tendency to blink rapidly when he is stressed. Despite having had some initial difficulties in school, both Crash and Walt are well educated, with degrees in aeronautical engineering and business administration, respectively. Finally, Mrs. McCool, an accountant by training, is something of a cleanliness nut, keeping her entire house (minus Mickey's room) spotless, vacuuming all the carpets in the house and washing the kitchen floor every day of the week. This practice contributes to the friction with Mickey who, if permitted, would file his socks under the bed and his shoes all over the house.

During your assessment of Mickey you notice that he is fidgety, constantly crossing and recrossing his legs and scratching his scalp, particularly when listening to his mother describing his wall-breaking proclivities. He still has the occasional tic in the form of a rapid head flick to the right. Throughout the visit he is very pleasant and cooperative, often quite amusing and insightful. His general physical and neurological examinations are completely normal; there are no dysmorphic features.

Objectives

- To learn the principal anatomical components of the central nervous system involved in the maintenance of attention, impulse control, learning, and mood regulation
- To review the neuromodulatory transmitter systems implicated in memory, learning, mood, and behavior
- To develop the ability to distinguish the various causes of long-standing brain dysfunction presenting with learning and behavioral disturbances
- To learn the main criteria for the diagnosis of some common developmental behavior disorders
- To review the clinical management of selected disorders of attention, impulse control, mood, and emotion

12.1 Clinical Data Extraction

At this point, please proceed to the DVD and complete the worksheets. Once you have developed hypotheses concerning the region of the nervous system implicated, the probable disease categories, and a possible diagnosis or diagnoses, you are ready to continue with the text.

12.2 The Main Clinical Points

From Mrs. McCool's perspective, the chief complaints would be inattention, learning problems, obsessiveness and violent behavior, all of long duration. She might also mention nervous tics or "habit movements" but, a decade after their decline, may well have forgotten about the problem.

From Mickey's standpoint, the chief complaint, assuming that you remember to ask him, is that he is "stupid and can't do anything right!"

The key findings in Mickey's history are as follows:

- A long-standing problem with attention span and impulse control, associated with hyperactive behavior, dating back to age four
- Stereotypic uncontrolled facial and limb movements, as well as vocalizations, dating from age 8, peaking at around age 12, then declining but never entirely disappearing
- Complex stereotypic compulsive behaviors, beginning in the early teens, and persisting
- Selective learning difficulties, particularly for mathematics, despite normal to superior intelligence, dating from the kindergarten years and persisting
- Obsessive thinking and disabling complex rituals, beginning in the mid-teens
- Attacks of uncontrollable rage triggered by arguments over trivial matters, or by the parents attempts to stop his rituals, also dating from the mid-teens
- Poor self-esteem, long-standing
- Positive family history of hyperactivity, tics, and obsessive-compulsive behaviors
- Except for the occasional motor tic, a completely normal neurological examination

12.2.1 Characterizing Mickey's Disorder

We can begin by carefully analyzing the main features of Mickey's history and attempting to categorize his main symptoms:

1. *Impaired selective attention*—Mickey has trouble concentrating on one task to the exclusion of others. He is thus easily distracted, witness the difficulty he had doing his classroom work when city buses went by the windows.
2. *Poor impulse control*—this is really a corollary of impaired attention in that, if you are unable to ignore extraneous stimuli while attempting a task, you are more likely to respond to the stimuli and proceed to do something irrelevant to the task at hand. Thus, Mickey might abandon the solution of a math problem, get out of his chair and go to the classroom window, disrupting classroom activities in the process. Persistent disruptive, irrelevant activity of this type is what is meant by the rather vague but popular term *hyperactivity*.
3. *A tic disorder*—a tic is defined as a stereotypic complex movement or vocalization involving the synchronous, organized action of many muscle groups, over which the individual has imperfect control. Tic-associated movements such as jaw-opening or head-turning would be considered normal purposeful movements under certain circumstances but, as tics, are inappropriate and irrelevant to the existing circumstances, for example, being required to give an oral presentation in class. Tics can be voluntarily suppressed, on request, but tend to resume as soon as the person is distracted—an obvious linkage to impaired selective attention. Given that tics can be suppressed, they are clearly distinct from choreic movements (encountered in Chapter 7), which cannot. In addition, choreic movements are more disorganized—chaotic—and would not be considered as "normal" movements under any circumstances.
4. *A learning disorder*—defined as the state of being two or more academic years behind age level in learning specific material (e.g., reading, mathematics) despite normal intelligence and appropriate opportunities to learn. We have seen that Mickey has measured intellectual ability in the high average range, but has great difficulty learning math. Obviously Mickey's poor attention span has contributed to his learning difficulties but this problem alone would not explain why Mickey cannot do math problems despite being able to read at his age level.
5. *Compulsive behaviors*—whether shooting imaginary baskets, twiddling knobs, counting colored pencils or repeatedly checking the bedroom closet for interlopers, these behaviors are, in a sense, uncontrolled actions akin to motor tics, but far more complex. They could be construed simply as manifestations of poor impulse control, but stem from an inner, dysfunctional drive rather than an uninhibited response to an irrelevant external stimulus.
6. *Obsessive thinking*—in essence, this term refers to unwanted, recurrent thoughts that cannot be suppressed and therefore interfere with normal thinking and reasoning. In being unwanted and

poorly suppressed they are the "thinking" equivalent of a motor or vocal tic. Far more than tics, however, obsessions, translated into the kinds of compulsive behaviors described above, interfere with the normal activities of daily life.

7. *Rage attacks*—as explosions of furniture smashing, these attacks are essentially a combination of unbridled anger and poor impulse control. The new element they introduce to Mickey's smorgasbord of behavioral problems is a disturbance of mood control. Mood is defined as a state of mind or feelings, the predominant examples being happiness, sadness, anger, anxiety, and tranquility. We all live with fluctuations in these feelings, some predominant one day, others the next; overall, however, they remain in a state of balance, without one mood predominating over time. In Mickey's case, anger is never far from the surface and, once in the foreground, cannot be resolved without resorting to antisocial and destructive behavior.

If we distill this list of symptoms further, we can postulate that Mickey's core "pathology" is an incontinence or instability of vigilance, thought, mood and behavior. This common thread of incontinence of mental states suggests the possibility of a single underlying mechanism that could produce, in varying combinations at different stages of brain development, inattention, tics, compulsions, obsessions, and rage. We will return to this hypothesis later in the chapter.

Attention, learning, thought, mood: these are all examples of the most complex functions of which the human brain is capable. This being the case, the localization of Mickey's neurological disorder would appear to be perfectly obvious: there has to be a problem involving the cerebral hemispheres—a process located in the brainstem or points below the foramen magnum would be illogical. Furthermore, one would be tempted to echo the fictional detective Hercule Poirot in localizing the problem to the "little gray cells" or, in neuroanatomical parlance, the cerebral cortex. As you will see, this conclusion is too simplistic, but does turn out to be half right.

Our next task, therefore, will be to review the neuroanatomical basis for selective attention, impulse control, learning, thought sequencing and mood.

12.3 Relevant Neuroanatomy and Neurochemistry

12.3.1 Neuroanatomical Substrates of Behavior

The brain structures involved in attention, learning and behavior are numerous, and comprise a widely distributed network. The main components of this network are as follows:

1. The prefrontal cortex
2. The basal ganglia, in particular the caudate nucleus and the ventral striatum (also known as the nucleus accumbens), and the thalamus, especially the dorsomedial and anterior nuclei
3. The cerebellar hemispheres
4. The limbic system
5. The monoaminergic and cholinergic diffuse projection systems

Obviously this is a formidable-appearing list and the interrelationships between the various components are exceedingly complex. Since this text is designed for novices in neuroanatomy, we will confine our discussion to the essential elements.

12.3.1.1 Prefrontal Cortex By this term we mean those portions of the frontal lobe anterior to the various motor cortices, that is, the precentral gyrus, premotor cortex, frontal eye fields, supplementary motor area, Broca's area (see Figure 12.1 and the relevant material in Chapters 7, 9 and 10). The prefrontal cortex is extensive and is distributed over the three surfaces of the frontal pole: the lateral frontal convexity, the midline parasagittal area, and the orbital-frontal region above the eyes (Figure 12.1).

Based in part on the study of patients with prefrontal area lesions, it is clear that the principal roles of the prefrontal cortex are selective attention, judgment, and working memory. Frontal lobe lesions are accompanied by inattention, distractibility, poor impulse control, and impaired ability to retain number sequences (as in looking up a telephone number and remembering it long enough to dial the required digits). Patients with frontal lobe damage (especially children) may be hyperactive or apathetic, sometimes both, by turns. As we noted in Chapter 1, it is

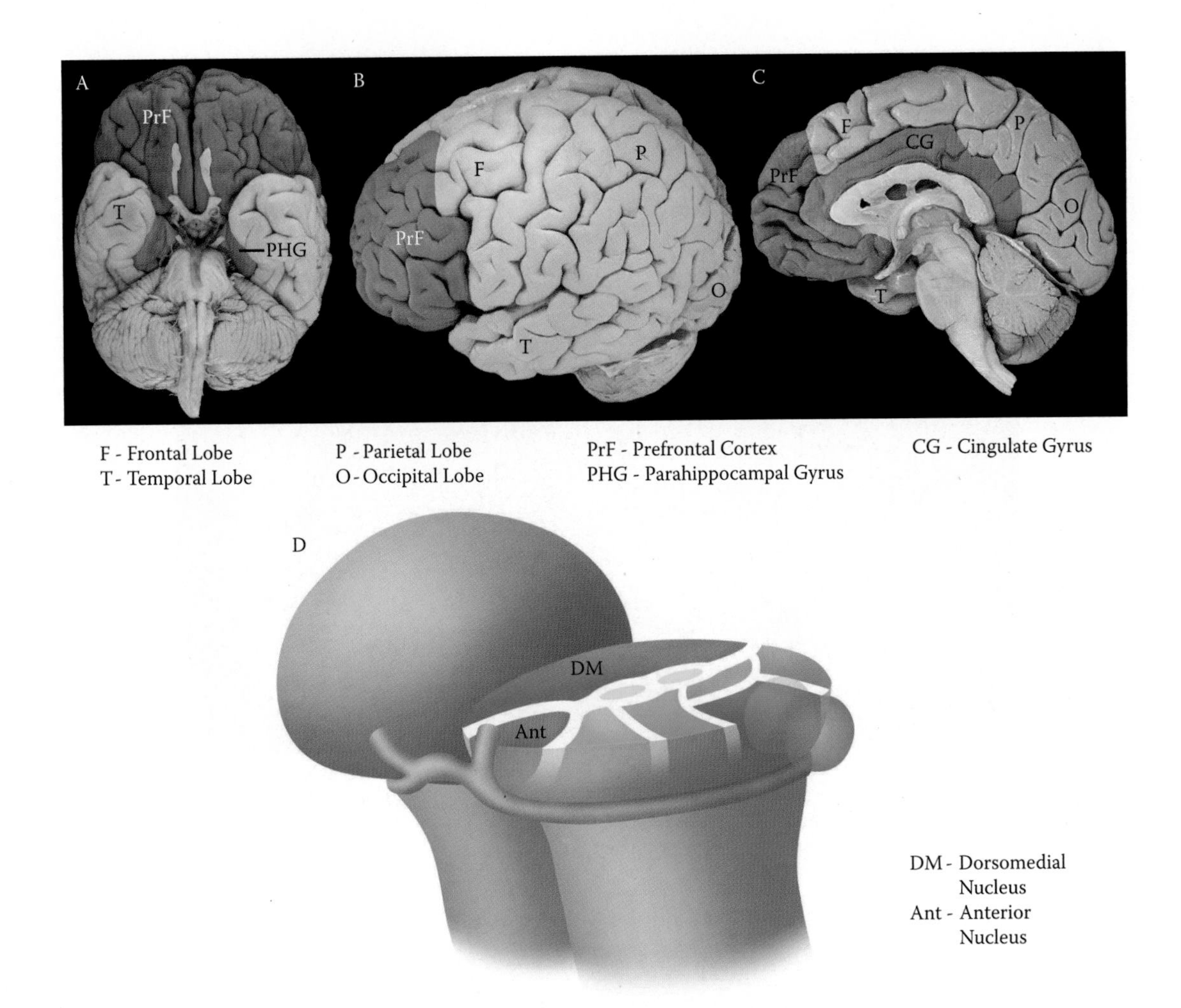

FIGURE 12.1 External view of the brain (upper panels) with the prefrontal and limbic cortices highlighted. A = ventral view; B = lateral view, left hemisphere; C = medial view, right hemisphere. The lower panel (D) shows the thalamus sectioned in the horizontal (axial) plane, demonstrating the thalamic nuclei most relevant to attention, memory and emotion (dorsomedial and anterior nuclei).

customary to refer to the prefrontal areas as the seat of executive function.

This is not to suggest that the other association cortices in temporal, parietal, and occipital lobes are unimportant in learning processes, as they clearly are, simply that the prefrontal cortex has an overarching, supervisory role in all forms of learning. The cellular architectural layering and modular functioning of the association cortices are reviewed in Chapters 9 and 10.

12.3.1.2 Basal Ganglia Circuit Just as the precentral gyrus cannot direct the performance of complex movements without the support of a variety of subcortical structures, the prefrontal area requires the collaboration of components of the same subcortical structures in deciding (for example) whether it is safe to chase an errant baseball into a busy street.

In Chapter 7 we noted that the initiation and cessation of movement appears to be centered in a looping circuit comprising the motor and sensory cortices, the putamen, the globus pallidus, the substantia nigra, the subthalamic nucleus, the lateral thalamus, and, finally, the motor cortex (see Figures 7.4 and 7.10). The prefrontal area, in the performance of its functions, participates in an analogous circuit that involves similar but geographically distinct portions of the striatum and thalamus (as well as the other structures just named). It is not necessary for you to know the details of this circuit. We will simply draw your attention to the two most important components of this prefrontal loop.

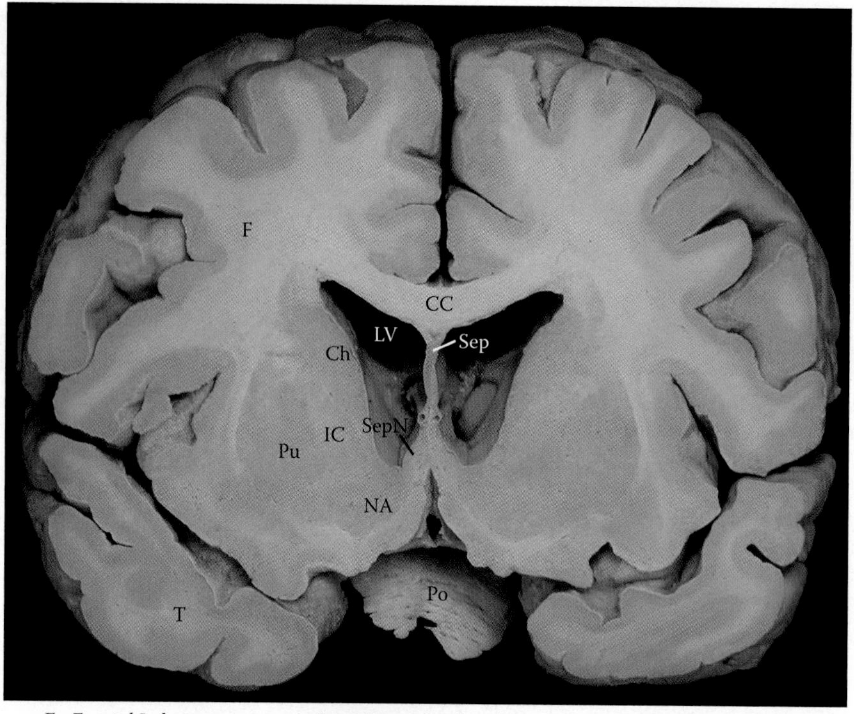

FIGURE 12.2 Coronal section through the frontal lobes at the level of the head of the caudate nucleus and the nucleus accumbens.

The first is the head of the caudate nucleus (see Figure 12.2); in the prefrontal loop it is the analogous structure to the putamen for the motor loop. The importance of the caudate head in executive functioning is revealed by the fact that its destruction (as, for example by hemorrhage or ischemic stroke) leads to deficits in attention and judgment similar to those seen in patients with lesions of the prefrontal cortex.

The second component is the dorsomedial thalamic nucleus (see Figures 12.1 and 1.8); it is analogous to the ventral anterior and ventral lateral thalamic nuclei for the motor loop. In the prefrontal loop, the dorsomedial thalamus projects back to the prefrontal cortex, having received input from the caudate head via distinct components of the globus pallidus, and indirectly from the subthalamic nucleus and substantia nigra. Neurobehavioral disturbances similar to those seen with prefrontal cortex and caudate head lesions may also be present in patients with isolated damage to the dorsomedial thalamic nucleus.

In the course of this text we have repeatedly made reference to the thalamus, its various nuclei and their projections to specific cortical areas. Previous thalamus-cortex illustrations have focused on the particular thalamocortical function(s) relevant to the clinical problem. Figure 12.3 summarizes these anatomical relationships for the thalamus as a whole.

12.3.1.3 Cerebellar Circuit In Chapter 7, we saw that the coordination of movement required the participation of an integrative, feedback loop involving the motor cortex, the contralateral cerebellar hemisphere (via the pontine nuclei) and the ventral lateral thalamic nucleus (thence back to motor cortex). Suffice it to say that there is an analogous circuit from the prefrontal area to different elements of the contralateral cerebellar hemisphere, returning via the dorsomedial thalamic nucleus. In effect, therefore, the cerebellar hemispheres represent the "steering wheel" for executive function.

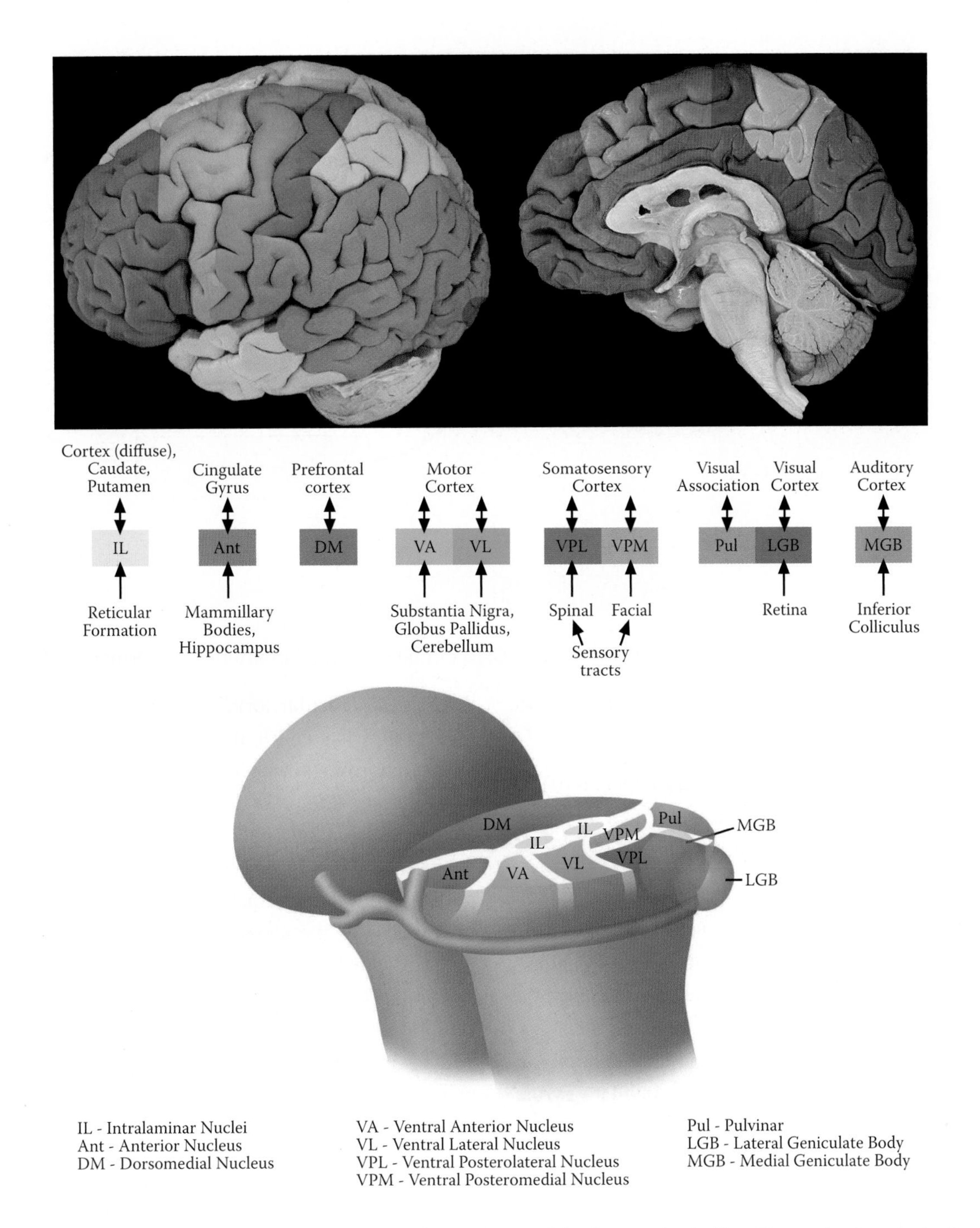

FIGURE 12.3 Reciprocal connectivity between the cortex and thalamus, including the major inputs to the various thalamic nuclei.

The combined basal ganglia and cerebellar circuits for the prefrontal area are outlined in a box diagram (Figure 12.4).

12.3.1.4 Limbic System The limbic system comprises a group of cerebral hemispheric structures located at the junction of the hemisphere with the diencephalon. Embryologically, they form a ring, or fringe (*limbus*, in Latin) around the foramen linking the lateral and third ventricles (see Chapter 11) at the origin of the primordial cerebral hemisphere. During the evolutionary process, the limbic elements of the cerebral hemispheres were the first to appear; indeed, in reptiles the cerebral hemisphere is almost entirely "limbic" in function.

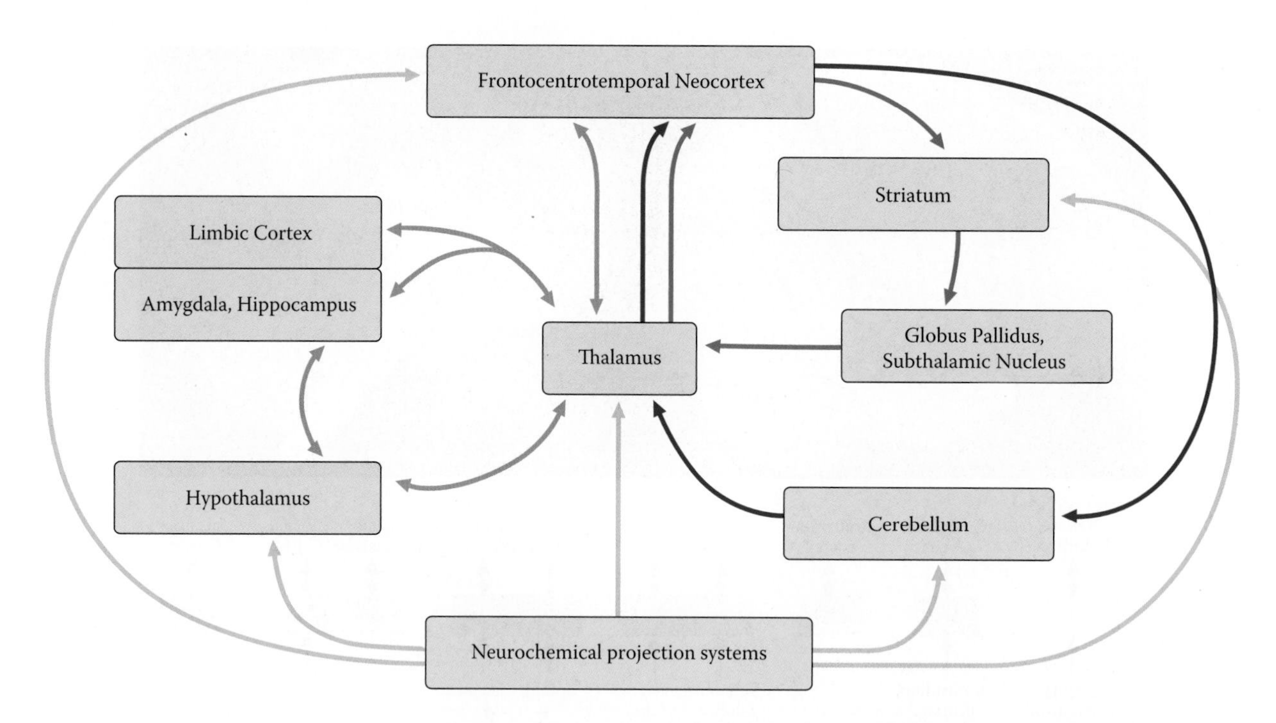

FIGURE 12.4 **Schematic diagram illustrating the main CNS circuits involved in attention, learning and emotion.**

There are both cortical and subcortical components to the limbic system, the most important of which are

1. Cortical—cingulate gyrus, parahippocampal gyrus, hippocampal formation, septal area
2. Subcortical—fornix, mammillary body, anterior thalamic nucleus, amygdala, ventral striatum (nucleus accumbens)

Of these structures, the nucleus accumbens is illustrated in Figure 12.2, in relation to the caudate head, while the remaining structures are illustrated in Figure 12.5. Figure 12.5 shows the "ring"-like configuration of the limbic cortex, the ring consisting of the cingulate gyrus, septal area and parahippocampal gyrus. The hippocampal formation is located medial to and is partially concealed by the parahippocampal gyrus.

The functions of the limbic system are only partially understood. The most important functions are memory registration (in which the hippocampal formation is a crucial element), social interaction (including sexual interaction), the response to external threats, and mood control (in which the amygdala is particularly important). In animal species the limbic system is implicated in the ability to distinguish friend from foe and in the emotional reaction to and ability to cope with an external threat—the so-called "fight or flight" response.

Like the reticular formation and nonspecific thalamic nuclei (see Chapter 10), the limbic system receives extensive input from peripheral pain receptors—witness the characteristic motor and emotional responses to being stung on the foot by a bee. The limbic system also has constant sensory input from visceral pain receptors, as is supported by the predominant emotions of anxiety and fear in response to the squeezing chest pain associated with myocardial ischemia.

Output from the limbic system projects diffusely to the cerebral cortex, diencephalon and brainstem, in keeping with the importance of emotional states in helping to direct motor function, decision making, language, learning and sensory perception. Output to the adjacent hypothalamus results in the characteristic autonomic nervous system reactions to emotional states; an obvious example would be the tachycardia, tachypnea, elevated blood pressure and intense perspiration that immediately follow a near-miss high speed collision at a traffic intersection.

As might be expected, many of the components of the limbic system are functionally linked by sweeping loops of fibers with reciprocal connections; they are analogous to those already described for the motor system (Chapter 7). Although limbic circuitry is of great interest to neuroanatomists and neurophysiologists, the subject is beyond the scope of an introductory, clinically based text.

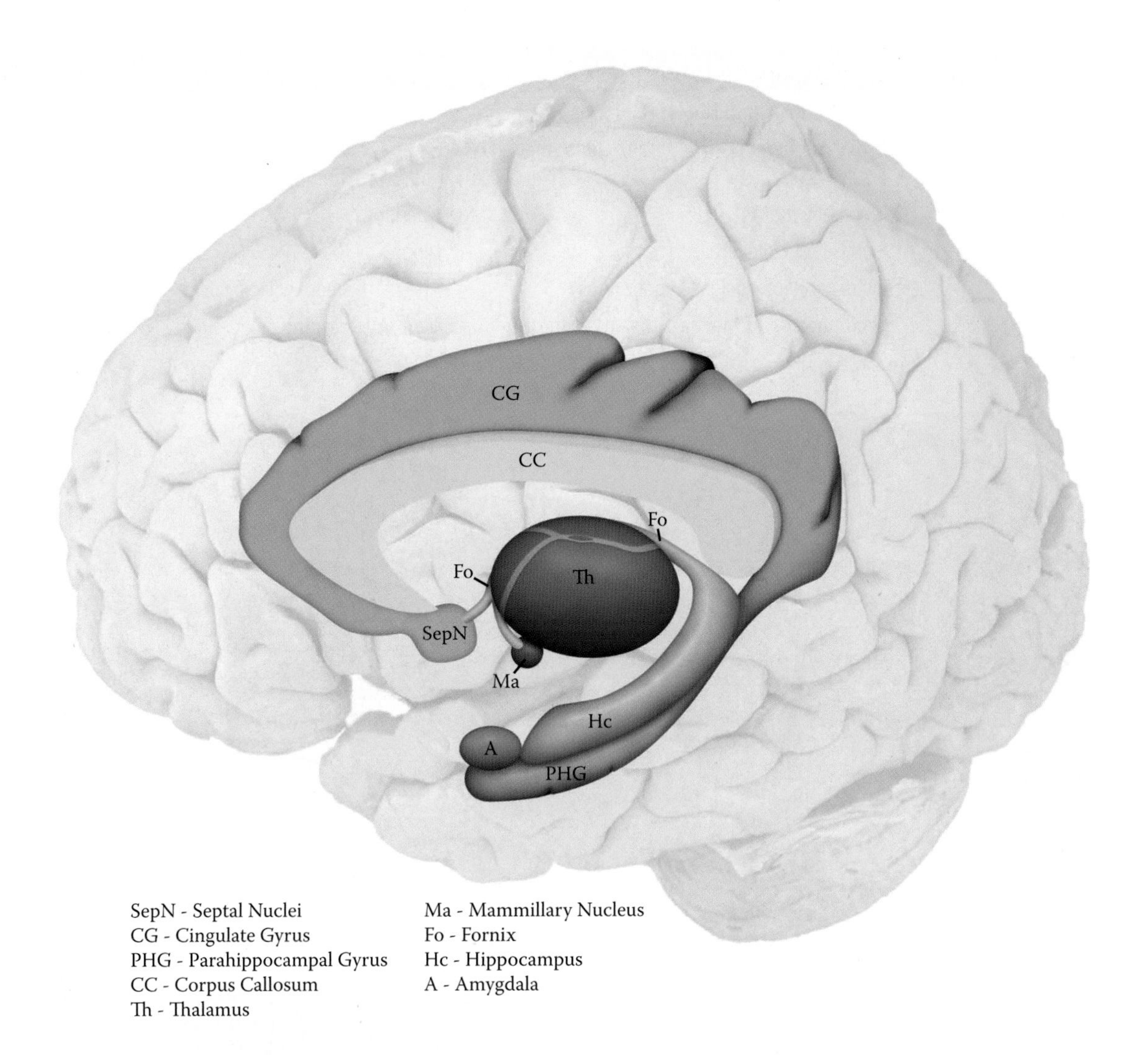

FIGURE 12.5 Illustration of the location and relationship of the main components of the limbic system, both cortical and subcortical.

12.3.1.5 Monoaminergic and Cholinergic Diffuse Projection The final components of the network are diffuse neuronal projection systems from brainstem and basal forebrain; their function is to modify the level of activity in the remaining components. Activity may be stimulated or inhibited depending on the type of balance between the various transmitters at a given moment.

The four most important transmitter systems are serotonin (5-hydroxytryptamine), dopamine, norepinephrine and acetylcholine. Of these, the cells of origin of the first three (all monoamines) are located variously in the pons and midbrain while those of the cholinergic system are in the basal forebrain inferior to the ventral striatum. Although the nuclei containing the monoaminergic and cholinergic cell clusters are relatively small, their axons project widely to the cerebral cortex, basal ganglia, thalamus and limbic system.

Table 12.1 outlines the location of the various nuclei participating in the diffuse projection system and the main functions regulated by each neurotransmitter. Figure 12.6 illustrates the main cerebral regions to which the transmitter systems project (panels A to D).

As can be seen in Figure 12.6, the serotonergic and noradrenergic systems project to the entire cerebral cortex (including the limbic cortex), the basal ganglia, diencephalic structures, and the cerebellum. The dopaminergic system, whose striatal projections we reviewed in Chapter 7, largely confines its cerebral cortical projection to the frontal lobes and limbic system; these relatively restricted projections are nevertheless crucial for the regulation of

Table 12.1 Main Neurotransmitter Projection Systems for Attention, Memory, and Behavior

Transmitter	Principal Nuclei (Location)	Functions
Serotonin (5-OH-tryptamine)	Median raphe nuclei (midbrain, pons, medulla)	Sleep–wake cycle, mood (pain inhibition via spinal cord projections)
Dopamine	Substantia nigra	Attention
	Ventral tegmental area	Mood
Norepinephrine	Locus ceruleus nucleus (upper pons)	Attention, sleep–wake cycle
Acetylcholine	Basal nucleus of Meynert (basal forebrain)	Memory, learning

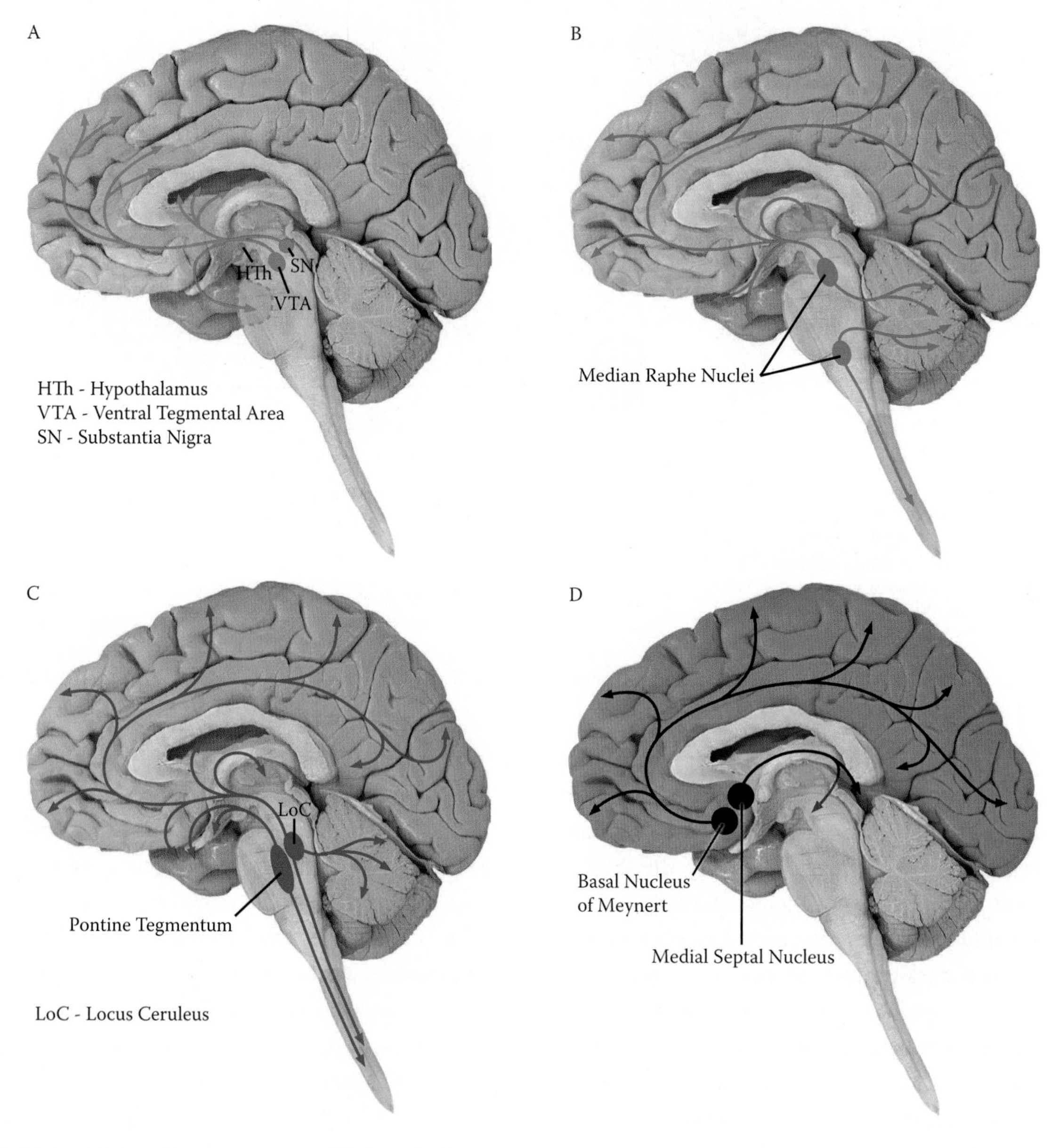

FIGURE 12.6 Mid-sagittal section of human brain showing the nuclei of origin and regions served by the monoaminergic and central cholinergic projection systems. (A) Dopamine system, from the midbrain; (B) serotonin system, from the raphe nuclei of the brainstem; (C) norepinephrine system, from the locus ceruleus and other brainstem nuclei; (D) cholinergic system, from the basal forebrain nuclei.

attention and mood. The cholinergic system originating in the basal forebrain projects primarily to the entire cerebral cortex, the hippocampus and the amygdala.

12.3.2 *Neurochemical Substrates of Behavior*

Having seen how the brainstem monoaminergic systems and the basal forebrain cholinergic system project variously to neocortical association areas, limbic cortex and central cerebral gray matter structures—all implicated in attention, learning, mood and behavior—we will now briefly review the mechanisms by which these projection systems alter neuronal function.

In Chapter 10, in the context of epileptic disorders, we discussed how the amino acid neurotransmitters glutamate and GABA excite or inhibit neurons by opening excitatory (Na^+, Ca^{2+}) or inhibitory (Cl^-) ion channels in neuronal membranes. These neurotransmitter actions are extremely rapid, the excitation or inhibition lasting only milliseconds. The excitatory effects of acetylcholine on nicotinic-type cholinergic receptors in the peripheral nervous system are equally rapid in onset, and short lasting.

In contrast, the effects on neuronal activity of the monoaminergic projection and basal forebrain cholinergic (muscarinic) systems are relatively slow in onset and long lasting, hundreds of milliseconds or even several seconds. The term used for such prolonged modifications in neuronal activity is *neuromodulation*.

Norepinephrine, dopamine, serotonin and forebrain-derived acetylcholine do not act by directly opening ion channels (so-called ligand-gated channels), but through action on specialized neuronal membrane proteins known as G-proteins (Figure 12.7, also mentioned in Chapter 10). Glutamate and GABA may also act through G-proteins via metabotropic and GABA-B receptors, respectively. As demonstrated in Figure 12.7, G-proteins may then act to open adjacent ion channels or may, in turn, activate or inhibit so-called second messenger molecules such as cyclic AMP (adenylate cyclase). Once activated, second messengers may initiate long-lasting changes in cell chemistry, such as stimulation of protein kinase activity or upregulating gene expression. Gene transcription and associated protein synthesis are believed to be essential components in long-term memory registration and, thus, in the learning of new material such as neuroanatomy!

In Chapter 7, during the discussion of the central mechanisms involved in the initiation of movement, we introduced the concept of a single transmitter chemical having either excitatory or inhibitory actions, depending on the type of receptor present. The example given was dopamine receptors in the striatum, D1 receptors being excitatory to the downstream neuron, D2 receptors inhibitory. The same is true for norepinephrine and serotonin.

Thus, for example, there are four main norepinephrine receptor subtypes ($\alpha 1$, $\alpha 2$, $\beta 1$, $\beta 2$), the β subunit types predominating in the central nervous system. The list of known serotonin (5-hydroxytryptamine) subtypes is long, the most

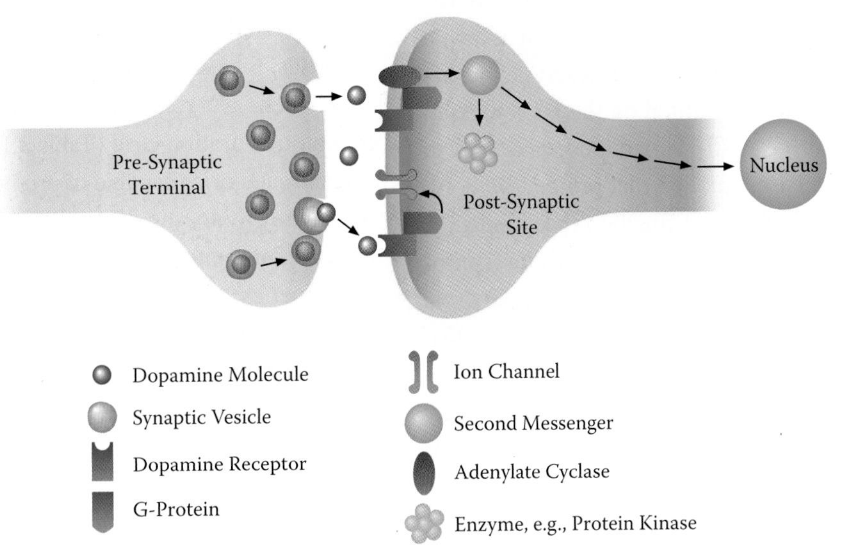

FIGURE 12.7 Schematic of a CNS synapse showing neuromodulation via G-proteins.

important being 5HT1A, 5HT1B, 5HT1C, 5HT1D, 5HT2A, 5HT2B, 5HT3; of these, the 5HT2 receptor subtypes appear to be the most important ones in modulating cortical and limbic function, thus modifying mood states.

The importance of normal functioning of the many neurotransmitter receptor subtypes in the regulation of behavior is illustrated by the fact that the current pharmacological management strategies for the treatment of inattention, hyperactivity, depression, rage, obsessive-compulsive disorder and memory disintegration largely involve the manipulation of central nervous system receptor functions. Thus, hyperactivity and inattention may be treated with dopamine upregulation, depression with 5HT2 receptor stimulation, and Alzheimer-related memory loss with central muscarinic cholinergic receptor activation.

12.4 Localization Process

Fortified with an expanded fund of information concerning the array of anatomical structures—cortical, subcortical, limbic and even brainstem—you can begin to appreciate that the "seat" of Mickey's behavior disorders could, in principle, reside in any or all of these sites. Indeed, it is conceivable that some or all of his symptoms may originate not so much from a structural lesion at one anatomical site, but from a generic defect in cell-to-cell interaction involving neurons of the cortex, limbic system and the basal ganglia—neurons that all participate in circuits mediating attention, learning, mood, and so forth.

Earlier in this chapter we speculated that, in a sense, nervous tics, obsessive thinking and rage outbursts could all share a common mechanism: all are stereotypic behavioral phenomena that are not wanted by the person experiencing them, cannot be easily suppressed, and lead to major problems in social interaction that profoundly compromise the individual's ability to function independently. *Incontinence* was the term used; this word seems appropriate given that bowel or bladder incontinence refers to a failure to develop or to a loss of an essential function that allows us to participate normally in society. The development of urinary incontinence, for example, is exceedingly embarrassing to the patient, significantly curtails activities outside the home, and may have devastating effects on self-esteem. The development of compulsive touching of strangers or of rage attacks (both forms of behavioral incontinence) is no different in these respects.

In other words, it is possible that stereotyped involuntary movements, compulsive thinking and uncontrollable rage outbursts may all involve a similar mechanism, possibly at the level of receptor function, residing in separate, but parallel cortical-basal ganglia-thalamocortical circuits originating in supplementary motor, prefrontal and limbic cortices, respectively. With this hypothetical conclusion, we have gone about as far as we can go in the complex task of trying to localize Mickey's disorder.

12.5 Etiology: The Nature of Mickey's Disorder

You may have noticed that, in this chapter, we have been using the term *disorder* in reference to Mickey's collection of behavioral disturbances rather than the term *disease*. This is a somewhat arbitrary distinction but it serves to point out that, thus far, Mickey's symptom complex is virtually lifelong (thus chronic), but has shown no sign of relentless progression such as one would expect with many ongoing "disease" processes. True, his symptoms have fluctuated over time, some waxing and others waning, but, on the basis of your neurological examination two decades into the process, Mickey appears to be remarkably "intact." This statement by no means abrogates the fact that the effects of Mickey's behavior disorder have had a catastrophic impact on his development as an independent adult human being, and on his entire family. In the psychiatric literature, mental health dysfunctions are typically referred to as *disorders* rather than *diseases*; for our purposes, the two terms are interchangeable.

If, then, we refer back to the Neurological Disease Symptom Duration Grid (Table 3.13), we see that the main disease categories capable of producing a process evolving over many years (as is the case with Mickey) are metabolic, inflammatory/autoimmune, neoplastic, degenerative and genetic.

In addition, however, there is the possibility that one of the other more acute disease categories could have injured Mickey's brain early in life, leaving him with a static functional deficit. Such an injury could have resulted from a traumatic, vascular or infectious process; if so, however, the process would likely have occurred before birth, given that there is no history of any significant postnatal illness or injury. The term typically used to describe such a process is a *static encephalopathy of prenatal origin*, of which a classic example is cerebral palsy.

Given that, in one way or another, we will have to consider most of the 10 disease categories, we will work through the whole list, as we did in most of the previous chapters.

12.5.1 Paroxysmal

Paroxysmal disorders of the nervous system such as epilepsy and migraine may be accompanied by significant behavioral dysfunction. Quite apart from the transient, dramatic behavior changes that occur during complex partial seizures, the interictal personality of patients with localization-related epilepsy may be chronically disturbed: mood disorders, learning disabilities and aggressive behavior are common. In the case of migraine attacks, there may be striking mood changes for hours prior to and following the headache symptoms. For both of these paroxysmal disorders, however, the cause of the behavior disturbance is obvious.

12.5.2 Traumatic

Cerebral contusions and hemorrhages following a head injury are a frequent cause of acquired, long-standing behavioral disturbances. The prefrontal and anterior temporal cortices are particularly vulnerable to traumatic injury. It is not surprising, therefore, that head injury patients often demonstrate poor impulse control, inattention, impaired judgment, apathy and rage attacks. Such behavioral disturbances may be seen whether the brain injury occurred at birth in the context of a traumatic delivery or in the twenties following an automobile accident or sports-related injury.

12.5.3 Vascular

Chronic behavioral problems are also common following ischemic or hemorrhagic strokes. In Mickey's case, there is obviously no history of such an event.

Prenatal arterial and venous strokes, however, if sufficiently small, may also lead to learning and behavioral disorders with preservation of normal intelligence. In most instances there is an obvious motor disability, usually a hemiparesis. Frontal lobe infarcts, however, may produce no symptoms other than features of impaired executive function: distractibility, difficulty remaining on task, hyperactivity and poor impulse control. Likewise, dominant hemisphere temporal lobe infarcts may lead to difficulty with reading comprehension.

While these deficits may be found on the list of Mickey's symptoms, prenatal focal infarcts have not been typically associated with tic disorders or obsessive-compulsive behaviors. The development of epileptic seizures is also relatively common in children with prenatal stroke, particularly if the stroke is large; when present, such seizures usually appear before the age of six.

Finally, perinatal hypoxic-ischemic brain injuries often lead to significant learning and behavior problems later in life, usually but not invariably associated with motor deficits. Since neonates with such brain pathology are clearly very ill in the neonatal period, and Mickey was not, we can eliminate this possibility.

12.5.4 Toxic

Long-standing behavior changes may occur following a toxic injury to the brain, as in methanol poisoning or chronic alcoholism. The brain may also be damaged prenatally following exposure of the mother to neurotoxic agents; examples include ionizing radiation and, again, ethanol. In the latter instance, there is a typical combination of facial dysmorphic features and severe disturbances of learning, attention and behavior that carries the name fetal alcohol syndrome. Indeed, Mickey shares many of the behavioral characteristics seen in children and adults with *fetal alcohol syndrome*; tics and obsessions, however, are not typically seen in this disorder.

Toxic compounds that impede the normal development of the brain or other organs are called *teratogens*.

12.5.5 Infectious

Chronic behavioral problems are one of many neurological complications of acute viral encephalitis or bacterial meningitis; these diagnoses, however, are not compatible with Mickey's history. Prenatal encephalitides may have the same behavioral consequences but are usually accompanied by varying degrees of global intellectual impairment, epilepsy, and (frequently) congenital deafness. Examples of pathogens responsible for prenatal encephalitis include rubella and cytomegalic inclusion viruses, syphilis, and toxoplasma, a unicellular parasite.

12.5.6 Metabolic

In adults, metabolic disorders are a common cause of behavioral and personality changes. Chronic hepatic failure may be accompanied by confusion and disinhibition, hypothyroidism by lethargy and depression, hyperthyroidism by a manic state. Consequently an unexplained behavior change in an older child or an adult should prompt a search for a metabolic disorder: liver function tests, urea, electrolytes, TSH (thyroid stimulating hormone), vitamin B12 level. None of these disorders would be likely to produce a neurobehavioral disorder lasting two decades without some other features of a systemic disease.

12.5.7 Inflammatory/Autoimmune

Some autoimmune disorders, most notably systemic lupus erythematosus (SLE), may produce a chronic, fluctuating encephalopathy characterized by behavioral and mood abnormalities, even frank psychosis, without any clear-cut motor or sensory findings. It would be highly unlikely, however, that a patient with CNS lupus would have neurological symptoms persisting for two decades without the development of other classical manifestations of the disease: arthropathy, nephritis, facial rash, cardiac valvulopathy.

The only autoimmune disorder worth considering in Mickey's case is PANDAS, an acronym for pediatric autoimmune neurobehavioral disorder associated with streptococcus. This behavioral syndrome is thought to be a cousin of Sydenham's chorea, a streptococcus-induced disease originally recognized in the seventeenth century by Thomas Sydenham. In Sydenham's chorea, a streptococcal infection (usually pharyngeal) is followed by the subacute appearance of chorea (whose distinction from tics has already been elucidated), sometimes lasting for months. Patients with this disorder may also develop hyperactivity, inattention, mood swings and obsessive-compulsive behaviors. In PANDAS, whose existence as a discrete disease entity is not universally accepted, the children develop, following a streptococcal infection, any or all of the behavioral problems seen in Sydenham's chorea, but do not have chorea. They may, however, develop a tic disorder. While all of these symptoms have a startling resemblance to Mickey's, they typically appear in a relatively explosive fashion, following the infection, last for weeks or even months, in a fluctuating fashion, then subside. Thus, as tempting as the diagnosis of PANDAS might be, the time course of Mickey's disorder does not fit.

12.5.8 Neoplastic/Degenerative

By their very natures, neoplastic and degenerative diseases produce a slow, relentless deterioration with the gradual accumulation of new and increasingly severe neurological deficits. Thus, while they may produce major behavioral alterations along with their many other manifestations, these categories do not need to be considered any further.

12.6.9 Genetic

Genetic disorders may have behavioral consequences at all stages in life, from fetal development to old age.

As a rule, genetic disorders that compromise fetal brain development (trisomy 21 or Down syndrome, fragile X syndrome, Rett syndrome, among many) also produce significant degrees of mental insufficiency—termed *mental retardation* in North America and *learning disability* in Europe, expressions that are, respectively, pejorative and unnecessarily vague.

Syndromes with X chromosome aneuploidy may be accompanied by learning difficulties but normal intelligence. These include Turner syndrome (phenotypic females with a single X chromosome: XO) and Klinefelter syndrome (phenotypic males with two X chromosomes: XXY). In Mickey's case, Turner syndrome is clearly not an option, while Klinefelter syndrome is typically accompanied by dysmorphic features including tall stature with excessively long arms and legs.

The genetic disorders considered thus far all seem unlikely explanations for Mickey's problem. Given the family history of similar, if less dramatic behavioral disturbances, however, we must look at the possibility of an entirely different type of genetic disorder, one whose only manifestation is behavioral.

In Section 12.4, we introduced the possibility of a global neurochemical dysfunction, possibly a neurotransmitter receptor malfunction, compromising several parallel cortical-basal ganglia–thalamocortical circuits and leading to a variety of behavioral abnormalities, including tics, obsessive-compulsive behaviors, rage attacks, and so forth. Since an inherited mechanism is a logical explanation for such a scenario, it is time to revisit it.

12.5.9.1 Genetic Neurobehavioral Disorders As a complex, nonprogressive behavior disorder in an adult with normal intelligence and a normal neurological examination, Mickey's disorder straddles the completely artificial separation between the clinical disciplines of neurology and psychiatry. The etiology of most psychiatric disorders is not presently known; in consequence, psychiatrists have had to define the disorders they encounter by using strict clinical criteria. As was formerly the case with many neurological disorders, it is hoped that a rigorous descriptive approach to psychiatric symptom complexes will clearly separate one disorder from another and eventually lead to the establishment of precise etiologies.

Leaders in psychiatry have collaborated in the publishing of generally accepted clinical criteria for specific behavioral syndromes. These criteria appear in the *Diagnostic and Statistical Manual of Mental Disorders*, usually abbreviated as *DSM*, with periodic updates. The most recent version is the *DSM-IV-TR*. Since the listed criteria for each disorder are those that the experts in the field agree upon, it is not surprising that, with accumulating clinical experience and analysis, the criteria for each disorder are modified somewhat over time. This process will continue to evolve until precise neurochemical and neurogenetic mechanisms are uncovered.

With this caveat in mind, we can look again at Mickey's collection of behavioral abnormalities in the light of *DSM-IV-TR* published criteria for generally accepted psychiatric disorders. When we do this, we find that Mickey has many features of several apparently discrete disorders:

1. Attention-deficit/hyperactivity disorder (criteria listed in Table 12.2)
2. Tourette's disorder (Table 12.3)
3. Mathematics disorder (Table 12.4)
4. Obsessive-compulsive disorder (Table 12.5)

If you review the criteria listed in Tables 12.2 through 12.5, you will note that Mickey meets all the criteria for attention-deficit/hyperactivity disorder, Tourette's disorder (i.e., his long-standing motor and vocal tics) and mathematics disorder, and most of the criteria for obsessive-compulsive disorder. Faced with this plethora of simultaneous psychiatric diagnoses, you may well ask: Are the four diagnoses distinct disorders that are usually present in individual patients in isolation, and, if so, is it possible that all four could occur in a single patient having one unitary diagnosis? The answer to both questions is "yes!"

You will have noted that the *DSM-IV-TR* criteria for these psychiatric disorders include the important proviso that potentially causative underlying "medical" conditions must have been excluded. This is a very important statement that obliges the physician to consider such conditions in any child or adolescent with any combination of inattention, hyperactivity, impulsivity, and motor or vocal tics.

Attention-deficit/hyperactivity disorder, for example, is a common behavioral complex with a strong hereditary predisposition that appears to affect 5 to 10 percent of the pediatric population. In some suspected cases, however, it turns out that the real cause of the symptoms is a sleep disorder: poor sleep hygiene, obstructive sleep apnea due to airway obstruction from enlarged tonsils and adenoid, or periodic limb movements in sleep. In other instances, the cause is found to be excessive consumption of antihistamine medications for respiratory allergies. These are just some examples among many.

Correction of the underlying cause will alleviate the inattention and hyperactivity. When faced with a patient having these symptoms, therefore, it is important to ask the parents such questions as:

1. Does your child snore?
2. Does he or she get up frequently during the night?
3. Does your child get out of bed in the morning complaining of feeling tired or having a headache?
4. What medications does your child take on a regular basis?

In the case of Tourette's disorder, the diagnosis cannot be made if the patient has a history of encephalitis, traumatic brain injury or other acquired neurologic pathology. We have already considered and excluded the hypothetical Tourette's disorder look-alike, PANDAS.

Returning to the apparent conundrum that Mickey has symptoms that reasonably satisfy the criteria for four mental disorders, we may ask whether three of the apparently discrete disorders may, in Mickey's case, be subsumed under the diagnostic umbrella of the fourth. Again, we can respond to this question in the affirmative, as Tourette's disorder patients frequently also meet the accepted criteria for attention-deficit/hyperactivity disorder, one or more learning disorders, and obsessive-compulsive disorder. While some Tourette's disorder patients only have tics, others initially present with all of the features of attention-deficit/hyperactivity disorder then, several years later, begin to demonstrate motor and vocal tics—just as Mickey did. Learning difficulties are common in Tourette's disorder, but not invariable; they may be the result of an

Table 12.2 Diagnostic Criteria for Attention-Deficit/Hyperactivity Disorder

A. Six or more symptoms of either inattention or hyperactivity-impulsivity
Inattention
• failure to give close attention to details; careless mistakes in schoolwork
• difficulty sustaining attention in tasks or play
• not appearing to listen when spoken to directly
• not following through on instructions or completing assignments
• difficulty organizing tasks or activities
• avoidance or reluctance to engage in tasks requiring sustained mental effort
• tendency to lose things necessary for tasks and activities
• easily distracted by extraneous stimuli
• forgetful in daily activities
Hyperactivity–impulsivity
• fidgeting with hands or feet; squirming in seat
• leaving seating in classroom or elsewhere when remaining seated required
• running about or climbing in inappropriate circumstances
• difficulty playing quietly
• acting as if "driven by a motor"
• talking excessively
• blurting out answers before questions completed
• difficulty awaiting turn
• interrupting or intruding on others, in conversations or games
B. At least some of the above symptoms were present prior to age 7 and were sufficient to cause impairment
C. Impairment from symptoms is present in two or more settings (e.g., school, home)
D. Clear evidence of clinically significant impairment in social, academic or occupational functioning
E. Symptoms are not present exclusively in the context of another disorder such as pervasive developmental disorder (autistic spectrum), schizophrenia or other psychosis

Table 12.3 Diagnostic Criteria for Tourette's Disorder

A. Both multiple motor and one or more vocal tics present at some time during the illness, although not necessarily concurrently
B. Tics occur many times daily, nearly every day or intermittently throughout a period of at least one year; no tic-free period of more than three consecutive months
C. Onset before age 18 years
D. Symptoms not due to the direct physiological effects of a medication (e.g., stimulants) or to a medical condition such as viral encephalitis or Huntington disease

impaired concentration span but also may result from a specific learning disability that persists after the attention span has improved.

Motor and vocal tics are, in a sense, forms of compulsive motor behavior. Many tic patients maintain that the reason they tic is an ill-defined internal pressure that requires the performance of the tic in order to release "tension." The same sense of pressure is present with compulsive hand-washing, counting and checking, all features of obsessive-compulsive disorder. Indeed, this psychological "pressure" is so compelling that attempts by well-meaning relatives to prevent or interfere with compulsive

Table 12.4 Diagnostic Criteria for Mathematics Disorder

A. Mathematical ability (as measured by standardized tests) substantially below the level expected for age, measured intelligence and appropriate opportunities to learn
B. Learning disturbance significantly interferes with academic achievement or activities of daily living requiring mathematical ability
C. In case of a sensory deficit (e.g., visual), the mathematical difficulties are in excess of what might be expected from the deficit
Reprinted with permission from the *Diagnostic and Statistical Manual of Mental Disorders*, Text Revision, Fourth Edition, (Copyright 2000). American Psychiatric Association.

Table 12.5 Diagnostic Criteria for Obsessive-Compulsive Disorder

A. Either obsessions or compulsions
Obsessions
• recurrent or persistent thoughts, impulses or images that are intrusive and inappropriate; cause marked anxiety or distress
• thoughts, impulses, images; not simply excessive worry about real-life problems
• individual attempts to ignore or suppress thoughts, impulses, images
• individual recognizes that the thoughts, impulses, or images are a product of one's own mind (not imposed from without)
Compulsions
• repetitive behaviors (e.g., hand-washing, ordering, checking) or mental acts (e.g., praying, counting, repeating words silently) the individual feels driven to perform in response to an obsession
• behaviors or mental acts designed to prevent or reduce distress but are either excessive or not related to a realistic problem
B. Individual recognizes at some point that the obsessions or compulsions are excessive or unreasonable
C. Obsessions or compulsions cause marked distress, are time-consuming and/or significantly interfere with normal routines and functions (school, occupation, social activities)
D. Obsessions or compulsions are not confined to another specific disorder (e.g., preoccupation with food in the presence of an eating disorder)
E. Disturbance not due to physiological effects of a substance (e.g., drug abuse) or a general medical condition
Reprinted with permission from the *Diagnostic and Statistical Manual of Mental Disorders*, Text Revision, Fourth Edition, (Copyright 2000). American Psychiatric Association.

behaviors may trigger an angry response, or even a rage attack. Thus, rage attacks, as we have already suggested, are really a secondary behavior disturbance derived from a combination of obsessive thinking, impulsivity and pronounced instability of mood.

Thus, we see how Tourette's disorder, as illustrated by Mickey's characteristic presentation, ties together the various symptoms and deficits he presents, all of which may be seen in isolation in other patient scenarios.

12.6 Case Summary

Mickey has a chronic, complex behavioral disorder that has troubled him, in various overlapping forms, for most of his life. The principal symptoms are poor attention span and impulse control, motor and vocal tics, complex compulsive behaviors, obsessive thinking, learning difficulties, and rage attacks. Given his otherwise normal developmental status, his completely normal neurological examination, and the positive family history of chronic motor tics and obsessive behaviors, the most likely diagnosis is Tourette's disorder.

12.7 Additional Information

Originally described in the late nineteenth century by Gilles-de-la-Tourette, a pupil of the legendary French neurologist Charcot, Tourette's disorder (or Tourette syndrome, as it is usually termed in the neurological literature) is a common behavioral disorder, with a prevalence of about 1/200–300 in the general population. Its presumptive genetic origin is illustrated by the fact that there is often a positive family history of one or more of the Tourette behavioral spectrum: attention-deficit disorder, chronic motor tics and obsessive-compulsive disorder, each in different individuals. Tourette's disorder and chronic motor tics are, for reasons unknown, more common in boys and men; isolated obsessive-compulsive disorder is more common in women.

The pathophysiologic basis of Tourette's disorder is unknown. While clinical evidence suggests a disturbance in the function of several neuromodulatory receptors, this may well be a second-order phenomenon. Although the fundamental defect may be a mutation or polymorphism in a gene or genes, it is clear that many other factors help determine phenotypic expression. Gender differences in rates and types of symptoms suggest an important hormonal effect. Environmental factors are also important, as revealed by the fact that, in identical twins with Tourette's disorder, the twin who was smaller at birth is usually the more symptomatic.

At least two neuromodulators are implicated in Tourette's disorder. Dopaminergic synapses are thought to be affected because drugs that upgrade dopaminergic function (such as methylphenidate) improve attention span and reduce impulsivity in Tourette patients (see Table 12.6). On the other hand, dopaminergic antagonists (such as haloperidol) reduce the number of motor and vocal tics. Thus, whatever the defect may be in dopaminergic transmission in Tourette's disorder, it is more complex than simple over- or under-expression.

Serotonergic synapses are also implicated in Tourette's disorder because serotonergic agonists, usually in the form of selective serotonin reuptake inhibitors (SSRIs), are used successfully to treat obsessive-compulsive behaviors, as well as to stabilize mood swings.

Finally, it is important to recognize that, just as Mickey's three other psychiatric "diagnoses" have been included under the umbrella of Tourette's disorder, that diagnosis itself may overlap with other relatively distinct syndromes. The most notable example is autistic spectrum disorder, a developmental syndrome with multiple etiologies whose main features include impaired expressive and receptive language, poor social interaction, impoverished imaginative play and obsessive fascination with moving objects or parts of objects. Many but not all autistic individuals are mentally handicapped. Some autistic patients with normal intellectual ability and good vocabulary may have marked obsessive traits as well as prominent motor and vocal tics; they meet all of the existing criteria for Tourette's disorder.

Table 12.6 Pharmacologic Interventions for the Symptoms of Tourette's Disorder

Symptom / Drug
Attention-deficit/hyperactivity
Methylphenidate
Dextro-amphetamine
Atomoxetine
Clonidine
Tics
Haloperidol
Pimozide
Risperidone
Clonidine
Obsessive-compulsive symptoms
Clomipramine
Fluoxetine
Paroxetine
Fluvoxamine
Venaflaxine
Rage attacks
Risperidone
Olanzapine
Clonidine

These variable overlapping neurobehavioral symptom complexes, with Tourette's disorder as the centerpiece, are illustrated in Figure 12.8 as multiple overlapping circles. This conceptualization of the relationship between the disorders will have to suffice until their exact physiological mechanisms are known. Ironically, once that achievement has been made, Tourette's disorder, as a clinical entity, may well disappear as the tic disorder will simply be seen as but one symptom of a fundamental neurochemical dysfunction.

12.8 Investigating and Treating Mickey's Disorder

As already suggested, the diagnosis of Tourette's disorder depends on the recognition of the classic clinical criteria; no supporting laboratory confirmation exists at present. MR imaging of the brain in Tourette's syndrome appears normal, with no consistent abnormalities. Hence, an MRI study in Tourette's disorder is not required unless the history or physical findings suggest

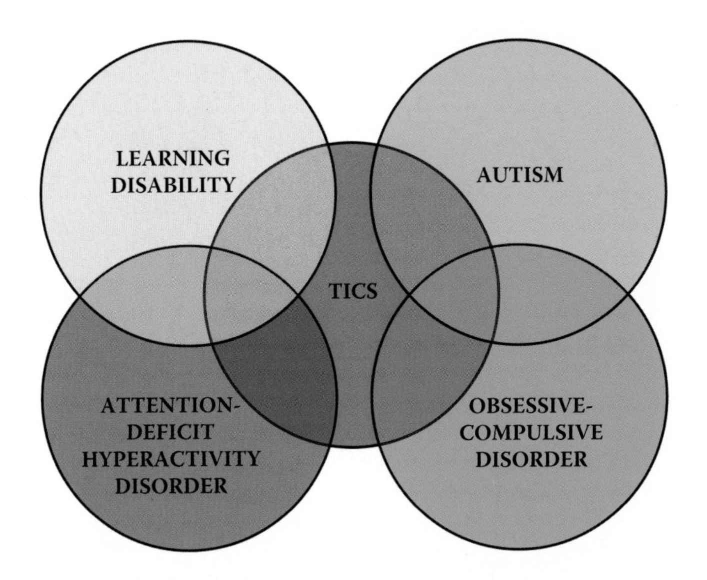

FIGURE 12.8 Schematic of the overlap between tic disorders and other developmental neurobehavioral syndromes.

the presence of a remote brain injury, whether traumatic, vascular or infectious.

Similarly there are no laboratory test abnormalities, whether biochemical or otherwise, to support the diagnosis. Given the evidence for a disturbance in monoaminergic transmitter function, it might be expected that measurement of levels of transmitters or their metabolites would reveal a characteristic pattern. In fact, however, no consistent abnormalities have been detected in large groups of patients with Tourette's disorder, whether the tissue sampled is serum, cerebrospinal fluid or brain.

Management of the various symptoms of Tourette's disorder remains largely supportive, especially educational, and pharmacologic. Some commonly used pharmacologic interventions are listed in Table 12.6. Clearly the main thrust of medical management for Mickey will be directed toward his obsessive-compulsive behaviors and his rage attacks. In all probability he would require a combination of an SSRI agent other than paroxetine and an atypical antipsychotic agent such as risperidone.

Related Cases

There are additional cases relevant to this chapter available for review on the text DVD and Web site.

Suggested Readings

Dooley, J. Tic disorders in childhood. *Semin. Pediatr. Neurol.* 13 (2006): 231–242.

Faridi, K., and O. Suchowersky. Gilles de la Tourette's syndrome. *Can. J. Neurol. Sci.* 30, Suppl. 1 (2003): S64–S71.

Robertson, M.M. Mood disorders and Gilles de la Tourette's syndrome. An update on prevalence, etiology, comorbidity, clinical associations, and implications. *J. Psychosom. Res.* 61 (2006): 349–358.

Shavitt, R.G. et al. Tourette's syndrome. *Psychiatr. Clin. North Am.* 29 (2006): 471–486.

Singer, H.S. Tourette's syndrome: From behaviour to biology. *Lancet Neurol.* 4 (2005): 149–159.

Web Sites

American Tourette Syndrome Association: www.tsa-usa.org

National Institute of Neurological Diseases and Stroke: www.ninds.nih.gov/disorders/tourette/detail_tourette.htm

Section III

Supplementary Considerations: Rehabilitation and Ethics

Chapter 13

Neurorehabilitation

Objectives

- To understand the definition of rehabilitation
- To become familiar with members of a rehabilitation team and their roles in the rehabilitation process
- To be able to explain the rehabilitation approach in common neurological conditions with functional deficits
- To be aware of differences in the approach to rehabilitation in the pediatric population in comparison with that in adults
- To know useful online resources that can help in clinical practice

13.1 Introduction

Rehabilitation is an active, goal-oriented process maximizing resumption of function. It represents hope of recovery. In 2001, the World Health Organization (WHO) developed the *International Classification of Functioning, Disability and Health* (ICF) (Figure 13.1). This important framework reminds us of the complexity of human function and allows dissection of factors that limit maximal performance. It therefore facilitates clear goal setting and treatment.

Common neurological diagnoses that require rehabilitation services include acquired brain injury, spinal cord injury and stroke. In children, all of these diagnoses exist but in smaller numbers. Rehabilitation principles and services are also utilized in childhood-acquired disabilities such as cerebral palsy, spina bifida, and neuromuscular diseases. In genetic and perinatally acquired conditions, the concept of resumption of function, or *rehabilitation,* is replaced by the process of maximizing function during growth and development through enhancement of the environment, or *habilitation.*

13.2 Team

Resuming function is often examined within three spheres: physical, cognitive and behavioral. It frequently requires a "*team*" (Figure 13.2), with the patient and family at the center. The size and composition of this team depends on the challenges faced by the individual participating in the rehabilitation program.

Following the onset of a newly acquired injury where rehabilitation is required, a physical medicine and rehabilitation assessment is frequently requested. This process involves:

1. A medical review of diagnosis, investigation and treatment
2. Definition of the neurological deficit and resultant alteration in function
3. Determination of rehabilitation readiness

Early involvement of rehabilitation specialists is important, as even prior to rehabilitation readiness, principles for early preventative care are frequently recommended. These may include muscle and joint stretching, medication or bracing to facilitate maintenance of range of motion, the use of pressure-relieving surfaces to prevent pressure sores,

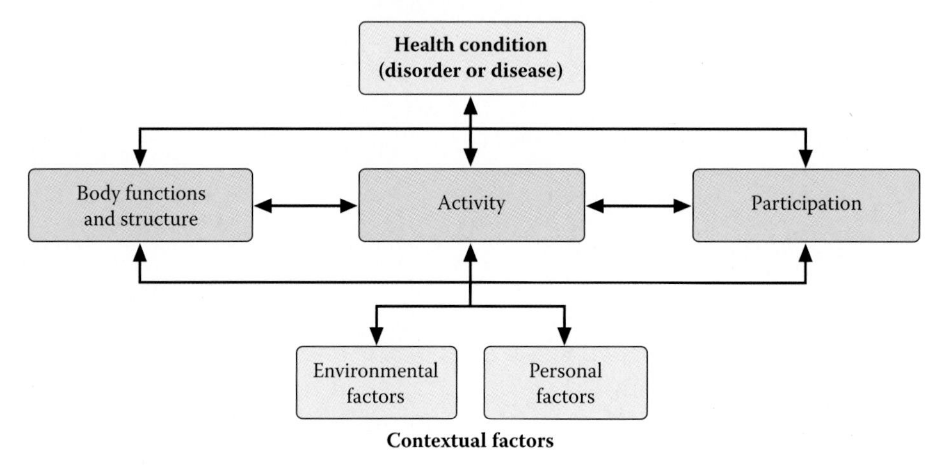

FIGURE 13.1 World Health Organization (International Classification of Functioning Disability and Health). Geneva: World Health Organization, 2001. (Used with permission.)

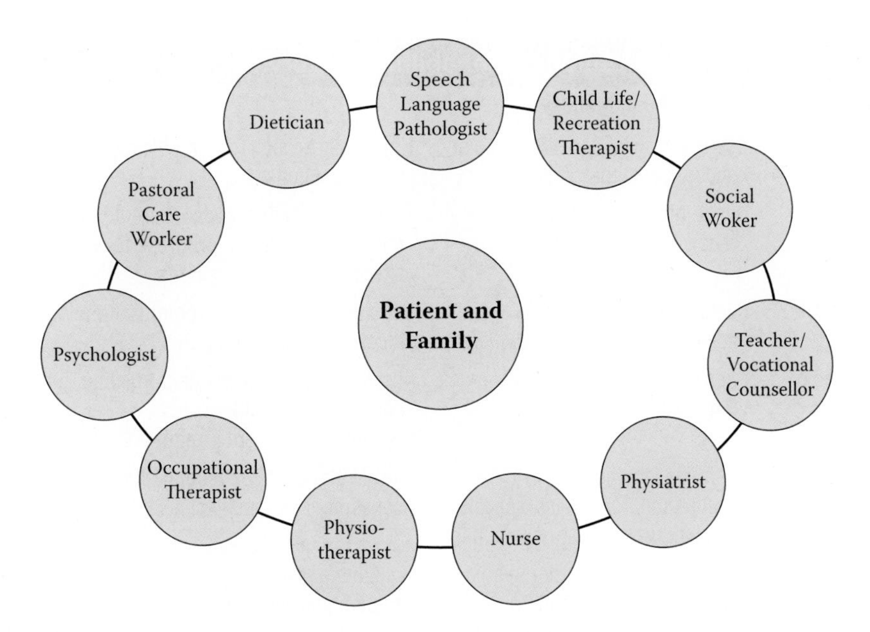

FIGURE 13.2 Rehabilitation team schematic. (Note: A physiatrist is a physical medicine and rehabilitation specialist.)

enhancement of nutrition, teaching early communication strategies, and provision of specialized family support. The goal in early intervention is to minimize complications and maximize health and well-being, so the patient is in the best possible condition to undertake rehabilitation.

Centers providing rehabilitation services may have varying admission criteria. Quite frequently these will include medical stability of the patient, patient ability to actively participate in therapy two to three hours daily, patient ability to lay down new memories and learn new information, and a willingness of the individual with the disability to work with the rehabilitation team. When the clients are at this stage of recovery, they can often work with their health care providers to define rehabilitation goals and actively embrace the rehabilitation process. Examples of general goal areas, team members involved and therapeutic interventions frequently utilized are outlined in Table 13.1.

Once general goal areas are set, each goal may be made more specific and tailored to each patient. For example, in the area of mobility, the goal may be to achieve 10 meters of independent ambulation. Specific care plans can be developed, timelines are approximated, and hard work begins! Team meetings occur regularly to allow clear communication, problem solving as required and maximal team function.

Outcome measures are available to assess patients and their progress in rehabilitation. Certain tools are very general and look at global rehabilitation. The Functional Independence Measure is one such tool that is completed with input from the whole team. Other tools are very specific and client focused, such as goal attainment scaling. An example is the Community Balance and Mobility Score, a tool utilized often used by physiotherapists to look at balance. These tools can be extremely helpful in defining needs for caregiver support, providing feedback to clients about progress and assessing utility of specific interventions. Each tool utilized, however, must be studied carefully with respect to its strengths and limitations.

Rehabilitation does not end on discharge from the hospital. A rehabilitation stay may range from quite short (weeks) to months and in unusual cases a year. On average, however, most rehabilitation stays are two to three months. As patients transition to outpatient care, they frequently continue to work their way through the hierarchy of rehabilitation-related tasks (Figure 13.3).

13.3 Specific Conditions

13.3.1 Brain Injury

The learner will:

- Know common causes and significant impacts of brain injuries
- Appreciate the unpredictability of brain injury recovery

Table 13.1 The Planning and Organization of Rehabilitation Services

Goal Area	Team Members	Intervention
1. Maximize range of motion	Client, physician, physiotherapist, occupational therapist	Heat, daily stretch, oral medication, bracing, botulinum toxin injections
2. Mobilization	Client, physiotherapist, rehabilitation assistant	Range of motion exercise, core strengthening, increase bed mobility and sitting, progress to walking when ready, explore alternative mobility options
3. Maximizing nutrition and initiation of oral feeding	Client, speech therapist, occupational therapist, dietician	Bedside assessment, oromotor stimulation, videofluoroscopy if concerns re: aspiration
4. Assessment of activities of daily living	Client, nurse, occupational therapist	Assess dressing, grooming, toileting, bathing, feeding, community mobility Prescribe equipment if required
5. Cognitive assessments with determination of beneficial learning strategies	Client, psychologist, occupational therapist, speech and language therapist, teacher	Formal standardized testing when able to participate Assessment in a small classroom setting
6. Community reintegration	Client, occupational therapist, nurse, child life, recreation therapist, teacher, social worker	Facilitate passes to home, school and community venues Problem solve around reintegration issues
7. Family support	Social worker, psychologist, pastoral care, entire team	Counseling sessions re: catastrophic injury, dealing with loss, coping with stress
8. Discharge planning	Client, nurse, social worker, entire team	Establish outpatient resources and transition plan

- Recognize the assessment tools utilized to evaluate the severity of the injury
- Appreciate how treatment goals are altered depending on level of consciousness
- Be aware of an excellent clinical resource for brain injury management

To put the material in this section in perspective, it would be helpful for the reader to work through problem 12e-4 (Eddie) on the text DVD.

Injuries to the brain are acquired in many ways and represent a major health concern worldwide. Common causes in North America include motor vehicle collisions, violent assaults, and sporting accidents. Financial impacts are profound, with medical and rehabilitation costs, as well as loss of income.

This diagnosis presents the greatest prognostic challenge. It is extremely difficult to predict outcome. When counseling families the term *predictably unpredictable* is useful. There is no single evaluation tool or even combination of tests that can predict ultimate outcome. Measures such as the Glasgow Coma Scale, the Glasgow Outcome Scale, and the durations of loss of consciousness and post-traumatic amnesia all evaluate the severity of initial injury; nevertheless, it is difficult to reassure families with

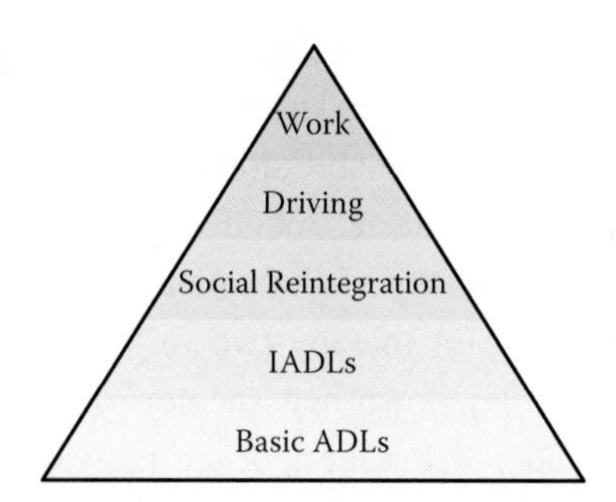

FIGURE 13.3 Hierarchy of rehabilitation goals. ADLs: activities of daily living (for example, mobility, feeding, dressing and grooming); IADLs: instrumental activities of daily living (more complex activities not always done daily but important in daily living, for example, banking and grocery shopping). (Courtesy of Dr. S. Marshall.)

certainty. Neuroradiological studies and neuropsychological evaluations help to predict deficits and define function but there is a lack of precision. It is simply amazing to watch individuals who have been in a coma for months suddenly start talking and eventually to watch them walk out of the hospital. It is equally frustrating to watch people stall and never regain meaningful communication with their environment.

It is notable that many individuals will quote neuropsychological evaluations at two years as being prognostic of

Table 13.2 Rehabilitation Principles Guided by Ranchos Los Amigos (RLA) Scale of Cognitive Functioning

RLA Level	Functional Description	Approach	Goals
Level 1	No response	Stimulation—orientation	Prevent sensory deprivation, elicit responses, improve attention
Level 2	Generalized response • Inconsistent and nonpurposeful	Provide heightened sensory input in a structured way Present one sensory stimulus at a time preceded by a brief explanation Sessions lasting 15 to 20 minutes	
Level 3	Localized response • Specifically but inconsistently to stimuli		
Level 4	Confused—agitated Heightened state of activity with severely decreased ability to process information	A quiet, highly structured environment Constant supervision for safety Calm reassurance and orientation information Physical or medical constraint if danger to self or others	Decrease intensity, duration and frequency of irritability/agitation Increase focus and attention to environment
Level 5	Confused—inappropriate • Appears alert and is able to respond to simple commands • Easily confused with complex information and frequent inappropriate responses	Team goals can be established (physical, cognitive and behavioral) Tasks highly structured to approach goals. Environment remains as constant as possible. Home passes should be initiated (careful supervision)	Establish focused transdisciplinary goals Introduce home environment
Level 6	Confused—appropriate • Starts to show goal-directed behavior but dependent on external cues/environment • Little carryover for new learning	Passes are short then lengthened to allow transition to home	
Level 7	Automatic—appropriate • Appropriate behavior within routine structured environment • Begins to show carryover for new learning • Judgment impaired	Slowly increase complexity of goals in the home and in the community Transition patients from dependent in a familiar environment to independent in society	Community reintegration
Level 8	Purposeful—appropriate • Alert, oriented • Evidence of new learning • Responsive to diverse environments within society • Cognitive and behavioral challenges may still be present		

Modified from Buck H. Woo, Ph.D., and Shanker Nesathurai, M.D., eds. *The Rehabilitation of People with Traumatic Brain Injury.* Boston: Boston Medical Center, 2000. (Used with permission.)

long-term cognitive function. Although neuropsychological testing is one of the most useful guides, it is to be used carefully. There are many examples of slow-to-progress cases that make significant improvements outside this timeline. As well, it is important to remember that results of one-on-one structured cognitive tests do not always predict function within a group or social setting.

Treatment of an individual with brain injury varies depending on level of consciousness; utilization of the Ranchos Los Amigos Score can be useful in directing therapy (Table 13.2).

It must be emphasized that physical, cognitive and behavioral realms all require evaluation and treatment plans. All team members are frequently essential. Therapist involvement is usually slowly decreased over two years but then may be required once again during times of transition (e.g., graduation from school, introduction into the workplace, independent living and marriage). Ensuring the appropriate intervention at the appropriate time is essential in maximizing treatment.

Keeping up with recommended treatment in rehabilitation is often challenging. Web-based clinical literature summaries and expert analysis can be extremely useful. Such a database exists for brain injury: Evidence-Based Review of Acquired Brain Injury (ERABI; see Web sites listing at the end of the chapter).

13.3.2 Spinal Cord Injury

The learner will:

- Know the difference between a complete versus incomplete spinal cord injury
- Appreciate the meaning of the level of injury and tests to determine this level
- Have knowledge of key aspects of teaching and care, including management of autonomic dysreflexia, neurogenic bladder, neurogenic bowel and skin care
- Gain an appreciation for timelines for recovery and reasons for hope

To help put this section in perspective, the reader is invited to revisit the story of Cletus (Chapter 5).

The first key to planning rehabilitation for an individual with a spinal cord injury is the designation of the injury as complete or incomplete. When the spinal cord reflexes have returned below the level of the lesion (signaled by return of simple sensory-motor pathways such as the anal wink or bulbocavernosus reflexes), the individual is out of spinal shock (see Chapter 5). Spinal shock can last from one day to three months post-injury, with an average of three weeks. The lesion can then be described as complete (with absence of sensory and motor function in the lowest sacral segment) or incomplete (having some degree of sensory and motor function in the lowest sacral segment). The level of the sensory and motor deficit can then be defined.

The American Spinal Injury Association (ASIA) scoring system is utilized to standardize documentation. The level of the lesion is defined as the most caudal segment that has a normal test of both sensory and motor innervation.

Table 13.3 Key Muscles to Test for Spinal Cord Level of Injury

Myotome	Index Muscle	Action
C5	Biceps brachii	Elbow flexors
C6	Extensor carpi radialis	Wrist extensors
C7	Triceps	Elbow extensors
C8	Flexor digitorum profundus	Finger flexors (e.g., FDP of middle finger, distal interphalangeal joint)
T1	Abductor digiti minimi	Small finger abductor
L2	Iliopsoas	Hip flexors
L3	Quadriceps	Knee extensors
L4	Tibialis anterior	Ankle dorsiflexors
L5	Extensor hallucis longus	Long toe extensors

There are certain key muscles, the evaluation of which helps define the motor level. Each key muscle must have a grade of 3/5 or full antigravity power to be intact (Table 13.3 and Table 2.2).

The most caudal sensory segment must be graded as normal for pinprick and soft touch. The face is used as the normal control point.

For pinprick testing, the patient must be able to differentiate the sharp and dull edge of a safety pin. Each segment is scored according to the individual's response (Table 13.4).

For light soft touch, a cotton tip applicator is compared to the face sensation (Table 13.5).

This information can then be inserted into a comprehensive chart provided by the American Spinal Injury Association (Figure 13.4). This clearly defines the injury.

Table 13.4 Sensory Testing by Pinprick

Scores:	0	Absent	Not able to differentiate between the sharp and dull edge
	1	Impaired	The pin is not felt as sharp as on the face, but the patient is able to differentiate sharp from dull
	2	Normal	Pin is felt as sharp as on the face

Table 13.5 Sensory Testing by Soft Touch

Scores:	0	Absent
	1	Impaired, less than on the face
	2	Normal, same as on face

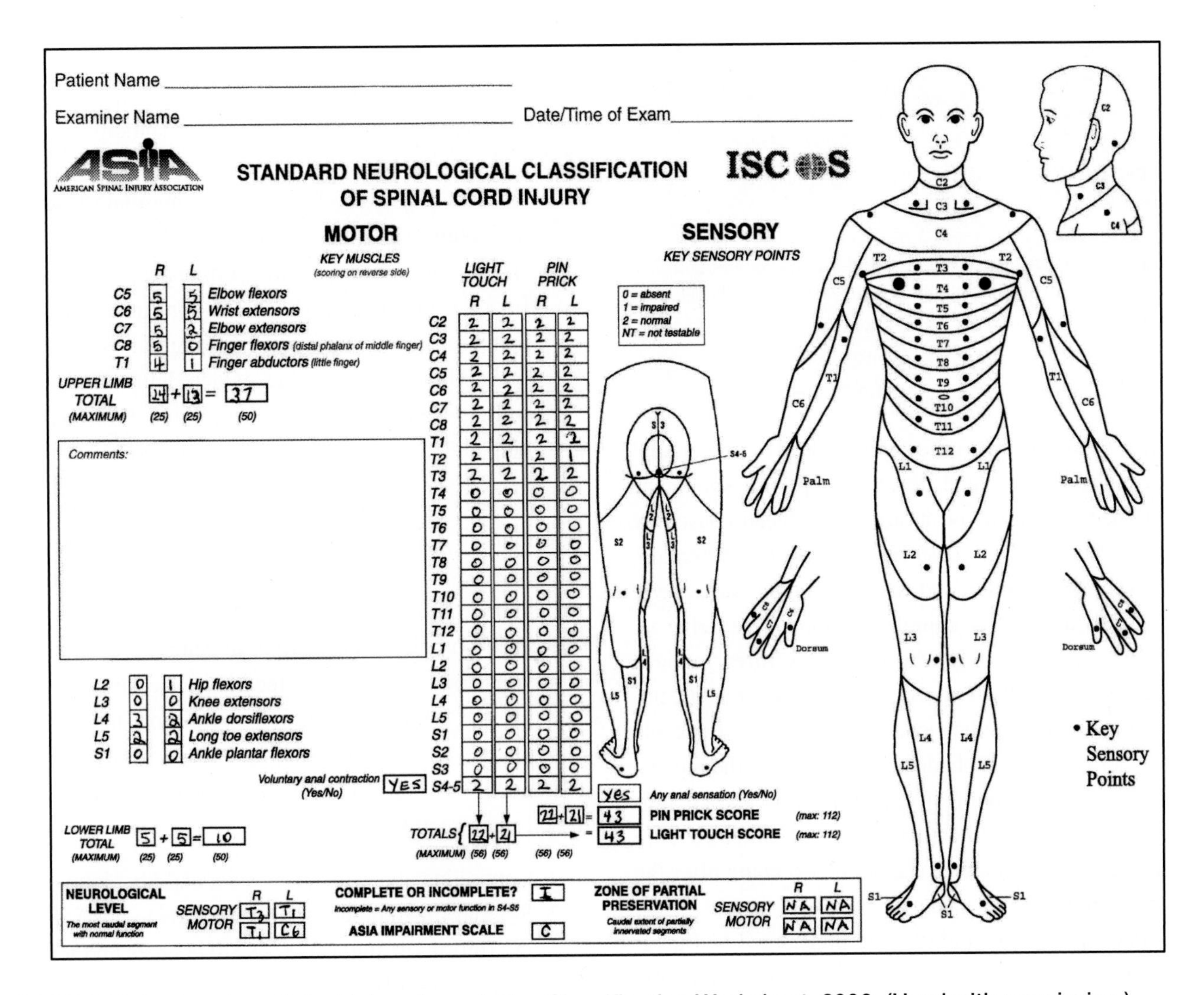

Patient Name ____________

Examiner Name ____________ Date/Time of Exam ____________

ASIA — AMERICAN SPINAL INJURY ASSOCIATION

STANDARD NEUROLOGICAL CLASSIFICATION OF SPINAL CORD INJURY

ISCoS

MOTOR

KEY MUSCLES (scoring on reverse side)

	R	L	
C5	5	5	Elbow flexors
C6	5	5	Wrist extensors
C7	5	2	Elbow extensors
C8	5	0	Finger flexors (distal phalanx of middle finger)
T1	4	1	Finger abductors (little finger)

UPPER LIMB TOTAL (MAXIMUM) 24 (25) + 13 (25) = 37 (50)

Comments:

	R	L	
L2	0	1	Hip flexors
L3	0	0	Knee extensors
L4	3	2	Ankle dorsiflexors
L5	2	2	Long toe extensors
S1	0	0	Ankle plantar flexors

Voluntary anal contraction (Yes/No) YES

LOWER LIMB TOTAL (MAXIMUM) 5 (25) + 5 (25) = 10 (50)

SENSORY

KEY SENSORY POINTS

0 = absent
1 = impaired
2 = normal
NT = not testable

	LIGHT TOUCH R	LIGHT TOUCH L	PIN PRICK R	PIN PRICK L
C2	2	2	2	2
C3	2	2	2	2
C4	2	2	2	2
C5	2	2	2	2
C6	2	2	2	2
C7	2	2	2	2
C8	2	2	2	2
T1	2	2	2	2
T2	2	1	2	1
T3	2	2	2	2
T4	0	0	0	0
T5	0	0	0	0
T6	0	0	0	0
T7	0	0	0	0
T8	0	0	0	0
T9	0	0	0	0
T10	0	0	0	0
T11	0	0	0	0
T12	0	0	0	0
L1	0	0	0	0
L2	0	0	0	0
L3	0	0	0	0
L4	0	0	0	0
L5	0	0	0	0
S1	0	0	0	0
S2	0	0	0	0
S3	0	0	0	0
S4-5	2	2	2	2

TOTALS (MAXIMUM) 22 (56) + 21 (56); 22 (56) + 21 (56)

Any anal sensation (Yes/No) Yes

PIN PRICK SCORE (max: 112) 43

LIGHT TOUCH SCORE (max: 112) 43

NEUROLOGICAL LEVEL — The most caudal segment with normal function: SENSORY R T3, L T1; MOTOR R T1, L C6

COMPLETE OR INCOMPLETE? I — Incomplete = Any sensory or motor function in S4-S5

ASIA IMPAIRMENT SCALE C

ZONE OF PARTIAL PRESERVATION — Caudal extent of partially innervated segments: SENSORY R NA, L NA; MOTOR R NA, L NA

FIGURE 13.4 American Spinal Injury Association Classification Worksheet, 2006. (Used with permission.)

If there is residual power or sensation below the level of the lesion there is hope for progressive recovery. Most recovery occurs in the first six months but it can take years to determine final functional capacity.

If the lesion is complete (motor and sensory), the focus is on strengthening the innervated muscles (up to two segments below the lesion) and teaching the client about maximizing independence. This includes intensive teaching about:

- The medical emergency of autonomic dysreflexia in lesions above T6 (see below)
- Spasticity management, including stretching, avoidance of provoking factors and medication options
- Care of insensate skin to avoid pressure sores, including pressure relief techniques and proper seating options
- Care of neurogenic bladder and bowel to avoid renal failure and social embarrassment

It is important to remember that very high levels of independence can be achieved in spinal cord injury, particularly in injuries affecting the cord below C6.

Autonomic dysreflexia (AD) is potentially life threatening (see Table 13.6). It occurs in individuals with a spinal cord injury at level T6 or above. Noxious stimuli to intact sensory nerves below the level of injury lead to unopposed sympathetic outflow and resultant dangerous blood pressure elevation. Due to the spinal cord injury, normal counterbalancing parasympathetic outflow has been disrupted. Parasympathetic outflow through cranial nerve X (vagus), can cause reflexive bradycardia. This does not, however, compensate for the severe sympathetic-induced vasoconstriction.

Table 13.6 Signs and Symptoms of Autonomic Dysreflexia

Slow pulse or fast pulse	Flushed (reddened) face
Hypertension (blood pressure greater than 20 mm Hg from baseline)	Nasal congestion
Pounding headache (caused by the elevation in blood pressure)	Red blotches on the skin above level of spinal injury
Restlessness	Sweating above the level of spinal injury
Nausea	Goose bumps

AD is a medical emergency and can lead to intracranial hemorrhage, seizures and death. The patient should be transitioned to a sitting position with the head constantly elevated. The bladder must be checked for distention, the bowel for impaction and the skin for pressure areas. Any cause of noxious stimulation should be relieved. Blood pressure should be checked every three minutes and treated medically if systolic blood pressure is greater than 150 mmHg. Medications commonly used include nitroglycerine paste (applied above the level of the injury), nifedipine capsules (immediate release form) and IV anti-hypertensives in a monitored setting. The blood pressure will need to be monitored for at least two hours after the resolution of an AD episode.

Bladder function is important to understand and manage (Figure 13.5). The bladder and its internal sphincter (bladder neck) have dual innervation. The stimulation of the cholinergic (parasympathetic) system allows normal voiding, causing contraction of the bladder musculature and relaxation of this sphincter. The sympathetic system allows storage with relaxation of the bladder and contraction of the sphincter. It is the balance between these two systems that determines if the bladder is storing or emptying. Injury to the spinal cord can disturb this balance and result in a bladder that stores excessively and overflows or fails to store due to excessive bladder contraction or sphincter failure.

From a practical perspective, when neurological injury has interfered with bladder function, it is important to remove a continuous catheter as soon as clinically possible. This will limit the possibility of infection. Often there is a combination of poorly coordinated contractions and incomplete emptying. Management of the bladder during recovery frequently includes:

1. Asking patients to try to void.
2. Checking to see if there is urine remaining in the bladder (with ultrasound or catheterization).
3. If the remaining urine is greater than 75 to 100 cc, catheterizing every four hours to maintain bladder volumes less than 350 cc.
4. If this does not achieve continence, anticholinergic medication or sympathomimetics can increase storage and prevent irregular contractions. Urodynamic studies, which define bladder filling, bladder pressures, and urine flow, will help define the need for these medications.

If recovery is not expected, catheterizing every three to four hours to maintain volumes less than 350 cc and normal pressures is recommended to maintain renal health.

In spinal cord injury, there is often evidence of upper motor neuron bowel dysfunction. In general, there is loss of spontaneous control of defecation. The key to treatment of the neurogenic bowel is training individuals to stimulate reflex evacuation and maintaining normal amounts of soft stool. This can be done using the following clinical treatments:

1. Counseling about adequate water and fiber intake
2. Regular timed toileting with attention to positioning (hip-thigh angle at 90 to 110 degrees and feet well supported)
3. Utilizing the gastrocolic reflex by toileting 15 to 20 minutes after a meal with digital stimulation of the anal sphincter if required

When these have been tried and there are still difficulties, stool softeners can be used if the stool is too firm. Oils (mineral oil or lansoyl) may be employed to help transit of the stool and bulk-forming agents may help as further distention of the bowel aides in emptying. In addition, suppositories may be used to stimulate emptying. As a last resort, bowel stimulants can be used. Usage of the latter may result in dependence over the long term. Enemas may be required initially, particularly if constipation has been prolonged.

Alteration in sexual function is also commonly experienced in patients following spinal cord injury. Individuals may be hesitant to speak of these issues and medical professionals may need to ask directly about these concerns.

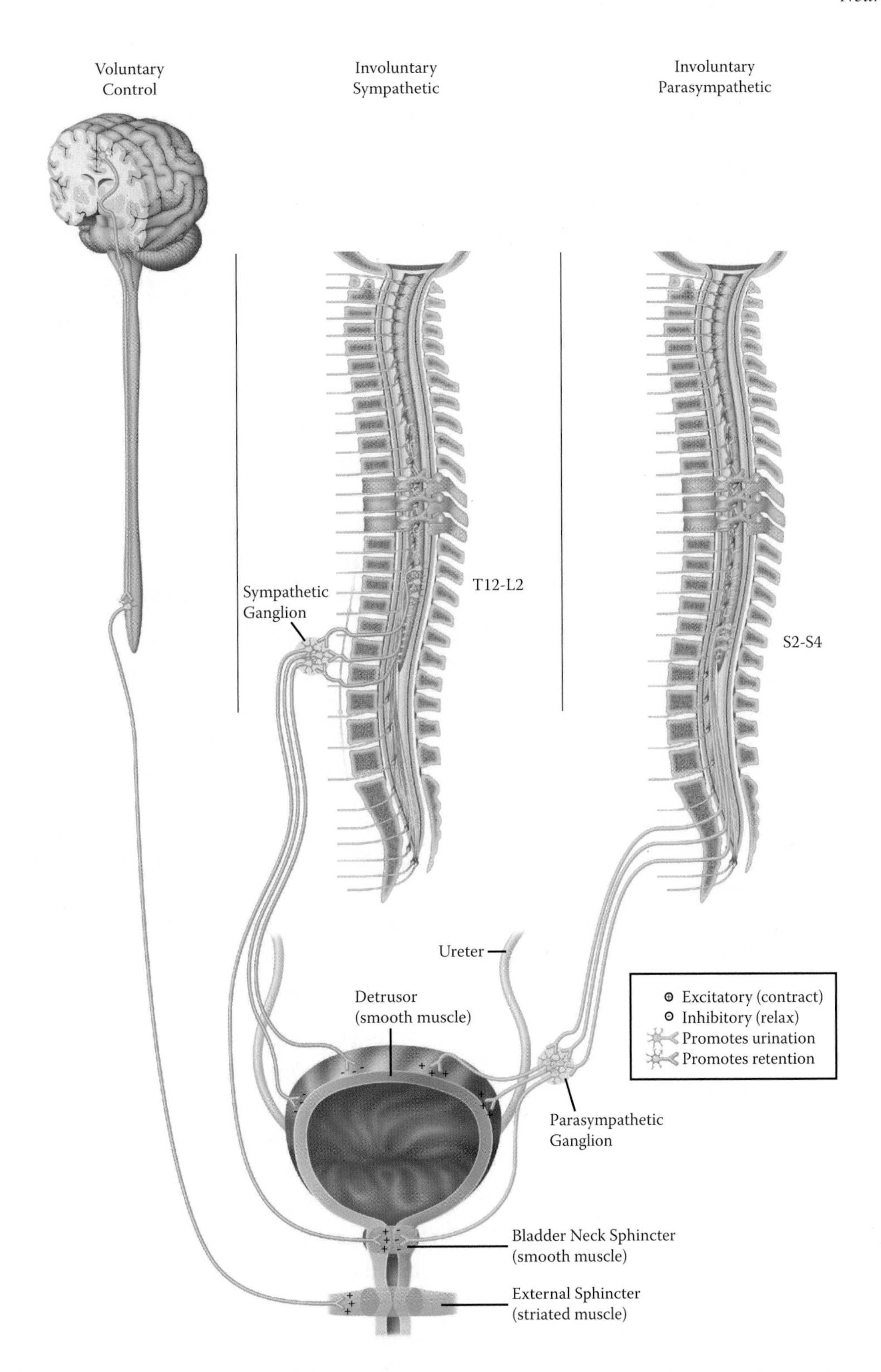

FIGURE 13.5 Bladder innervation: The effect of each system on the bladder (detrusor) muscle and the internal (smooth muscle) and external sphincters (striated muscle) is indicated by + = excitatory and - = inhibitory. The parasympathetic system (S2, 3, 4) promotes emptying of the bladder (indicated in green). The sympathetic system (T12, L1, L2) promotes urinary storage (indicated in red). The cortical voluntary pathway goes to lower motor neurons in the sacral cord (the conus medullaris) and these neurons control the external voluntary sphincter; destruction of the descending influence may lead to urinary incontinence.

In male patients, difficulty with initiating or sustaining firm erections is frequently reported. In general terms, there are two regions of the spinal cord implicated in these difficulties. Outflow from T12–L2 (sympathetic nervous system) is primarily responsible for the psychogenic component of the erection. Spinal cord injuries above T12, therefore, often result in difficulties in erections stimulated by arousing thoughts originating from the brain. Reflex erections stimulated by tactile input are mediated through S2–S4 (parasympathetic nervous system via the cauda equina). Lesions below S1, therefore, limit input into this reflex and diminish response. A lesion between S1 and L3 decreases neuronal communication between both centers. This communication is important for increasing the strength and sustainability of the erectile response.

Patients need to know that these problems exist and that there are a number of possible treatments for erectile dysfunction related to spinal cord injury, including physical techniques, medications and surgical procedures.

In female patients concerns include diminished responses to tactile stimulation, decreased lubrication and increased spasticity during sexual activity. Each of these issues can also be addressed to help individuals improve their quality of life.

Fertility may also be a concern. Commonly female fertility is unchanged when regular menses are reestablished. This usually occurs approximately one year post-injury. There are, however, some increased risks reported with pregnancy and delivery in this population. These include increased risk of autonomic dysreflexia, urinary infections, pressure sores and premature deliveries. Specialized obstetrical attention is recommended.

In males, decreased fertility is common. Although erection can frequently be attained, ejaculation occurs in only a small percentage of individuals with spinal cord injury and sperm quality is often diminished. There is hope for fertility but treatment may be required through a fertility clinic and a center that has experience in treating individuals with spinal cord injury.

Whether an injury to the cord is complete or incomplete, there is certainly reason for optimism. There are many promising areas of research, including neuroprotection, nerve regrowth and stem cell work. Excellent care of an injured patient is essential, for both present and future opportunities.

13.3.3 Stroke

The learner will:

- Appreciate the different etiologies for stroke, the variability in presentation of neurologic deficits post-stroke and thus the variation in needs from a rehabilitation perspective
- Gain knowledge about indicators that assist in the choice of appropriate candidates for intensive rehabilitation programs
- Be aware of the urgency of timely swallowing assessment post-stroke
- Appreciate the complexity of communication-based difficulties after neurologic injury
- Be aware of possible timelines for recovery
- Be introduced to an excellent evidence-based database for stroke rehabilitation

To help put this section into perspective, the reader is invited to revisit the story of Etienne (Chapter 8).

A stroke is the neurologic injury that occurs as a result of one of a number of vascular disease processes. Stroke is classified into two major types:

- Brain ischemia due to thrombosis, embolism or systemic hypoperfusion (approximately 80 percent)
- Brain hemorrhage due to direct intracerebral bleeding or to an adjacent subarachnoid hemorrhage (approximately 20 percent)

The challenges a stroke survivor experiences are highly dependent on the location and extent of the vascular insult. This, in turn, determines the patient's needs from a rehabilitation perspective.

The first issue that often faces a rehabilitation clinician is whether a patient is a good candidate for rehabilitation interventions. Initial cognitive assessments are very important. Individuals must learn throughout the rehabilitation process. Evidence of severe memory problems or an inability to follow instruction significantly limits one's ability to participate in rehabilitation. Other poor prognostic indicators include previous stroke, advanced age, prolonged urinary and bowel incontinence and severe visuospatial deficits.

Swallowing problems need careful assessment early in the course of this condition. Speech and language pathologists and occupational therapists will complete a bedside assessment. If concerns are noted, a videofluoroscopic assessment of the patient's swallowing in the radiology suite, with therapist involvement, may be required. This will assess coordination of swallowing and check for aspiration of solids or liquids into the lungs. Aspiration may produce surprisingly few signs and symptoms; subsequent pneumonias can be life threatening. Careful attention to swallowing issues is therefore imperative to good care.

Speech problems associated with stroke are also common and can be complex. They may be receptive, expressive or both. Frequently, the Boston classification system is utilized. The problems with language are separated into those that are fluent (with smooth flow of word production, Table 13.7) or nonfluent (without such flow, Table 13.8). An experienced speech and language pathologist can assist in the diagnosis of this stroke-related problem and develop a treatment plan for communication. The rehabilitation team that is working with the client will then follow these recommendations to maintain consistency and promote learning in this area. If there is evidence of receptive abilities but significant language output problems, communication tools including newly designed computerized output devices may be quite useful.

Table 13.7 Fluent Aphasias

Wernicke	Frequent word substitutions
	Limited comprehension
Anomia	Word finding difficulties
	Good comprehension and language use
Conduction	Word substitutions with similar sounding syllables
	Difficulty with repetition
	Good receptive function/comprehension
	Bursts of fluent language
Transcortical sensory	Repetition possible
	Poor receptive function/comprehension
	Language frequently incomprehensible

Table 13.8 Nonfluent Aphasias

Broca	Speech with mostly nouns and verbs (like a telegram)
	Associated with apraxia
	Comprehension intact except for more abstract tasks
Transcortical motor	Naming and repetition possible
	Auditory comprehension intact
	Low quantity of verbal language output
Global	Severe language dysfunction (receptive and expressive)
Mixed transcortical	Repetition possible
	Decreased auditory comprehension
	Low quantity of verbal output

From a motor perspective, the majority of recovery occurs in the first three to six months following the stroke. Multiple hands-on techniques are used by therapists to facilitate recovery. As a clinician, it is important to support weak upper extremities with the goal of preventing subluxation of the shoulder and pain syndromes. When high tone interferes with function, stretching, avoiding nociceptive input, bracing, botulinum toxin injections and oral tone-mediating agents can be extremely useful. This will allow therapists to focus on strengthening antagonist muscles and maximizing function.

There are summaries of critically appraised data for the clinician now available on the Internet and in written form. The Evidence-Based Review of Stroke Rehabilitation (see Web sites section at the end of this chapter) is an ongoing review and synthesis of data that examines practices in stroke rehabilitation. This comprehensive information can guide best practices. The new treatments and aspects of prevention described are constantly evolving; this gives hope to survivors of stroke.

13.3.4 Pediatrics

The learner will:

- Appreciate the differences in rehabilitation of children (versus adults) that result from ongoing growth and development
- Understand the concept of "growing gap"
- Gain knowledge about unique features of common childhood acquired disabilities, such as cerebral palsy and spina bifida
- Be aware of advances in pediatric rehabilitation, including classification of cerebral palsy and spasticity management

To help put the material in this section in perspective, it would be helpful for the reader to work through problem 7e-4 on the text DVD.

Rehabilitation in the pediatric population has many unique features. A child is rapidly changing in size, cognitive skill and behavior. Physically, children with motor deficits are at great risk for rapidly developing soft tissue contracture or orthopedic complications, such as scoliosis or hip subluxation. Appropriate stretching, bracing and spasticity management are essential to keep muscles growing along with rapidly growing bones. Frequent equipment changes are also required as needs change with development and growth.

From a developmental and behavioral perspective there is also rapid change! Learning is a very active process; children are adaptable and studies regarding neuroplasticity hold great hope. From another perspective, children who have physical, cognitive or behavioral challenges secondary to neurological insult are at risk of developing a gap between their skills and the skills of normally developing peers. As rapid development occurs, this gap may progressively enlarge. This is referred to as the concept of *growing gap*. The role of many clinicians in the field of pediatric rehabilitation is to promote development and limit the size of this gap. Frequent reassessments and therapeutic interventions are required, particularly during times of change and transition, for example, school entry, entry to elementary school, transfer to high school, planning for after-school training or moving out of a parent's home. Each of these transitions may require environmental alterations, new training and changes to resources or equipment. Rehabilitation in children must match the rate of change.

Diagnostically, the patient population requiring rehabilitative intervention in the pediatric group is quite different from that in adults. There is a high percentage of individuals with congenital, neonatal or perinatally acquired causes of disability, such as cerebral palsy, spina bifida and neuromuscular disorders. In these cases, the lack of a premorbid baseline of normal function may present a challenge. Early intervention is the standard of care, with the provision of an enriched environment and age-appropriate habilitative treatment plans. The children are not trying to regain function that was already established, but are trying to maximize the potential they have.

Cerebral palsy is the most common cause of complex physical disability in childhood, with an incidence of 1.5 to 2.5 per 1000 live births. CP is defined as "a group of disorders of *development of movement and posture* causing activity limitation that are attributed to *nonprogressive* disturbances in the fetal or infant *developing brain*" (see Bax et al. in the Suggested Readings). The underlying motor deficit is frequently accompanied by abnormalities in sensation, cognition, communication, perception and behavior, as well as seizures. The clinical picture is therefore complex.

In the last decade, researchers have established a motor classification for cerebral palsy* that allows prediction of whether a child will eventually walk independently or will require the support of a walker or a wheelchair for independent mobility. This tool, the Gross Motor Function Classification System, can allow motor prediction at the age of two and the classification appears to be stable into the second decade of life. Any clinician working with individuals with this diagnosis should be aware of this scale, as this knowledge will help in the setting of reasonable motor goals.

Management of high tone states (spasticity, rigidity) has always been a challenge in the area of cerebral palsy. Over the past two decades, management of spasticity, as well as involuntary movements (dyskinesias) has changed. Newer options such as botulinum toxin injections in spastic muscles, intrathecal placement of baclofen, and surgical cutting of dorsal nerve root fibers have all been investigated and present alternatives for those individuals who are good candidates for such interventions. As well, oral anti-spasticity medications such as baclofen, dantrolene and tizanidine can be considered after careful consideration of the benefit versus side-effect ratio. The clinical key is careful assessment of patients to match the proper candidate with the appropriate intervention. Specialty clinics have been developed to facilitate this decision making.

There have also been improvements in the management of infection, nutrition and musculo-skeletal deformities. These improvements, coupled with early evidence of functional changes in the brain on fMRI studies following therapeutic intervention, provide reasons for hope. There is great need for clinicians to work hard to promote health and well-being in these patients throughout their life spans.

Spina bifida, though decreasing in incidence, continues to be a significant cause of childhood acquired disability. The complexity of this condition should not be underestimated. This population is most challenging

* (See suggested readings: Palisano et al., 1997.)

from a rehabilitation perspective. The combination of an abnormality in spinal cord development (with motor deficits, sensory deficits, neurogenic bladder and bowel) and concomitant cerebral malformations (with cognitive limitations) results in the need for a flexible individualized habilitation program.

13.4 Conclusion

In general, ongoing advancements in the field of rehabilitation lead to hope for improvement. The use of functional MRI in training, botulinum toxin and intrathecal baclofen in the treatment of spasticity, electrical stimulation to prevent muscle wasting, new bracing techniques to treat contractures, as well as the use of body weight support systems and robotics to support gait training, allow highly skilled team members to develop challenging treatment protocols. Internet databases with expert evaluation of literature are now becoming available to assist the clinicians in dealing with the ever-growing information explosion. Clinicians, clients and families can therefore be actively involved in the rehabilitation process as they strive for maximal outcome.

Suggested Readings

Bax, M. et al. Proposed definition and classification of cerebral palsy. *Dev. Med. Child Neurol.* 47 (2005): 571–576.

DeCorwin, S. *Life after Spinal Cord Injury.* Montreal: Quebec Paraplegic Association (QPA), 1997.

DeLisa, J.A., ed. *Rehabilitation Medicine Principles and Practice,* 3rd ed. Philadelphia: Lippincott-Raven, 1998.

Molnar, G.E., and M.A. Alexander, Eds. *Pediatric Rehabilitation,* 3rd ed. Philadelphia: Hanley & Belfus, 1999.

Palisano, R. et al. Development and reliability of a system to classify gross motor function in children with cerebral palsy. *Dev. Med. Child Neurol.* 39 (1997): 214–223.

Woo, B.H., and S. Nesathurai, Eds. *The Rehabilitation of People with Traumatic Brain Injury.* Malden, MA: Blackwell Science, 2000.

Web Sites

Evidence-Based Review of Moderate to Severe Acquired Brain Injury (ERABI): www.abiebr.com

The Evidence-Based Review of Stroke Rehabilitation: www.ebrsr.com

Chapter 14

Neuroethics

Introduction

In the previous chapters various clinical neurological situations have been presented that challenge physicians, be they specialists or students, in their problem-solving abilities and mirror those found in the setting of bedside, office or consulting room. In real life there are almost always issues that complicate the management of medical problems because patients are people in an environment where the illness impacts on their personal lives, on their family, on their loved ones and, not infrequently, on society. In addition, medical decisions must take into account issues of ethics. It should be emphasized that ethics is an integral part of medical care and that good ethical practice makes for good medicine and good medicine is ethical medicine. In this chapter we will present another neurological case but one where the emphasis will be on some of these other features of medical practice.

While ethics is involved in all of medicine, there is a particularly close association between ethics and neurology. The relationship may be considered as a reciprocal one.

Ethical decisions are frequently required in the management of patients suffering from neurological diseases. There are many neurological diseases that invoke life and death situations where intervention calls ethics into play. These involve people both at the beginning of life and at the end of life, but also in the most active and productive stages of life. Examples include the treatment of anencephalic newborns, patients with fatal conditions such as amyotrophic lateral sclerosis, muscular dystrophy, multiple sclerosis and persistent vegetative states.

The understanding of the source of ethics may have its underpinnings in neurology, and neurological studies may contribute to our understanding of the basis of ethics. Humans seem to be ethical creatures and ethical behavior is dependent on functioning of the brain. Indeed, some recent work suggests that there are areas in the brain whose function is necessary for moral and ethical decision making on the part of the individual.

The format to be followed is the presentation of a theoretical medical narrative in which there are many ethical issues. The story is presented as a vignette. The reader is encouraged to think about the narrative as it unfolds and to try to see what ethical issues may be inherent in the situation. Questions will be raised at the end of each section. Try to answer them yourself before proceeding to the discussion of these questions. Superscripts refer to discussion of specific ethical dilemmas that are printed at the end of each section. Suggestions for further reading are given at the end of the chapter.

Case Study Part A

You are a consultant neurologist and have been practicing several decades.

A man comes to see you with regard to his mother, whom you have been treating for many years. Mr. C. Ewan Court is a 45-year-old lawyer who has developed a successful practice and has become partner in an important, well-established law firm. He comes to see you for information on the health of his mother, Mary Court,[1] *who has been suffering from a peculiar neurological disease for more than a decade.*

Mrs. Court, now in her mid-60s, has been living at home alone since the death of her husband. When she was in her 40s she developed some problems with balance and coordination, and she exhibited peculiar involuntary movements of her hands. She developed mannerisms where she would bring her hand to her face and then sweep it along the side of her head as if patting down her hair. Occasionally she would shrug her shoulders, first the right and then the left. These peculiar movements were considered as nervous habits or tics as she was under a great deal of stress because of her husband's poor health. He was being treated for bipolar disorder and at the time was in a depressed mood. Later she suffered from many falls and developed a lurching gait. Initially, her condition was quite puzzling to you and you have had to do a considerable amount of reading to make a diagnosis. You remain uncertain although you have some ideas.[2, 3]

She is now showing not only abnormal erratic involuntary movements but seems to be getting very "absent minded." It is clear that in addition to the movement disorder, she has a dementing illness.[4] *She has been leaving the stove on, the doors unlocked and even forgetting to close them at night. Recently, the neighbors called the fire department because they saw smoke coming out of an open kitchen window. She had apparently left the stove on and, when a frying pan caught fire, she opened the window to let the smoke out but did not turn off the burner or remove the frying pan from the stove. The family physician had advised her son that his mother should be hospitalized for her own protection and that he should consider making plans to apply to long-term care facilities on his mother's behalf. However, she is most insistent that she remain in the house that has been her home for the past 30 years and flatly refuses to consider a move.*[5]

While Mr. Court came in to ask you about his mother, it is apparent that he is also concerned about his own health and takes advantage of the visit to consult you about his concerns.[6]

Questions to Be Considered

1. Does Mr. Court have a right to know about his mother's medical condition, or is this a private matter? What would you do? Is there any ethical way in which you feel free to discuss her health with her son? Are there situations in which confidentiality can be set aside?
2. What are the most likely causes for these symptoms?
3. How do you proceed when you are uncertain of the diagnosis? When should you consult colleagues? Is it necessary to get the patient's permission before discussing it with colleagues?
4. Do you know what the diagnosis is in this case? Is this a condition that is familiar to you? If not, what do you tell the patient? How do you handle the situation where you don't know what you are dealing with or are uncertain about the diagnosis?
5. At what point do you consider it necessary to recommend that a patient be hospitalized or placed in a long-term care facility? What rights does a patient have to refuse such advice?
6. Should a specialist-consultant demand that all patients be referred by a family physician before they are seen?

Discussion

1. At this point the patient is Mrs. Court and not her son. The physician owes the patient privacy and confidentiality. **Privacy** and **confidentiality** are important ethical issues in medical practice. Every patient expects that the physician will treat any information given as being confidential and not to be divulged to any third party. This includes family members. Ethical considerations depend on certain basic principles. One of the most important of basic principles in ethics is **respect for the autonomy** of the individual. This is the basis underlying the respect for privacy and confidentiality.

The observance of the rule regarding confidentiality should include family members.

There are some exceptions to this rule. A child before the age of majority will usually be accompanied by a parent or guardian who is responsible for the care of the child. The physician will discuss the child's medical problem with the parent who is charged with the care of the child. The exact age where this applies varies from one jurisdiction to another and from province to province and state to state. The greatest difficulty arises in cases of sexual activity or birth control where a young person may seek medical advice and does not wish to have parental involvement. This is a very delicate situation and requires great sensitivity on the part of the physician. As in many areas of medical ethics, there may not be an absolute right answer, but a good rule is to consider the best interests of the patient, while still realizing that the family should be treated with respect.

Another exception occurs when the patient is not mentally or otherwise capable or competent and a family member is involved and responsible for the person's welfare. This may be formalized where a person is appointed as having power of attorney for health care; more often there is an informal understanding that a close family member or members are looking after a sick and incapable family member or loved one.

In this age of multiculturalism these matters of privacy and confidentiality, as observed in much of the Western world, may have to be modified somewhat. In certain societies it is accepted that the family plays a much more intrusive role in the affairs of all its members: sometimes an elder or the dominant female or male assumes decision-making responsibility for the other members. A husband may make decisions for his wife's medical care. While this may be offensive to Canadian practitioners, it is important that the doctor be aware of the traditions of their patients and handle them with diplomacy and sensitivity.

In this case under discussion, the son has assumed the responsibility for the care of his mother, and it may be obvious to you who have been consulting in this woman's care for many years that the son has a legitimate interest in the health of his mother and is a responsible family member. This may not always be the case, and it may be appropriate to request that the person who is your patient give formal permission before you even speak to family members.

Occasionally a person may come into your office and ask casually, "I saw my neighbor, Bill Jones, in your waiting room. Doc, how's my old friend, Bill, doing?" It may be tempting to respond in an equally casual manner, but you should reply: "I'm sorry, but I don't talk about other patients. I know that you wouldn't want me to discuss your health with Mr. Jones. Why don't you ask Bill yourself? I am sure he would be glad to know that you are interested in his health."

2. Loss of balance and incoordination commonly occur together, and when they do, they may point to certain functional areas of the nervous system. Balance or equilibrium involves appreciation of position sensory information from the lower extremities, mechanisms in the inner ear and a disturbance of monitoring and modification of movement by cerebellar pathways (discussed in Chapters 6 and 7). Involuntary movements may arise when there is pathology in the circuitry involved in the basal ganglia or may arise from pathology in the motor cortex. Abnormal movements that seem involuntary, such as tremor or tremulousness, most commonly are related to anxiety, but are also seen in hypermetabolic states such as hyperthyroidism and excessive use of stimulants such as caffeine and nicotine.

Changes of behavior are often early evidence of a dementia. Such changes, however, are not always due to organic disease; anxiety and depression may cause an "absent-mindedness" or distractibility that has been described as "pseudo-dementia." However people with depression may also show reduced activation of certain prefrontal areas of the brain on SPECT scan independent of their dementia.

3. One of the most important ethical issues in medical practice is that of **fidelity** or **faithfulness**, which may sound like old-fashioned terms and perhaps are better described as responsibility or trust, of which there are many aspects. Among them is the implication that the physician has certain duties to the patient who will have certain expectations of his or her physician. One expectation is that the physician is competent and knows what she or he is doing. The doctor is expected to be knowledgeable about diseases from which the patient may be suffering. The study of medicine is a never ending educational process, and physicians may have to spend time updating their knowledge of conditions with which they may not be familiar or have not thought about since medical school. Wise physicians know the limits of their competence. Physicians should be honest with patients and not deceive them by implying that they know more than they do. Most patients will feel even more confident in physicians who tell them of their uncertainty but assure them that they will make every effort to establish the diagnosis and to treat them appropriately.

Where the physician feels incapable about handling a specific condition or situation, she or he has a duty to refer the patient to a more knowledgeable or more experienced colleague or specialist. More often one may simply discuss the case with a colleague, but taking great pains to hide the identity of the patient. These "corridor consults" may save time, but often the consultant is put on the spot and may not be able to give appropriate advice. Furthermore some of the particular skills that a specialist has acquired include specific history taking and experience in a specific aspect of the physical examination. These provide information that cannot be easily transmitted from one physician to another. It is preferable to ask for a formal consultation. You should never let personal pride get in the way of providing the best service to your patient.

4. What do you think may be the disorder afflicting Mrs. Court?

What conditions present with the constellation of symptoms of movement disorder and cognitive decline? Your observations of the patient are most important. What kind of abnormal movements do you see? Indeed, are they abnormal or are they merely restlessness associated with anxiety? Would you describe them as tremors? Are the movements sudden, tic-like, or do they fit the description of chorea (dancing) or athetotic writhing movements. If tremor, essential tremor or the resting tremor of Parkinson's disease come first to mind, but metabolic conditions with accelerated metabolism such as hyperthyroidism may also present with tremor. Cognitive decline may be an accompaniment of Parkinson's disease but is usually not apparent until late in the course of the disease. Parkinson's disease with dementia is sometimes referred to as Lewy body disease because of the accumulation of abnormal material, in the form of Lewy bodies, inside neurons (see Chapter 9).

Also, there are a number of obscure conditions to be considered. The two most likely, although themselves rare, are Wilson's disease (hepatolenticular degeneration) and Huntington's disease (HD, Huntington's chorea). Other possibilities include the once common, but now rare, Sydenham's chorea or the even rarer dentato-rubro-pallido -luysian atrophy and one of the many variants of spino-cerebellar atrophy. Neuroacanthocytosis should not be difficult to diagnose with simple blood examination. Suspected Wilson's disease can be confirmed by the finding of corneal pigmentation (Kayser–Fleischer rings) and blood tests in which abnormally low levels of the copper-binding protein, ceruloplasmin, are found, and there is evidence of abnormal copper metabolism. Huntington's disease can be confirmed by genetic testing, which will also reveal a carrier status in 90 percent of cases; however, the diagnosis is easily established on a clinical basis alone. In both conditions neuroimaging will show abnormalities in the parts of the brain most affected. In Wilson's disease there may be atrophy of the putamen and in Huntington's, caudate atrophy will be apparent. Although helpful, these findings are not specific. The picture presented by Mrs. Court fits best with that of Huntington's disease. More information on Huntington's disease may be found in the appendix to this chapter on the text DVD and Web site.

5. There is reasonable concern that Mrs. Court may harm herself by setting her house on fire or through other inadvertent events. She plays down these dangers and resists all coaxing to move to a safer environment. While one must respect autonomy, there is another basic principle in medical ethics that may seem to be in conflict with autonomy, **beneficence**. This means that the physician should always consider what is best for the patient. She or he should always treat the patient with beneficence. In cases where the patient is not capable of acting independently, a choice may have to be made that would seem to override the autonomy of the patient in order to serve the patient's best interests. In such situations we say that "beneficence trumps autonomy." In this case, although Mrs. Court wants to remain in her home, it may be necessary to override her autonomy by admitting her to a residence where she would not be in danger of harming herself. In the past physicians assumed the controlling role and were presumed to always know what was best for a patient. This has been referred to as **paternalism** (parentalism would be a better gender-neutral term). Now patients are granted more autonomy, in part because medical knowledge is now more widespread. Consequently medical decision making should be a shared activity between caregivers and those cared for, with involvement of other interested parties such as family members and significant others as agreed upon by the patient (further discussed under 13, below).

Someone, usually a family member, will have the power of attorney to ensure that the patient, if considered incompetent, will be given the best care, which may include commitment against the person's will. Before one can force a patient to go into a hospital or chronic care facility, one must be assured that he or she is incapable of appreciating the facts of the situation. Competency adjudicators are established in many jurisdictions. The family may be well advised to seek legal counsel in questionable situations.

Commitment to a chronic care facility may be temporarily deferred by making some compromise arrangements for Mrs. Court at home. For instance, measures to ensure safety in the kitchen such as disconnecting the stove and allowing her to prepare meals with a microwave and a kettle with an automatic cut-off may be a reasonable solution in the early stages of her dementia.

6. While Mr. Court is supposedly coming in to ask about his mother, it is apparent that he is concerned about his own health and future. While most consultants request that a formal request be made from the family physician, it sometimes happens that a new patient will come in this way without formal referral. This is especially true where there is a familial or hereditary condition. While there may be concerns about "queue jumping," it is often advantageous to treat the family members as a group if there is a familial or related condition present. It is important, however, that the physician keep patients separate, using separate sets of records and utilizing treatments according to each individual's needs. It may be difficult for the doctor to avoid giving confidential information about one family member to another member. It is sometimes preferable to refer one of the patients to a colleague.

Case Study Part B

Mrs. Court finally agrees to go into hospital, but only for some tests. You obtain a CT scan of the brain and this shows considerable bilaterally symmetrical atrophy of the caudate nuclei. You now feel confident that this lady does suffer from Huntington's disease.

The following week, you receive a phone call from one of the senior partners in Mr. Court's law firm Didley, Didley, and Skwatt. He says that he and some of the other partners have noted some personality changes in Mr. Court; the caller is aware that Mr. Court has consulted you. He says that Mr. Court had been a good partner and had made major contributions to the firm, and they were even considering making him a full senior partner. He was very hard working and often stayed late at the office. Recently, however, he has on occasion gone to a bar after work with some of the secretaries. Furthermore, he has been seen at a disco bar dancing with women who appear to be decades younger than he. Some of his partners are concerned about the appropriateness of some of this socializing with younger office staff. Although he didn't say so, you sense that he was more concerned that such behavior would reflect badly on the firm; nevertheless, you also sense that there is some genuine concern over his colleague's wellbeing. You are rather surprised to receive this call and acknowledge that you have seen Mr. Court as a patient. However, you ask whether Mr. Court has authorized the caller to make these enquiries. The lawyer is somewhat taken aback and says that, although he is well aware that doctors are not allowed to divulge private information to third parties, he is concerned that Mr. Court may be mentally ill and his colleagues wanted to help. Furthermore, he adds, Mr. Court is the key person in a most important deal with their major real estate client. You try to be courteous and say that you appreciate the situation but you are not allowed to divulge any information without your patient's express permission.[7]

Questions to Be Considered

7. How would you respond to this situation? What are your obligations to the patient? What are your obligations to significant others on whom your actions may impact? What obligations does the physician have to society? When may your obligations to your patient be secondary to your obligations to other parties?

Discussion

7. This brings up again the important medical ethical issue of privacy and confidentiality as has been discussed already. One can appreciate that the partners of the law firm have legitimate concerns and are anxious to protect their interests. This situation presents a dilemma for the physician. He must be ever mindful of the confidentiality that he owes his patient but he has also some responsibility to society. The situation sometimes arises in which withholding privileged information may allow harm to come to a third party or even to a large number of people. One can think of situations where a physician might become aware of planned terrorist activities. The physician must make a decision as to which is the greater good and act accordingly. There may not be one right answer to the dilemma. A famous landmark case is that of *Tarasoff vs. Regents of University of California* (1976), where a university psychologist became aware of his patient's professed intent to murder a college student but was hesitant to report it; after the murder took place, he was charged with criminal negligence. A somewhat similar case occurred more recently in Canada where an emotionally disturbed man shot his estranged wife with a crossbow and had previously disclosed his intentions to his psychiatrist. Unfortunately the physician did not feel that the intentions were sufficiently certain to report his patient to authorities. In these cases the duty to warn is perceived as overriding the duty to confidentiality.

While our case is hardly one with such potentially fatal results the same principle applies. One way of dealing with this situation would be to discuss the work situation with Mr. Court and inquire whether he feels confident that he is able to carry on the heavy responsibility that he has to the firm. He may well be unaware of the partner's concerns. Since he has expressed his own self doubts previously to you, this may be seen as an opening for further discussion.

Case Study Part C: What about Mr. Ewan Court?

As already mentioned Mr. Court had expressed concerns about his own health. He admits that he has difficulty concentrating and finds that he is very impatient. He confides that an important client assigned to him involves a complicated legal battle regarding land claims. He feels unsure of his ability to meet the demands of the contract; this makes him anxious and has led to disturbed sleep. You notice that he does seem quite anxious. While he is talking to you he is fidgeting with his pen and with the buttons on his jacket. During the functional enquiry he confides that he has had some emotional difficulties in the past and at college had a bout of depression during which he had contemplated harming himself. He was seen by the campus physician and the diagnosis of bipolar disorder was considered at that time but he was not prescribed any medication. You carry out a thorough neurological examination but do not find any obvious abnormalities other than his nervous fidgeting. However, you start to wonder if he may be developing Huntington's disease himself.

You ask yourself whether you should suggest that he have genetic testing.[8] *You also think about the implications of genetic testing and who, other than Mr. Court, might be affected by the results: who might benefit and who might be harmed.*

Before considering genetic testing you decide that you should get more information about the family. In fact, in situations such as Mr. Court's, a detailed family history is essential.

You find out the following: Ewan Court's father died under rather mysterious circumstances. He had suffered from a bipolar disorder all his life but had been a successful businessman, in part because of his periods of untiring energy during the manic phases of his illness. Just before his death he had suffered a significant financial loss related to an unwise business decision and had become very depressed. At the time of this decision, he was in an ebullient mood and spoke about how he expected to "make a killing." Despite advice from his partners against this decision he went ahead and invested a large sum of money in what turned out to be a scam. He was on the verge of personal bankruptcy. He told the family that he wanted to be alone to consider what he should do and went to the family cottage. His body was found there several days later. The cause of death was undetermined, but there was no evidence of foul play. It was suspected that he had taken an overdose of medication that had been prescribed for depression. A probable further contributor to his disturbed emotional state was his concern over his wife's deteriorating health and the question of whether she had inherited Huntington's disease. This all took place before you had been asked to see Mrs. Court.

Ewan has a younger sister who has been living on the west coast for several decades; they have had little communication over the years. He presumes his sister is in good health. She is aware of her mother's illness. Along with her brother, she has joint power of attorney for their mother.[9]

Ewan Court has been married for more than 20 years but he admits that the stress of looking after his mother has caused considerable tension in their marital relationship. He has two teen-aged sons; he says that he would like to spend more time with them but that his job leaves little free time. He also admits that his wife has done most of the parenting.

As you have been the consulting physician for the senior Mrs. Court, Ewan's mother, you already are quite aware of her family history. She was born in western Canada and her family came from Ireland. Her mother was said to be rather queer and may have taken her own life. She had spent much time in mental institutions before she died. Her father was a farmer and was killed in a farm accident. Mrs. Court was the only member of her family in whom a definite diagnosis of Huntington's disease had been made. However, it may well be that her mother was also affected.

Questions to Be Considered

8. What are the potential harms and benefits to be considered in ordering genetic testing? Why do other family members have to be considered as well as the patient?

9. What is meant by *power of attorney*? What is implied by the terms *substitute*, *proxy* or *surrogate decision maker*?

Discussion

8. **Genetic testing** is readily available for people at risk for developing Huntington's disease. It is important that the physician be well informed about the accuracy and implications of positive or negative test results for a disease. It is also most important to be aware of the implications of

these tests for family members, not only the person being tested. Ewan's sons are potentially at risk if their father tests positive but are very unlikely to develop the disease if he tests negative. Should they be told if their father has a positive test? How would this affect them emotionally? Even if their father is positive, they have a chance that they will not develop the disease. What do you do if one of them asks to be tested, even if the father declines to be tested? Testing children raises a whole new set of ethical questions. While tests can be obtained by simply sending a sample to a lab, it is usually advisable, because of these very complex ethical issues, to refer the patient to a genetics specialist who will be prepared to make counseling available to the patient and his or her family. These matters will be discussed in more detail in the appendix in the particular context of Huntington's disease.

Various organizations have issued policy statements on genetic testing.

Several points are particularly important and are common to all genetic testing: Adult patients have a right to know whether they carry the gene. Testing must be done only on persons who are fully informed and aware of the implications of the testing (informed consent). There must be no coercion by family members or other parties for the person at risk to be tested (autonomy). Testing should not be done on minors. The results must be kept in strict confidence and not be revealed to third parties (privacy and confidentiality). However, the physician is obligated to make full disclosure of results of tests and should avoid using euphemistic terms or creating false impressions that may confuse or deceive the patient. Truth-telling is one of the most important cornerstones of a good physician-patient relationship. At the same time the skilled physician will be sensitive to the patient's feelings and anxieties and be ready to provide reassurance and support. Preferably, counseling services should be provided by people who have special expertise and experience in genetic counseling prior to the testing being done and the results revealed by the same counselors.

On occasion, complicated situations may arise in which the need of one family member to know his or her genetic status may contravene the autonomy rights of another family member. An example would be the granddaughter of an affected person wishing to know her own status, whereas her parent may refuse to have testing, even though that person is at risk. A positive test in the grandchild would obviously implicate the at-risk parent. In such a situation it is difficult to honor the rights of both the parent and the adult child.

Another reason why pretesting as well as posttesting counseling is advisable is that patients may be overwhelmed by the result if positive. Some patients have attempted suicide on learning that they may well end up like their affected parent. However, it has been shown that most patients benefit from having the test results, whether positive or negative, and the benefits outweigh the potential harm. Each case must be carefully assessed. In the case of our patient, Mr. Court, this risk is increased by the history of past propensity toward depression and thoughts of harming himself. The diagnosis of bipolar illness in his father adds another risk factor.

9. **Power of attorney (POA) substitute decision maker, surrogate decision maker, proxy decision maker,** depending on the jurisdiction, have various definitions and are not necessarily interchangeable terms allows individuals to appoint another person or organization to handle their affairs while they are unavailable or unable to do so. Durable power of attorney for medical care allows individuals to appoint another person to make decisions for medical care when they are deemed unable to decide for themselves. The legal definitions of these terms and what is entailed in their application vary from one jurisdiction to another, and physicians should make themselves aware of the legislation governing their particular area. It is presumed that persons who assume POA will act in accordance with the presumed wishes and best interests of the patients they are representing and will be careful to not act in their own interests. Advanced directives, commonly called a "living will," allows people to lay out a directive for their subsequent medical care should they become incompetent or unable to communicate their wishes. Most physicians take as a matter of duty the adherence to these expressed wishes but there are certain caveats. The wishes may not be followed when they require the doctor to carry out treatment that is inconsistent with the physician's conscience (such as assisted suicide), that is against the policy of the health care institution or that is deemed to be ineffective in treating the condition (see 15. Futility, below). In general advanced directives are a useful guide for the health care team and for the surrogate decision makers.

Case Study Part D

After considerable debate it is decided that Mrs. Court must be transferred to a nursing home, even against her own wishes. She is assessed by a qualified neuropsychologist and is declared incompetent to decide about her own placement. She seems quite depressed and you are concerned that not only might she be a harm to herself and others through negligence but that she might entertain the idea of taking her own life.[10] *You go to visit her at the nursing home and find that she is extremely unhappy with her surroundings: Most of the other patients have severe dementia and many are in a terminal state. She tells you that she knows that she soon will be one of them. She says she remembers how her mother was before she died and does not want to "go that way." She pleads with you to help her to end her life while she is still capable. You feel very sympathetic to her plight. She has almost continuous abnormal movements, cannot stand by herself, and cannot feed herself. She has lost a lot of weight. The situation makes you feel extremely uncomfortable and your immediate reaction is to brush aside her request and change the subject. You mutter something like, "Now, Mrs. Court, you really don't mean that!" although you know she does. You look at your watch and say you have to leave. You feel even more uncomfortable as you do.*[11]

Questions to Be Considered

10. Under what circumstances can a patient be forced to enter hospital? Should a patient be hospitalized against her will when you suspect that she is a suicide risk?

11. How would you respond to a patient who asks for assistance to end his or her life? What are the legal, ethical and moral implications?

Discussion

10. Hospitalization against a patient's will. Where a physician has a good reason to suspect that the patient is at risk, one is legally obliged to commit a patient to a hospital for his or her own protection. Usually a family member who has power of attorney will request that the physician invoke the law. The issue of committing a suicidal patient to hospital against that person's will is not as simple as it might seem. Members of the medical and legal professions are far from unanimous as to the ethics of the situation. This is particularly evident when it comes to a patient who has an incurable disease such as Huntington's disease and who may still be competent. One recent survey of medical and law students showed an almost even split on the subject (Elger and Harding, 2004). Mental health acts in most jurisdictions allow for forced commitment for a limited period—pending assessment—if the individual, in the view of the committing physician, is a danger to himself or others.

11. **Assisted suicide** and **euthanasia** present major ethical dilemmas for those caring for patients with terminal or incurable illness. A family history of suicide seems to be an added risk factor. Mr. Court has had risk factors on both sides of the family. However, Mrs. Court does not have that background and perhaps is at less risk. However, as she sees herself deteriorating and is haunted by her own mother's demise, she does contemplate ending her life while she is still capable of it. She pleads with you to assist her. There are few more difficult ethics issues than those surrounding the question of assisted suicide. From an ethics standpoint it involves truth telling, disclosure, fidelity, respect for autonomy, nonmaleficence, beneficence, patient's best interest, justice, public policy and public danger. In addition there is the medical-legal or, perhaps more bluntly, legal matter.

Volumes have been written on the subject and the present discussion can only highlight a few of the issues. The laws vary from jurisdiction to jurisdiction and may also change over time. Furthermore, enforcement of the law may also be variable according to the personal beliefs of the authorities. Physicians must be aware of the current legal status in their area of practice. In general, the Canadian Federal law prohibits assisted suicide.

In the United States, the law varies from state to state, and there is some tension between federal and state legislation. The state of Oregon has passed legislation that would allow assisted death under very particular circumstances. In actual fact there have been only a small number of cases where assisted suicide has actually been carried out under the law. Australia also has different laws in each state. European countries have considerable variation in their approach, with the Netherlands, and more recently Switzerland, having the most permissive legislation. In general it is emphasized that more emphasis should be placed on palliative care for patients who are in a terminal state rather than legitimizing assisted suicide.

The ethics stance is also variable. Those who support euthanasia cite autonomy of the individual as the overriding principle to be followed. Those who might consider assisting a person to end his or her life may refer to beneficence. However, there is a fear that allowing assisted suicide may lead to a general devaluation of life and of individual autonomy.

The religious authorities who oppose assisted suicide also oppose suicide in any form as they believe that it is thwarting God's will. They also emphasize "sacredness of life." Groups who represent the disabled fear that allowing assisted suicide may start society on a slippery slope toward eventually accepting assisted suicide for those who do not wish it, or cannot state their preferences, but may be perceived as a drain on society.

Various medical organizations have studied these issues and many have developed policy statements to guide their membership. The World Medical Association has outlined its stance in the *WMA Medical Ethics Manual* (Williams, 2005). In the manual, *euthanasia* is defined as "knowingly and intentionally performing an act that is clearly intended to end another person's life and that includes the following elements: the subject is a competent informed person with an incurable illness who has voluntarily asked for his or her life to be ended; the agent knows about the person's medical condition and desire to die, and commits the act with the primary intention of ending the life of that person, and the act is undertaken with compassion and without personal gain" (pp. 57–58).

"Assistance in suicide means knowingly and intentionally providing a person with the knowledge or means or both required to commit suicide, including counseling about lethal doses of drugs, prescribing such lethal doses or supplying the drugs" (p. 58).

The WMA concludes its review with a strong statement on euthanasia:

> Euthanasia, that is the act of deliberately ending the life of a patient, even at the patient's own request or at the request of close relatives, is unethical. This does not prevent the physician from respecting the desire of a patient to allow the natural process of death to follow its course in the terminal phase of sickness.

Medical people are often in the position of wanting to relieve suffering but the methods of doing so may also cause death. There is a doctrine in ethics (curiously originated in the Roman Catholic Church itself and probably first annunciated by St. Thomas Aquinas) that is referred to as *the law of double effect*. This was applied to situations where the intention is to benefit the person but that an unintentioned harmful effect may result. This would allow a pregnant woman to receive chemotherapy for cancer even though it might cause loss of the fetus. This would also allow a physician to administer large, potentially fatal, doses of narcotics to alleviate pain, as long as the intent was to assuage suffering and not to cause death. The justice system may respond variously to these situations.

In the situation in this anecdote, Mrs. Court is not in severe or uncontrollable pain, although she may well be in severe mental and emotional anguish. She is not in danger of imminent death. Undoubtedly, the law would deal severely with a physician prescribing fatal doses of medication in such a case. Mrs. Court may well recall with horror the situation she experienced with her own mother and may also have suspicions about the events surrounding her demise. The physician's inclination may well be to simply dismiss the request with a firm "no" and try to change the subject, pretending that the question was never asked. However, it would be preferable to discuss the situation with the patient and allow her to express her anxieties. The discussion may be continued on further occasions. It is important to reassure the patient that you will do everything to help her otherwise and will not abandon her. Fear of abandonment and the feeling of loss of control are some of the major factors leading to a person's committing suicide or asking for assistance to end his or her life.

Case Study Part E: At the Nursing Home

During the next several years, Mrs. Court's condition deteriorates progressively until she seems unable to recognize or respond to those about her. She lies in bed, now immobile as the choreic movements no longer occur. Sometimes her eyes are open and she stares blankly at the ceiling, but occasionally she seems to look toward a stimulus, such as a nurse entering the room. At other times she seems to be in a deep sleep from which she cannot be aroused. She has had a feeding tube inserted into her duodenum. Her state is described as "vegetative" and a diagnosis of "persistent vegetative state" is written in the doctor's notes in the chart.[12] *She develops pneumonia but responds to antibiotic therapy. In the past a physician might act in a paternalistic fashion*[13] *and decide whether to prescribe antibiotics, but in this case the family members were consulted and it was their wish to give antibiotics. The physician argued against it but eventually gave in to the family's wishes. It has already been established that no attempts at resuscitation will be made if Mrs. Court has a cardiac or respiratory arrest and DNR (Do Not Resuscitate) orders are written in her chart.*

However, after a few more months, it is apparent that she is not showing any improvement but is still not in a state where death seems imminent. The family asks whether the tube feedings should be discontinued and active treatment withdrawn. They also ask whether aggressive treatment should be withheld, such as antibiotics, should she develop another bout of pneumonia. The social worker at the nursing home calls a family conference that is attended by the staff nurses and the attending physician. Early in the conference, someone mutters that the nursing home is very short of beds and that society cannot afford to maintain people in a vegetative state forever. This individual suggests that we should be more careful with scarce resources and that decisions have to be made on the basis of distributive justice.[14] *Others quickly dismiss this argument.*

As the conference proceeds, several issues are raised. What treatments, if any, are appropriate? If Mrs. Court were to develop pneumonia, should antibiotics be given? When is treatment considered to be a matter of futility?[15] *The problems of withdrawal or withholding treatment are discussed.*[16] *The family, composed of her son and her daughter, who flew in from the west coast, have joint power of attorney for health care. They agree that they would be honoring their mother's wishes and serving her best interests by not prolonging her life under such circumstances. She then developed pneumonia, which, untreated, led to her death. An autopsy confirmed the diagnosis of Huntington's disease.*

Questions to Be Considered

12. What is meant by "persistent vegetative state"? What is the neuropathological basis of this clinical condition?

13. What is meant by *paternalistic*?

14. What is *distributive justice*?

15. What is meant by *futility*? Is it a useful term?

16. What is meant by *withdrawal* or *withholding of treatment*?

Discussion

12. **Persistent vegetative state** is a neurological condition in which the patient has diffuse cortical damage and is believed to have lost all conscious awareness of his or her environment. "Vegetative" functions remain relatively intact and cardiovascular function and respiration are maintained. Some brainstem function is preserved and chewing and swallowing may occur. Random eye movements may be seen and at times there may seem to be a following of objects moving in the field of vision. With hydration, feeding, and attention to intervening medical conditions, the patient may be kept in this state for months or years. Needless to say, this is a most distressing situation for both family and caregivers.

13. **Paternalism**, or better, **parentalism**, is a term that was applied to physicians' tendency to act as if they had all the answers and treat the patient as they would a little child. The physician was the parent, the patient the child. It was the accepted attitude that prevailed until the past few decades, but is less common now since patients are better educated and have access to medical information that allows them to share decision making concerning their own health and medical care. Physicians, however, are not obligated to carry out a treatment that they feel may be detrimental to patients' well-being or to institute a course of action felt to be futile (see 15, Futility, below).

14. **Justice (distributive justice)** is one of the four principles in the oft-quoted "Georgetown Mantra,"

promulgated by the ethics group at Georgetown University in Washington, D.C. The others are autonomy, beneficence, and nonmaleficence. In medical care, justice demands the greatest good for the greatest number. In the real life situation where there are never unlimited resources, the individual physician is often in the difficult position of having to decide between the needs of his or her patient and those of the myriad of other patients who are equally deserving of provision of health care. Most physicians feel the strongest loyalty to their individual patient and will advocate first for that patient. Justice principles may impact on public policy if not on formalized legislation.

15. The term **futility** is not much used anymore but refers to a treatment that is not considered to be of benefit or serving the aims of treatment. The problem is that, like beauty being in the eye of the beholder, the definition of *futility* varies with the point of view of the user. A physician may not consider a treatment as futile when it prolongs life for a few more months. However, the patient may feel it is futile, and therefore undesirable, if the months of extended life are spent in the hospital and require tolerating adverse effects of the medication. Of course the reverse may be the case and some patients and families may want prolongation of life at any cost. The issue of futility arises many times in the situation of patients in a persistent vegetative state.

16. Withdrawing and withholding treatment are situations that evoke great emotional as well as practical problems for the caregivers. In some patients in whom the quality of life is very low or where they are in a state of chronic suffering, it may seem that prolonging suffering is not beneficent and indeed may be considered maleficent or harmful. This is especially so in patients who are in a terminal state. Family members may request that aggressive medical treatment be stopped. Most empathetic caregivers would comply especially if the patient has previously expressed (such as in a living will) the desire not to have his or her life prolonged in such a state. While there may be a consensus that aggressive medical treatments or surgical interventions should be avoided, there is ongoing debate as to what constitutes a medical act. Is hydration and feeding a medical act or is it merely caregiving? Is there a difference between feeding or hydrating through a feeding tube and spoon feeding a patient who is able to swallow?

Indeed, whether to initiate and when to discontinue feeding and hydration is one of the more controversial issues in medical ethics. The issue often is centered on what is natural and what is extraordinary. Some authorities, including religious authorities, say that tube feeding is unnatural and a medical act and therefore may well be refused or terminated. Others maintain that nourishment and hydration are simple supportive acts and the tube is only another way of carrying these out. It is often most difficult for the nursing staff and those caring directly for a patient to see the person fade away when they feel they could be doing something seemingly helpful.

Physician's opinions are widely divided on these issues and a survey of physicians only a decade ago revealed that there are even those who would go beyond routine medical care and provide more extensive treatments, such as flu shots or even breast or prostate examinations for patients who are in a persistent vegetative state.

As in most of medical ethics there is no absolute right answer. However, the underlying principles of doing the best for the patient must remain uppermost in the physicians' minds.

Suggested Readings

General Resources

Beauchamp, T. L., and J. F. Childress, eds. *Principles of Biomedical Ethics*, 5th ed. Oxford: Oxford University Press, 2008.

Bernat, J. L. *Ethical Issues in Neurology,* 3rd ed. Philadelphia: Lippincott Williams & Wilkins, 2008.

Hebert, P. *Doing It Right: A Practical Guide to Ethics for Medical Trainees and Physicians,* 2nd ed. Don Mills, Ont.: Oxford University Press, 2009.

Illes, J., Ed. *Neuroethics: Defining the Issues in Theory, Practice, and Policy*. Oxford: Oxford University Press, 2006.

Kluge, E. W. *Biomedical Ethics in a Canadian Context*. Scarborough, Ont.: Prentice-Hall Canada, 1992.

Roy, D. J., J.R. Williams, and B.M Dickens, Eds. *Bioethics in Canada*. Scarborough, Ont.: Prentice-Hall Canada, 1994.

Singer, P. A., Ed. *Bioethics at the Bedside: A Clinician's Guide*. Ottawa: Canadian Medical Association/Association Médicale Canadienne, 1999.

References

Canadian Medical Association. 2007. Euthanasia and assisted suicide. Retrieved June 5, 2009, from http://www.cpsns.ns.ca/publications/2007-euthansia-assisted-suicide-cma.pdf

Ethics Committee of the College of Family Physicians of Canada. 2009. Statement concerning euthanasia and physician-assisted suicide. Retrieved June 5, 2009, from http://www.cfpc.ca/English/cfpc/communications/health%20policy/2000%20statement%20concerning%20euthanasia/default.asp?s=1

Decruyenaere M. et al. Non-participation in predictive testing for Huntington's Disease: Individual decision-making, personality and avoidant behaviour in the family. *Europ. J. Human Genetics* 5 (1997): 351–363.

Downie, J. *Dying Justice: A Case for Decriminalizing Euthanasia and Assisted Suicide in Canada.* Toronto: University of Toronto Press, 2004.

Elger, B. S., and Harding, T. W. Should a suicidal patient with Huntington's disease be hospitalized against her will? Attitudes among future physicians and lawyers and discussion of ethical issues. *General Hospital Psychiatry* 26 (2006): 136–144.

Hayden M.R. et al. Positron emission tomography in the early diagnosis of Huntington's disease. *Neurology* 36 (1986): 888–894.

Lavery, J.V., et al. Bioethics for physicians: 11. Euthanasia and assisted suicide. *CMAJ* 156 (1997): 1405–08.

McTeer, M. A. *Tough choices: Living and Dying in the Twenty-First Century*. Toronto, Ont.: Irwin Law, 1999.

Practice Parameter: Genetic testing alert. Statement of the Practice Committee Genetics Testing Task Force of the American Academy of Neurology. *Neurology* 47 (1996): 1343–1344.

Sibbald R., J. Downar, and L. Hawryluck. Perceptions of "futile care" among caregivers in intensive care units. *CMAJ* 177 (2007): 1201–1208.

Somerville, M. *Death Talk: The Case against Euthanasia and Physician-Assisted Suicide*. Montreal: McGill-Queen's University Press, 2001.

Tarasoff v. Regents of the University of California, 17 Cal. 3d 425 (1976).

Williams, J. R. *Medical Ethics Manual*. Ferney-Voltaire, France: World Medical Association, 2005.

The World Medical Association. Medical Ethics Manual, 2nd ed. 2009. Available online at http://www.wma.net/e/ethicsunit/pdf/manual/ethics_manual.pdf.

Web Site

http://www.neuroethics.upenn.edu/

Glossary of Terms

NOTE TO READER: The Glossary contains neuroanatomical terms and terms commonly used clinically to describe neurological symptoms, as well as common physical findings detected on a neurological examination; some selected laboratory investigation techniques and clinical syndromes are also included.

Afferent: Sensory—conduction toward the central nervous system.

Agnosia: Loss of ability to recognize the significance of sensory stimuli (tactile, auditory, visual), even though the primary sensory systems are intact.

Agonist: A muscle that performs a certain movement of the joint; the opposing muscle is called the antagonist.

Agraphia: Inability to write owing to a lesion of higher brain centers, even though muscle strength and coordination are preserved.

Akinesia: Absence or loss of motor function; lack of spontaneous movement; difficulty in initiating movement (as in Parkinson's disease).

Alexia: Loss of ability to grasp the meaning of written words; inability to read due to a central lesion; word blindness.

Alpha motor neuron: Another name for the anterior (ventral) horn cell, also called the lower motor neuron.

Amygdala: Amygdaloid nucleus or body in the temporal lobe of the cerebral hemisphere; a nucleus of the limbic system.

Angiogram: Display of blood vessels for diagnostic purposes, using cineradiography, MRI or CT, usually by using contrast medium injected into the vascular system.

Anopia/Anopsia: A defect in the visual field (e.g., hemianopia—loss of half the visual field).

Antagonist: A muscle that opposes or resists the action of another muscle, which is called the agonist.

Antidromic: Relating to the propagation of an impulse along an axon in a direction that is the reverse of the normal or usual direction.

Apnea: Total interruption of breathing for 10 seconds or more. Apneas can be central or obstructive in origin.

Apnea test: This is a procedure conducted as part of the protocol to determine brain death. It is a test of the ability of the medulla to stimulate respiration in response to rising levels of pCO_2. With no other factors present (such as medication effects or collateral medical conditions) and a baseline pCO_2 of 40 mm Hg, failure of a patient to generate respiratory effort in response to pCO_2 levels of 55 to 60 mm Hg indicates lack of function of the medulla.

Aphasia: An acquired disruption or disorder of language, specifically a deficit of expression using speech or of comprehending spoken or written language; global aphasia is a severe form affecting all language areas.

Apoptosis: Programmed cell death, either genetically determined or following an insult or injury to the cell.

Apraxia: Loss of ability to carry out purposeful or skilled movements (such as combing the hair or brushing the teeth) despite the preservation of power, sensation and coordination.

Arachnoid: The middle meningeal layer, forming the outer boundary of the subarachnoid space.

Areflexia: Loss of reflex as tested using the myotatic stretch/deep tendon reflex.

Ascending tract: Central sensory pathway, for example, from the spinal cord to the brainstem, cerebellum or thalamus.

ASIA Scoring System: American Spinal Injury Association Scoring Scale used to document and evaluate spinal cord injury (see Chapter 13).

Association fibers: Fibers connecting parts of the cerebral hemisphere, on the same side.

Astereognosis: Loss of ability to recognize the nature of objects or to appreciate their shape by touching or feeling them.

Asterixis: A slow, flapping tremor of the outstretched hands; sometimes seen in patients with hepatic encephalopathy.

Astrocyte: A type of neuroglial cell with metabolic and structural functions; reacts to injury of the CNS by forming a gliotic "scar."

Asynergy: Disturbance of the proper sequencing in the contraction of muscles, at the proper moment, and of the proper degree, so that an action is not executed smoothly or accurately.

Ataxia: A loss of coordination of voluntary movements, often associated with cerebellar dysfunction; can also be caused by loss of proprioception or by motor weakness.

Athetosis: Slow writhing movements of the limbs, especially of the hands, not under voluntary control, caused by impaired function in the striatum.

Autonomic: Autonomic system; usually taken to mean the efferent or motor innervation of viscera (smooth muscle and glands).

Autonomic nervous system (ANS): Visceral innervation; sympathetic and parasympathetic divisions.

Axon: Efferent process of a neuron, conducting impulses to other neurons or to muscle fibers (striated and smooth) and gland cells.

Babinski response: Babinski reflex is not correct; stroking the outer border of the sole of the foot in an adult normally results in a flexion (downward movement) of the toes. The Babinski response consists of an upward movement (extension or dorsiflexion) of the first toe and a fanning of the other toes, indicating a lesion of the corticospinal (pyramidal) tract.

Ballismus: Involuntary, usually unilateral flinging motion of the arm or leg due to damage to the ipsilateral subthalmic nucleus.

Basal ganglia (nuclei): Gray matter structures in the deep cerebral hemispheres adjacent to the diencephalon: the caudate, putamen and globus pallidus; including functionally the subthalamus and the substantia nigra. Among many other functions, they are crucial components of the central apparatus of motor control.

Basilar artery: The major artery supplying the brainstem and cerebellum, formed by fusion of the two vertebral arteries.

Blink reflex: An electrical equivalent of the corneal reflex: it measures the latency of the input of an electrical stimulus to CN V and the resulting eye closure or blink output of CN VII on both sides. If the reflex response is slow or absent, this can help localize the lesion to CN V, the pons or CN VII on either side.

Boston Classification: Patient classification system designed for medical rehabilitation that predicts resource use and outcomes for clinically similar groups of individuals.

Bradykinesia: Abnormally slow initiation of voluntary movements (usually seen in Parkinson's disease).

Brainstem: Includes the medulla, pons, and midbrain.

Brainstem evoked potentials (BSEP): A neurophysiological technique involving testing the auditory portion of CN VIII using an auditory stimulus (clicking sound) and measuring the responses from CN VIII and the various brainstem structures that relay auditory information in the brainstem toward the thalamus and the cortex. This test might show a delay or slowing of the signal evoked by the sound on one or both sides, allowing for precise localization of a brainstem lesion.

Brodmann areas: Numerical subdivisions of the cerebral cortex on the basis of histological differences between different functional areas (e.g., area 4 = motor cortex; area 17 = primary visual area).

Bruit: French word meaning "noise"; refers to a swishing noise heard when listening over an artery, a vein, or over the head; results from turbulent blood flow in the area auscultated and may reflect critical narrowing of an arterial lumen heralding stroke, or of an abnormal fistulous connection between an artery and a vein.

Bulb: Referred at one time to the medulla but, in the context of "corticobulbar tract" refers to the whole brainstem in which the motor nuclei of cranial nerves and other nuclei are located.

CAT or CT scan: Computerized axial tomography; a diagnostic imaging technique that uses x-rays and computer-aided reconstruction of the brain to provide two-dimensional images of the brain and spine.

Calcarine cortex: Primary visual cortex, situated just above and below the calcarine fissure in the medial occipital lobe.

Carotid artery: Large artery in the anterior neck with an internal branch supplying the retina, basal ganglia, anterolateral and anteromedial cerebral hemispheres, and an external branch supplying the face and scalp on the same side.

Carotid siphon: Hairpin bend of the internal carotid artery within the skull.

Cauda equina: ("Horse's tail") The lower lumbar, sacral, and coccygeal spinal nerve roots within the subarachnoid space of the lumbar (CSF) cistern.

Caudal: Toward the tail or hindmost part of the neuraxis.

Caudate nucleus: Part of the neostriatum, of the basal ganglia; consists of a head, body, and tail (which extends into the temporal lobe).

Central nervous system (CNS): Cerebral hemispheres, including diencephalon, cerebellum, brainstem, and spinal cord.

Cerebellar peduncles: Inferior, middle, and superior; fiber bundles linking the cerebellum and brainstem.

Cerebellum: The little brain; an older part of the brain with important motor functions (as well as others), dorsal to the brainstem, situated in the posterior cranial fossa.

Cerebral aqueduct (of Sylvius): Aqueduct of the midbrain; passageway carrying CSF through the midbrain, as part of the ventricular system.

Cerebral peduncle: Descending cortical fibers in the "basal" (ventral) portion of the midbrain; sometimes includes the substantia nigra (located immediately behind).

Cerebrospinal fluid (CSF): Fluid in the ventricles and in the subarachnoid space and cisterns.

Cerebrum: Includes the cerebral hemispheres and diencephalon, but not the brainstem and cerebellum.

Cervical: Referring to the neck region; the part of the spinal cord that supplies the structures of the neck; C1 to C7 vertebral; C1 to C8 spinal segments.

Chorea: A motor disorder characterized by abnormal, irregular, spasmodic, jerky, uncontrollable movements of the limbs or facial muscles, thought to be caused by dysfunction or degeneration in the basal ganglia.

Choroid: A delicate membrane; choroid plexuses are found in the ventricles of the brain.

Choroid plexus: Vascular structure consisting of pia with blood vessels, with a surface layer of ependymal cells; responsible for the production of CSF.

Circle of Willis: Anastomosis between internal carotid and basilar arteries, located at the base of the brain, surrounding the pituitary gland.

Cistern(a): Expanded portion of subarachnoid space containing CSF, for example, cisterna magna (cerebello-medullary cistern), lumbar cistern.

Clonus: Abnormal sustained series of contractions and relaxations following stretch of the muscle; usually elicited in the ankle joint. Present following lesions of the descending motor pathways, and associated with spasticity. When elicted as a deep tendon reflex, implies a Grade of 4^+ (see Table 2.3).

CNS: Abbreviation for central nervous system.

Cog-wheeling: This is a clinical sign of altered tone usually at the wrist and elbows. When the wrist or forearm are moved in a circular motion, the resulting tone feels like a "ratcheting" or "on-off" movement.

Cold caloric response: Similar to the doll's eye response, this is a test of the slow component of eye movement to vestibular input in the absence of visual fixation. It requires normal function of CN VIII, the pons and the connections to CN III, IV and VI. The stimulus for this test consists of 150 cc of cold water injected slowly into the ear canal on one side (after verifying that the tympanic membrane is intact). The normal response would be slow conjugate movement of the eyes toward the side being irrigated. If performed on an awake patient with visual fixation, the fast component would follow the slow component with resulting nystagmus with the fast component away from the cold ear. The subject will complain of dizziness and often will have nausea and vomiting.

Colliculus: A small elevation; superior and inferior colliculi compose the tectum of the posterior midbrain.

Commissure: A group of nerve fibers in the CNS connecting structures on one side to the other across the midline (e.g., corpus callosum of the cerebral hemispheres; anterior commissure).

Concussion: Brain injury due to closed head trauma associated with brief loss of neurological function, usually consciousness, with no damage found by neuroradiological imaging.

Conjugate eye movement: Coordinated movement of both eyes together, so that the image falls on corresponding points of both retinas.

Connors Rating Scale Revised (CRS-R): Designed to assess attention-deficit/hyperactivity disorder (ADHD) and related problems in children.

Consciousness: The state of awareness and vigilance in which the individual has full function of sensory, motor and modulating systems to interact with the environment. This implies normal waking function of the brainstem, diencephalon and cerebral hemispheres.

Consensual reflex: Light reflex; refers to the bilateral simultaneous constriction of the pupils after shining a light in one eye.

Contralateral: On the opposite side (e.g., contralateral to a lesion).

Coordination: The integration of sensory, motor and modulating systems to provide smooth control of limb and trunk movements. Lack of coordination is ataxia.

Corona radiata: Fibers radiating from the internal capsule to various parts of the cerebral cortex—a term often used by neuroradiologists.

Corpus callosum: The main (largest) neocortical commissure of the cerebral hemispheres.

Corpus striatum: Caudate and putamen, nuclei inside the cerebral hemisphere, including tissue bridges that connect them (across the anterior limb of the internal capsule); part of the basal ganglia; the neostriatum.

Cortex: Layers of gray matter (neurons and neuropil) on the surface of the cerebral hemispheres (mostly six layers) and cerebellum (three layers).

Corticobulbar: Descending fibers connecting motor cortex with motor cranial nerve nuclei and other nuclei of the brainstem (including the reticular formation).

Corticofugal fibers: Axons carrying impulses away from the cerebral cortex.

Corticopetal fibers: Axons carrying impulses toward the cerebral cortex.

Corticospinal tract: Descending tract, from motor cortex to anterior (ventral) horn cells of the spinal cord (sometimes direct); also called pyramidal tract.

Cranial nerve nuclei: Collections of cells in the brainstem giving rise to or receiving fibers from cranial nerves (CN III to XII); may include sensory, motor or autonomic.

Cranial nerves: Twelve pairs of nerves arising from the cerebrum and brainstem and innervating structures of the head and neck (CN I and II are actually CNS tracts).

CSF: Cerebrospinal fluid, in ventricles and subarachnoid space (and cisterns).

CT angiogram: CT scan of the brain, neck and great vessels combined with the use of intravenous contrast material to outline the arterial and venous vascular anatomy. Often used in acute stroke to determine the location and degree of narrowing of a cerebral artery.

Cuneatus (cuneate): Sensory tract (fasciculus cuneatus) of the posterior column of the spinal cord, from the upper limbs and body; cuneate nucleus of medulla.

Decerebrate posturing: Characterized by reflexive extension of the upper and lower limbs in a comatose individual; lesion at the brainstem level between the vestibular nuclei (pons) and the red nucleus (midbrain).

Decorticate posturing: Characterized by reflexive extension of the lower limbs and flexion of the upper limbs in a comatose individual; lesion is located above the level of the red nucleus (of the midbrain).

Decussation: The point of crossing of CNS tracts, for example, decussations of the pyramidal (corticospinal) tract, medial lemnisci and superior cerebellar peduncles.

Dementia: Progressive brain disorder that gradually destroys a person's memory, starting with short-term memory, and loss of intellectual ability, such as the ability to learn, reason, make judgments and communicate; in addition, the inability to carry out normal activities of daily living; usually affects people with advancing age.

Demyelination: Degeneration and loss of the myelin sheath in a region or regions of the central or peripheral nervous system.

Dendrite: Receptive process of a CNS neuron; usually several processes emerge from the cell body, each of which branches in a characteristic pattern.

Dendritic spine: Cytoplasmic excrescence of a dendrite and the site of an excitatory synapse.

Dentate: Dentate (toothed or notched) nucleus of the cerebellum (intracerebellar nucleus); dentate gyrus of the hippocampal formation.

Dermatome: A patch of skin innervated by a single spinal cord segment (e.g., T1 supplies the skin of the inner aspect of the upper arm; T10 supplies the umbilical region).

Descending tract: Central motor pathway (e.g., from cortex to brainstem or spinal cord).

Diencephalon: Consisting of the thalamus, epithalamus (pineal), subthalamus and hypothalamus.

Diffuse axonal injury (DAI): Injury to white matter tracts caused by shearing forces produced by trauma; this injury leads to impaired white matter transmission and is often in combination with cortical injury.

Diplopia: Double vision; a single object is seen as two objects, can be horizontal, vertical or oblique.

Doll's eye response: This is a test of the slow component of eye movement to vestibular input in the absence of visual fixation. It requires normal function of CN VIII, the pons and the connections to CN III, IV and VI. This test involves moving the head, usually in the horizontal direction, and observing the conjugate movement of the eyes in the opposite direction. See *Cold caloric response*.

Dominant hemisphere: The hemisphere responsible for language; this is the left hemisphere in about 85 to 90 percent of people (including many left-handed individuals).

Dorsal column: Alternate term for the posterior column of the spinal cord.

Dorsal root: Afferent sensory component of a spinal nerve, between dorsal root ganglion and spinal cord.

Dorsal root ganglion (DRG): A group of peripheral neurons whose axons carry afferent information from the periphery; their central processes enter the spinal cord.

Dorsiflexion: Active elevation of the foot at the ankle joint; the opposite of plantar flexion.

Dura: Dura mater, the thick external layer of the meninges (brain and spinal cord).

Dural venous sinuses: Large venous channels for draining blood from the brain; located within the dura of the cranial meninges.

Dysarthria: Difficulty with the articulation of words.

Dysdiadochokinesia: Impairment of or inability to perform rapid alternating distal limb movements, such as alternating pronation/supination of the forearm or toe tapping.

Dyskinesia: Purposeless movements of the limbs or trunk, usually due to a lesion of the basal ganglia; also difficulty in performing voluntary movements.

Dysmetria: Disturbance of the ability to control the range of movement in muscular action, causing under- or over-shooting of the target (usually seen with cerebellar lesions).

Dysphagia: Difficulty with swallowing.

Dyspraxia: Impaired ability to perform a voluntary act previously well performed, in the presence of intact motor power, coordination and sensation.

Dystonia: Impaired control of limb and trunk posture, often asymmetric, with twisted postures of one or more limbs, and of the neck and face; may be focal (e.g., one foot), segmental (one limb, several joints) or generalized.

EEG: Electroencephalography, a technique for recording cerebral cortical electrical activity via electrodes applied to the scalp in a standardized configuration. The EEG is a useful tool to detect the presence of seizures as well as assess the background rhythms generated by the interplay of the cortex and thalamus. Localized slowing on the EEG can reflect focal cortical damage, whereas diffuse slowing can represent widespread injury (such as in dementia) or metabolic/toxic injury, which affects the whole brain.

Efferent: Away from the central nervous system; usually means motor to muscles.

Electromyogram (EMG): A technique for investigating peripheral neuromuscular disease by recording electrical activity from muscle, either by the use of needle or surface recording electrodes.

Encephalitis: Inflammation of parts or all of the cerebral hemispheres; may be of infectious or autoimmune origin.

Encephalopathy: Generic term for any disease affecting the cerebral hemispheres.

ENG: This is a test of vestibular function performed either by movement in a rotating or by the infusion of cold water in the ear canal. The normal output of the stimulus is the production of a slow vestibular component of eye movement followed by a first corrective component.

Electronystagmogram (ENG): A diagnostic test performed to determine inner ear function of the semicircular canals; uses cold or warm water (caloric) stimulation or circular movement using a rotating chair. The velocity in degrees per second of the slow and fast eye movement components from each ear can be precisely measured. This test is used to distinguish central from peripheral causes of vertigo and ataxia.

Entorhinal: Associated with olfaction (smell); the entorhinal area is the anterior part of the parahippocampal gyrus of the temporal lobe, adjacent to the uncus.

Ependyma: Epithelium lining of the ventricles of the brain and the central canal of the spinal cord; specialized tight junctions at the site of the choroid plexus.

Equinovarus: Abnormal posture of the foot in which it is both flexed and inverted at the ankle; commonly seen in severe upper motor neuron cerebral lesions such as stroke.

Extrapyramidal system: An older clinically used term, usually intended to include the basal ganglia portion of the motor systems and not the pyramidal (corticospinal) motor system.

Falx: Dural partitions in the midline of the cranial cavity; the large falx cerebri between the cerebral hemispheres, and the small falx cerebelli.

Fascicle: A small bundle of nerve fibers.

Fasciculation: Spontaneous (uncontrolled) discharge of a motor unit, visible as transient, irregular focal muscle twitches.

Fasciculus: A large tract or bundle of nerve fibers.

Fasciculus cuneatus: Part of the posterior column of the spinal cord; ascending tract for discriminative touch, conscious proprioception and vibration from upper body and upper limb.

Fasciculus gracilis: Part of the posterior column of the spinal cord; ascending tract for discriminative touch, conscious proprioception and vibration from lower body and lower limb.

Fiber: Synonymous with an axon (either peripheral or central).

Flaccid paralysis: Muscle paralysis with hypotonia due to a lower motor neuron lesion; may also occur in the acute phase of an upper motor neuron lesion, e.g., spinal shock.

fMRI: Functional magnetic resonance imaging. A technique for assessing which areas of the brain are involved in specific functions, for example, language comprehension; based on the principle that, while such regions are in use, metabolic activity and regional blood flow are relatively increased.

Folium (plural = folia): A flat leaf-like fold of the cerebellar cortex.

Foramen: An opening, aperture, between spaces containing CSF (e.g., Monro, between lateral ventricles and IIIrd ventricle; Magendie, between IVth ventricle and cisterna magna).

Forebrain: Anterior division of the embryonic brain; cerebrum and diencephalon.

Fornix: The efferent (noncortical) tract of the hippocampal formation, arching over the thalamus and terminating in the mammillary nucleus of the hypothalamus and in the septal region.

Fourth (IVth) ventricle: Cavity between the brainstem and cerebellum, containing CSF.

Functional Independence Measure (FIM): A scale used in the care of patients undergoing rehabilitation. It is an 18-item ordinal scale, used with all diagnoses within a rehabilitation population. It is viewed as most useful for assessment of progress during inpatient rehabilitation.

Funiculus: A large aggregation of white matter in the spinal cord, may contain several tracts.

Ganglion (plural = ganglia): A collection of nerve cell bodies in the peripheral nervous system—the dorsal root ganglion (DRG) and autonomic ganglion. Also inappropriately used for certain regions of gray matter in the brain (i.e., basal ganglia).

Geniculate bodies: Specific relay nuclei of thalamus—medial (auditory) and lateral (visual).

Genu: Knee or bend; middle portion of the internal capsule.

Glial cell: Also called neuroglial cell; supporting cells in the central nervous system—astrocyte, oligodendrocyte, and ependymal; also microglia.

Glioma: Central nervous system tumor arising from glial cells (whether astrocytes, oligodendroglia, or ependymal cells).

Globus pallidus: Efferent part of the basal ganglia; part of the lentiform nucleus with the putamen; located medially.

Gracilis nucleus: Gracilis nucleus of the medulla. Nucleus at the termination of the sensory tract (fasciculus gracilis) of the posterior column of the spinal cord; from lower limbs and pelvic area.

Gray matter: Nervous tissue, mainly nerve cell bodies and adjacent neuropil; looks "grayish" after fixation in formalin (e.g., cerebral cortex, basal ganglia, thalamus).

Gyrus (plural = gyri): A convolution or fold of the cerebral hemisphere; includes cortex and white matter (e.g., precentral gyrus).

Hemiballismus: Violent jerking or flinging movements of one limb, not under voluntary control, due to a lesion of the ipsilateral subthalamic nucleus.

Hemiparesis: Muscular weakness affecting one side of the body.

Hemiplegia: Paralysis of one side of the body.

Hemorrhage: As applies to the nervous system, leakage of blood from an artery or vein into either the parenchyma of the muscle, nerve, spinal cord, brain, or anatomic spaces that overlay these structures. Hemorrhage, therefore, causes direct damage to the tissue affected and indirect damage to adjacent structures due to mass effect depending on the size and location of the hemorrhage.

Herniation: Bulging or expansion of the tissue beyond its normal boundary (e.g., uncal).

Heteronymous hemianopia: Loss of (different) halves of the visual fields of both eyes; bitemporal for the temporal halves and binasal for the nasal halves.

Hindbrain: Posterior division of the embryonic brain; includes pons, medulla and cerebellum (located in the posterior cranial fossa).

Hippocampus (or hippocampus "proper"): Involved in the acquisition of new memories for names and events. Part of the limbic system; a cortical area "buried" within the medial temporal lobe, consisting of phylogenetically old (three-layered) cortex; protrudes into the floor of the inferior horn of the lateral ventricle.

Homonymous hemianopia: Loss of the visual fields serving the visual space on one side, in both eyes; this involves the nasal half of the visual field in one eye and the temporal half of the visual field in the other eye.

Horner's syndrome: Miosis (constriction of the pupil), anhidrosis (dry skin with no sweat) and ptosis (drooping of the upper eyelid) due to a lesion of the sympathetic pathway to the head. Lesions can occur anywhere along the course of the sympathetic pathway such as in the descending portion in the brainstem, in the paraspinous sympathetic ganglia in the chest, and along the course of the carotid artery to Mueller's muscle in the upper eyelid.

Hydrocephalus: Enlargement of the ventricles due to excessive accumulation of cerebrospinal fluid. If the CSF pathway obstruction is within the ventricular system, the disorder is termed noncommunicating (obstructive) hydrocephalus; communicating hydrocephalus means that the obstruction is in the extracerebral subarachnoid space.

Hypo/hyperreflexia: Decrease (hypo) or increase (hyper) of the stretch (deep tendon) reflex.

Hypo/hypertonia: Decrease (hypo) or increase (hyper) of the tone of muscles, manifested by decreased or increased resistance to passive movements.

Hypokinesia: Markedly diminished movements (spontaneous).

Hypothalamus: A region of the diencephalon that serves as the main controlling center of the autonomic nervous system and is involved in several limbic circuits; also regulates the pituitary gland.

Infarction: Local death of an area of tissue due to loss of its blood supply.

Infratentorial: This refers to the space and structures contained in the area below the tentorium cerebelli or cerebellar tentorium such as the brainstem and cerebellum.

Infundibulum (funnel): Infundibular stem of the posterior pituitary (neurohypophysis).

Innervation: Nerve supply, sensory or motor.

Insula (island): Cerebral cortical area not visible from outside inspection and situated at the bottom of the lateral fissure (also called the island of Reil).

Internal capsule: White matter between the lentiform nucleus and the head of the caudate nucleus (anteriorly) and the thalamus (posteriorly); consists of anterior limb, genu and posterior limb; carries motor, sensory, and integrative fibers connecting the thalamus and cortex.

Internuclear ophthalmoplegia (INO): Loss of adduction of one eye due to damage to the medial longitudinal fasciculus connecting CN VI to CN III.

Ion channel: Pore in a neuronal cell membrane that selectively allows a specific ion to pass into or out of the cell; allows for the development of an electric charge within the neuron.

Ipsilateral: On the same side of the body (e.g., ipsilateral to a lesion).

Ischemia: A condition in which an area is not receiving an adequate blood supply.

Kinesthesia: The conscious sense of position and movement.

Lacune: The pathological small "hole" remaining after an infarct in central cerebral hemispheric structures such as the internal capsule; also irregularly shaped venous "lakes" or channels draining into the superior sagittal sinus.

Lateral ventricle: CSF cavity in each cerebral hemisphere; consists of anterior horn, body, atrium (or trigone), posterior horn, and inferior (temporal) horn.

Lemniscus: A specific pathway in the CNS; medial lemniscus for discriminative touch, conscious proprioception and vibration; lateral lemniscus for audition.

Lenticulo-striate arteries: Small vessels originating from the circle of Willis or the proximal middle cerebral arteries and supplying the basal ganglia; obstruction of or hemorrhage from one of these vessels is a common cause of stroke.

Lentiform: Lens-shaped; lentiform nucleus, a part of the corpus striatum; also called the lenticular nucleus; composed of putamen (laterally) and globus pallidus.

Leptomeninges: Arachnoid and pia mater, part of the meninges.

Lesion: Any injury or damage to tissue, vascular, traumatic.

Leukodystrophy: Diffuse degenerative disease of the central nervous system white matter, typically of hereditary origin; some types may be accompanied by peripheral nerve demyelination.

Limbic system: Parts of the brain, cortical and subcortical, that are associated with emotional behavior.

Locus ceruleus: A small nucleus located in the uppermost pons on each side of the IVth ventricle; contains melanin-like pigment, visible as a dark bluish area in freshly sectioned brain; source of noradrenergic projections to the cerebral hemispheres, cerebellum, and spinal cord.

Lower motor neuron: Anterior horn cell of the spinal cord and its axon, and also the cells in the motor cranial nerve nuclei of the brainstem; also called the alpha motor neuron. Its loss leads to atrophy of the muscle and weakness, with hypotonia and hyporeflexia; also fasciculations are often noted.

Lumbar puncture (LP): This test is also called a spinal tap. The purpose is to obtain samples of cerebrospinal fluid (CSF) to determine infection, inflammation, malignancy and bleeding within the subarachnoid space. The test is performed by inserting a spinal needle into the subarachnoid space of the extension of the dural sac, the lumbar cistern. Usually performed between the spinal (vertebral) interspaces at the L3–L4 level. The commonest complication is headache, or LP headache. This is caused by the reduction of the hydrostatic pressure in the subarachnoid space caused by the lumbar puncture. This headache is characteristic in that it is worse when the patient is sitting or standing and immediately better when recumbent.

Mammillary: Mammillary bodies; nuclei of the hypothalamus, which are seen as small swellings on the ventral surface of the diencephalon (also spelled *mamillary*).

Mass Effect: The process of displacement or shutting of normal brain structures due to swelling caused by a mass lesion pushing against adjacent brain tissue.

Medial lemniscus: Brainstem portion (crossed) of sensory pathway for discriminative touch, conscious proprioception and vibration, formed after synapse (relay) in nucleus gracilis and nucleus cuneatus.

Medial longitudinal fasciculus (MLF): A tract throughout the brainstem and upper cervical spinal cord which interconnects visual and vestibular input with other nuclei controlling movements of the eyes and of the head and neck.

Medulla: Caudal portion of the brainstem; may also refer to the spinal cord as in a lesion within (intramedullary) or outside (extramedullary) the cord.

Meninges: Covering layers of the central nervous system (dura, arachnoid and pia).

Meningismus: A condition in which there is irritation of the meninges causing extreme neck stiffness: Brudzinski's sign occurs with meningeal irritation; there is reflex flexion of the legs when the head is flexed on the neck by the examiner. Similarly, Kernig's sign is reflex flexion of the head when the lower leg is extended at the knee. Neither of these signs is specific but if present should guide the clinician to look for causes of meningeal irritation by performing a lumbar puncture (if there are no contraindications).

Meningitis: Inflammation of the cerebral and spinal leptomeninges, usually of infectious origin.

Mesencephalon: The midbrain (upper part of the brainstem).

Microglia: The "scavenger" cells of the CNS, that is, macrophages; considered by some as one of the neuroglia.

Midbrain: Part of the brainstem; also known as the mesencephalon.

MMSE: Mini-Mental Status Examination or Folstein Test. This is a standardized test of mental status that includes evaluation of attention, immediate and delayed recall, verbal and written language. The test has a maximum score of 30. Scores of less than 27 without intercurrent medical causes are considered to be abnormal. (See Chapters 2 and 9.)

MoCA: Montreal Cognitive Assesment. This is similar to the MMSE but is more extensive in terms of frontal lobe testing. It has a maximum score of 30. Scores of less than 27 without intercurrent medical causes are considered to be abnormal. (See Chapters 2 and 9.)

Motor: To do with movement or response.

Motor unit: A lower motor neuron, its axon and the muscle fibers that it innervates; includes the neuromuscular junction.

MRI: Magnetic resonance imaging. A diagnostic imaging technique that uses an extremely strong magnet, not x-rays. In clinical applications, this technique uses the electromagnetic echo returned by excited hydrogen ions to give anatomical, functional and spectral information of the brain and spinal cord. This technique is extremely sensitive to changes in water density caused by tumors and inflammatory processes such as multiple sclerosis.

Muscle spindle: Specialized receptor within voluntary muscles that detects muscle length; necessary for the stretch/myotatic reflex (DTR); contains muscle fibers within itself capable of adjusting the sensitivity of the receptor.

Muscular dystrophy: Hereditary degenerative disease of muscle tissue.

Myelin: Proteolipid layers surrounding nerve fibers, formed in segments; it is important for rapid (saltatory) nerve conduction.

Myelin sheath: Covering of a nerve fiber, formed and maintained by oligodendrocyte in CNS and Schwann cell in PNS; interrupted by nodes of Ranvier.

Myelitis: Inflammatory disease of the spinal cord, of infectious or autoimmune origin.

Myelography: Imaging after performing a lumbar puncture; an iodinated dye is injected into the subarachnoid space. The spread of the dye in the subarachnoid space is followed up to the neck as the patient is tilted head down to make the dye flow with gravity.

Myelopathy: Generic term for disease affecting the spinal cord.

Myopathy: Generic term for muscle disease.

Myotatic reflex: Deep tendon reflex (DTR) elicited by stretching the muscle, causing a reflex contraction of the same muscle; monosynaptic, from muscle spindle afferents to anterior horn cell (also spelled *myotactic* reflex).

Myotome: Muscle groups innervated by a single spinal cord segment; in fact, usually two adjacent segments are involved (e.g., biceps, C5 and C6).

Neocerebellum: Phylogenetically newest part of the cerebellum, present in mammals and especially well developed in humans; involved in coordinating precise voluntary movements and also in motor planning.

Neocortex: Phylogenetically newest part of the cerebral cortex, consisting of six layers (and sublayers) characteristic of mammals and constituting most of the cerebral cortex in humans.

Neostriatum: The phylogenetically newer part of the basal ganglia consisting of the caudate nucleus and putamen; also called the striatum, or corpus striatum.

Nerve conduction studies (NCS): A technique involving transcutaneous electrical nerve stimulation to determine the speed of conduction and amplitude of response from peripheral nerves. This aids in localization of lesions to specific peripheral nerves. Slowing of conduction suggests disruption of myelin; low amplitude suggests axonal dysfunction or destruction.

Nerve fiber: Axonal cell process, plus myelin sheath if present.

Neuralgia: Pain, severe, shooting, "electrical" along the distribution of a peripheral nerve (spinal or cranial).

Neuraxis: The straight longitudinal axis of the embryonic or primitive neural tube, bent in later evolution and development.

Neuroglia: Accessory or interstitial cells of the central nervous system; includes astrocytes, oligodendrocytes, ependymal cells, and microglial cells.

Neuron: The basic structural unit of the nervous system, consisting of the nerve cell body and its processes—dendrites and axon.

Neuropathy: Disorder of one or more peripheral nerves.

Neuropil: An area between nerve cells consisting of a complex arrangement of nerve cell processes, including axon terminals, dendrites, and synapses.

Neuropsychological testing: Complex evaluations of mental status usually performed by certified neuropsychologists. These tests are usually ordered when the MMSE or MoCA does not provide enough sensitivity to determine if a given individual has experienced a decrement in cognitive functions. Examples of these tests include the Wechsler Adult Intelligence Scale (WAIS), Wisconsin card sorting task (WCST), Minnesota Multiphasic Personality Inventory (MMPI), and others. The neuropsychologist will often choose the battery of tests that best suits the type of difficulty that the patient is reporting.

Neurotransmitter: Chemical compound released into a synaptic cleft by the terminal process of the presynaptic neuron; excites or inhibits the postsynaptic site, depending on the type of transmitter and the type of receptor.

Nociception: Perception of a potentially injurious stimulus, typically via small myelinated and unmyelinated sensory nerve fibers; may or may not be associated with the sensation of pain.

Node of Ranvier: Gap in myelin sheath between two successive internodes; necessary for saltatory (rapid) conduction.

Nucleus (plural = nuclei): An aggregation of neurons within the CNS; in histology, the nucleus of a cell.

Nystagmus: An involuntary oscillation of the eye(s), typically slow in one direction and rapid in the other; named for the direction of the rapid component.

Oligodendrocyte: A neuroglial cell; forms and maintains the myelin sheath in the CNS; each cell is responsible for several internodes on different axons.

Operculum (from the Latin: "to cover"): Regions of cerebral cortex along the upper and lower edges of the lateral (Sylvian) fissure.

Optic chiasm(a): Partial crossing of optic nerves—the nasal half of the retina representing the temporal visual fields—after which the optic tracts are formed.

Optic disc: Area of the retina where the optic nerve exits; also the site for the central retinal artery and vein; devoid of receptors, hence, the physiological blind spot.

Optic nerve: Second cranial nerve (CN II); special sense of vision; actually a pathway of the CNS, from the ganglion cells of the retina until the optic chiasm as the optic nerve, and its continuation, the optic tract.

Palmomental reflex: This is a reflex movement of the mouth and lips of the contralateral face that occurs when the palm is stroked briskly on one side. This indicates disinhibition of the frontal cortex due to injury from stroke, trauma, or a degenerative process.

Papilledema: Edema of the optic disc (also called a choked disc); visualized with an ophthalmoscope; usually a sign of increased intracranial pressure.

Paralysis: Complete loss of muscular action.

Paraplegia: Paralysis of both legs and lower part of trunk.

Paresis: Muscle weakness or partial paralysis.

Paresthesia: Spontaneous abnormal sensation (e.g., tingling; pins and needles).

Parkinsonism: Impaired motor control syndrome characterized by resting tremor, muscle rigidity, akinesia and postural dyscontrol; classic example is Parkinson's disease.

Paroxysmal depolarization shift (PDS): An initial decline in resting membrane potential of a cortical neuron followed by a burst of action potentials, then a return to a normal resting membrane potential.

Pathway: A chain of functionally related neurons (nuclei) and their axons, making a connection between one region of the CNS and another; a tract (e.g., visual pathway, posterior column—medial lemniscus sensory pathway).

Peduncle: A thick stalk or stem; a bundle of nerve fibers (cerebral peduncle of the midbrain; also three cerebellar peduncles: superior, middle, and inferior).

Penumbra: The penumbra is defined as an area of ischemic brain that has partially lost its blood supply but not to the degree that cell death has occurred. This is usually a circumferential area surrounding an area of brain that has been irreversibly damaged (an infarct). The goal of stroke treatment is to identify patients with areas of penumbra and to target treatment to these areas before irreversible damage occurs.

Perikaryon: The cytoplasm surrounding the nucleus of a cell; sometimes refers to the cell body of a neuron.

Peripheral nervous system (PNS): Nerve roots, peripheral nerves and ganglia (motor, sensory and autonomic) outside the CNS.

PET: Positron emission tomography. A radionuclear technique using the properties of positron emission. This technique uses radionuclear analogs of biological substrates such as glucose, oxygen or neurotransmitters such as dopamine to visualize areas of the brain where these substances are metabolically active.

Pia (mater): The thin innermost layer of the meninges, attached to the surface of the brain and spinal cord; forms the inner boundary of the subarachnoid space.

Plexus: An interweaving arrangement of vessels or nerves.

Pons (bridge): The middle section of the brainstem that lies between the medulla and the midbrain; appears to constitute a bridge between the two hemispheres of the cerebellum.

Posterior column: Fasciculus gracilis and fasciculus cuneatus of the spinal cord, pathways (tracts) for discriminative touch, conscious proprioception and vibration; dorsal column.

Praxis: Greek word meaning "do" or "act"; in neurological parlance refers to normal performance of complex procedural motor activities such as combing hair or brushing the teeth.

Projection fibers: Bidirectional fibers connecting the cerebral cortex with structures below, including basal ganglia, thalamus, brainstem, and spinal cord.

Proprioception: The sense of body position (conscious or unconscious).

Proprioceptor: One of the specialized sensory endings in muscles, tendons and joints; provides information concerning movement and position of body parts (proprioception).

Prosody: Vocal tone, inflection, and melody accompanying speech.

Ptosis: Drooping of the upper eyelid; can be unilateral or bilateral.

Pulvinar: The posterior nucleus of the thalamus; functionally, linked with visual association cortex.

Putamen: The larger (lateral) part of the lentiform nucleus, with the globus pallidus; part of the neostriatum with the caudate nucleus.

Pyramidal system: So-called because the corticospinal tracts occupy pyramid-shaped areas on the ventral aspect of the medulla; may include corticobulbar fibers. The term pyramidal tract refers specifically to the corticospinal tract.

Quadriplegia: Paralysis affecting the four limbs (also called tetraplegia).

Radicular: Refers to a nerve root (motor or sensory).

Ramus (plural = rami): The division of the mixed spinal nerve (containing sensory, motor and autonomic fibers) into anterior and posterior branches.

Rancho Los Amigos Levels of Cognitive Functioning (RLA): This scale was designed to measure and track an individual's progress early in the recovery period following brain injury. The RLA scale describes levels of functioning and is used to assess the efficacy of treatment programs.

Raphe: An anatomical structure in the midline; in the brainstem, several nuclei of the reticular formation are in the midline of the medulla, pons, and midbrain (these nuclei use serotonin as their neurotransmitter).

Red nucleus: Nucleus in the midbrain (reddish color in a fresh specimen).

Reflex: Involuntary movement of a stereotyped nature in response to a stimulus.

Reflex arc: Consisting of an afferent fiber, a central connection (synapse), a motor neuron and its efferent axon leading to a muscle movement.

REM behavior disorder (RBD): A sleep disorder characterized by what appears to be the acting out of dreams. This phenomenon occurs during REM sleep in individuals who have lost REM atonia; seen in various forms of synucleinopathies such as Parkinson's disease and multisystem atrophy.

Repetitive nerve stimulation: A technique by which a repetitive electrical stimulus is applied to a peripheral nerve and the output is measured over a muscle innervated by that nerve; changes in compound motor action potential (CMAP) amplitude over time are assessed. A decremental response (decreasing CMAP amplitude with repetitive stimulation) suggests a fatiguable neuromuscular junction (NMJ) and therefore a postsynaptic problem. An incremental response (increasing CMAP amplitude with repetitive stimulation) suggests a presynaptic problem.

Reticular: Pertaining to or resembling a net; reticular formation of the brainstem.

Reticular formation: Diffuse nervous tissue nuclei and connections in brainstem, quite old phylogenetically.

Rhinencephalon: Refers in humans to structures related to the olfactory system.

Rigidity: Abnormal muscle stiffness (increased tone) with increased resistance to passive movement of both agonists and antagonists (e.g., flexors and extensors), usually seen in Parkinson's disease; velocity independent.

Root: The peripheral nerves—sensory (dorsal) and motor (ventral)—as they emerge from the spinal cord and are found in the subarachnoid space.

Rostral: Toward the snout, or the most anterior end of the neuraxis.

Rubro (red): Pertaining to the red nucleus, as in rubrospinal tract and corticorubral fibers.

Saccadic (to jerk): Extremely quick movements, normally of both eyes together (conjugate movement), while changing the direction of gaze.

Schwann cell: Neuroglial cell of the PNS responsible for formation and maintenance of myelin; there is one Schwann cell for each internode segment of myelin.

Secretomotor: Parasympathetic motor nerve supply to a gland.

Sensory (afferent): To do with receiving information, from the skin, the muscles, the external environment or internal organs

Septal region: An area below the anterior end of the corpus callosum on the medial aspect of the frontal lobe that includes the cortex and the (subcortical) septal nuclei.

Septum pellucidum: A double membrane of connective tissue separating the frontal horns of the lateral ventricles; situated in the median plane.

Single fiber EMG: A test of the integrity of the neuromuscular junction. The ability of neuromuscular junctions to transmit impulses within a given motor unit is tested and expressed as a parameter called "jitter." The test is based on measuring the variance of jitter between two neuromuscular junctions within the same motor unit.

Somatic: Used in neurology to denote the body, exclusive of the viscera (as in somatic afferent neurons from the skin and body wall); the word *soma* is also used to refer to the cell body of a neuron.

Somatic senses: Touch (discriminative and crude), pain, temperature, proprioception, and the sense of "vibration."

Somatosensory evoked potentials (SSEP): A neurophysiological test that measures the response to an electrical stimulus applied to either upper or lower limbs in order to determine the speed of conduction from peripheral nerves through the spinal cord to the contralateral somatosensory cortex. An electrical stimulus is usually applied to the median nerves at the wrist or to the posterior tibial nerves at the ankles; each side and limb is done separately. This test assists in localization by assessing where there is slowing of transmission of the evoked potential in the various structures through which the signal passes.

Somatotopic: The orderly representation of the body parts in CNS pathways, nuclei, thalamus and cortex; topographical representation.

Somesthetic: Consciousness of having a body; somesthetic (somatic) senses are the general senses of touch, pain, temperature, position, movement and "vibration."

Spasticity: Velocity-dependent increased tone and increased resistance to passive stretch of the limb muscles; in humans, most often the elbow, wrist, knee and foot flexors as well as the hip adductors; usually accompanied by hyperreflexia.

Special senses: Sight (vision), hearing (audition), balance (vestibular), taste (gustatory) and smell (olfactory).

Sphingolipid: Complex glycolipid component of neuronal and glial cell membranes; may accumulate intracellularly in hereditary disorders of sphingolipid degradation and recycling.

Spinal shock: Complete "shut down" of all spinal cord activity (in humans) following an acute complete lesion of the cord (e.g., severed cord after a diving or motor vehicle accident), below the level of the lesion; usually up to two to three weeks in duration. This can also occur after indirect trauma such as a blast injury.

Spinocerebellar tracts: Ascending tracts of the spinal cord, anterior and posterior, for transmission of "unconscious" proprioceptive information to the cerebellum.

Spinothalamic tracts: Ascending tracts of the spinal cord for pain and temperature (lateral) and nondiscriminative or light touch and pressure (anterior).

Split brain: A brain in which the corpus callosum has been severed in the midline, usually as a therapeutic measure for intractable epilepsy of frontal lobe origin.

Stereognosis: The recognition of an object using the tactile senses and also central processing, involving association areas especially in the parietal lobe.

Strabismus (a squint): Lack of conjugate fixation of the eyes; may be constant or variable.

Striatum: The phylogenetically more recent part of the basal ganglia (neostriatum) consisting of the caudate nucleus and the putamen (lateral portion of the lentiform nucleus).

Stroke: A sudden severe attack of the CNS; usually refers to a sudden focal loss of neurologic function due to death of neural tissue; most often due to a vascular lesion, either infarct (embolus, occlusion) or hemorrhage.

Subarachnoid space: Space between the arachnoid and pia mater, containing CSF.

Subcortical: Not in the cerebral cortex, that is, at a functionally or evolutionary "lower" level in the CNS; usually refers to the white matter of the cerebral hemispheres, and also may include the basal ganglia or other nuclei.

Substantia gelatinosa: A nucleus of the gray matter of the dorsal (sensory) horn of the spinal cord composed of small neurons; receives pain and temperature afferents.

Substantia nigra: A flattened nucleus with melanin pigment in the neurons located in the midbrain and having motor functions—consisting of two parts: the pars compacta, with dopaminergic neurons that project to the striatum and which degenerate in Parkinson's disease, and the pars reticulata, which projects more diffusely to forebrain structures.

Subthalamus: Region of the diencephalon beneath the thalamus, containing fiber tracts and the subthalamic nucleus; part of the functional basal ganglia.

Sulcus (plural = sulci): Groove between adjacent gyri of the cerebral cortex; a deep sulcus may be called a fissure.

Supratentorial: The space and structures contained in the area above the tentorium cerebelli or cerebellar tentorium, such as the cerebral hemispheres and ventricular system.

Synapse: Area of structural and functional specialization between neurons where transmission occurs (excitatory, inhibitory or modulatory), using neurotramsmitter substances (e.g., glutamate, GABA); similarly at the neuromuscular junction (using acetylcholine).

Syringomyelia: A pathological condition characterized by expansion of the central canal of the spinal cord with destruction of nervous tissue around the cavity.

Tectum: The "roof" of the midbrain (behind the aqueduct) consisting of the paired superior and inferior colliculi; also called the quadrigeminal plate.

Tegmentum: The "core area" of the brainstem, between the ventricle (or aqueduct) and the corticospinal tracts; contains the reticular formation, cranial nerve and other nuclei, and various tracts.

Telencephalon: Rostral part of embryonic forebrain; primarily cerebral hemispheres of the adult brain.

Tensilon test: A procedure that temporarily increases the amount of acetylcholine present in the synaptic cleft to overcome a block at the neuromuscular junction. This test uses a medication called Tensilon or edrophonium, a mild reversible acetylcholinesterase inhibitor. The medication (usually a 9 mg bolus) is given intravenously (IV) after an initial dose of 1 mg is given to test for bradycardia. The test is positive if the Tensilon reverses the neurological deficit for 5 to 10 minutes. This test should always be performed with cardiac monitoring and with IV atropine drawn and available should bradycardia or hypotension develop.

Tentorium: The tentorium cerebelli is a sheet of dura between the occipital lobes of the cerebral hemispheres and the cerebellum; its hiatus or notch is the opening for the brainstem—at the level of the midbrain.

Thalamus: A major portion of the diencephalon with sensory, motor and integrative functions; consists of several nuclei with reciprocal connections to areas of the cerebral cortex.

Third (IIIrd) ventricle: Midline ventricle at the level of the diencephalon (between the thalamus of each side), containing CSF.

Tic: Brief, repeated, stereotyped, semipurposeful muscle contraction; not under voluntary control although may be suppressed for a limited time.

Tinnitus: Persistent intermittent or continuous ringing or buzzing sound in one or both ears.

Tomography: Radiological images, CT or MRI, done sectionally.

Tone: Referring to muscle, its firmness and elasticity—normal, hyper or hypo—elicited by passive movement, and also assessed by palpation.

tPA: Tissue plasminogen activator. Catalyzes the conversion of plasminogen to plasmin resulting in the lysis of newly formed clots. If given within three to four hours of the onset of symptoms, this treatment can lyse acute clots and reverse deficits due to acute cerebral ischemia. It is now considered the standard of care for acute ischemic stroke.

Tract: A bundle of nerve fibers within the CNS, with a common origin and termination (e.g., optic tract, corticospinal tract).

Transient ischemic attack (TIA): A nonpermanent focal deficit, caused by a vascular event; by definition, usually reversible within a few hours, with a maximum of 24 hours.

Tremor: Oscillating, rhythmic movements of the hands, limbs, head or voice; intention (kinetic) tremor of the limb commonly seen with cerebellar lesions; tremor at rest commonly associated with Parkinson's disease.

Two-point discrimination: Recognition of the simultaneous application of two points close together on the skin; varies with the area of the body (e.g., compare finger tip to back).

Uncus: An area of cortex—the medial protrusion of the rostral (anterior) part of the parahippocampal gyrus of the temporal lobe; the amygdala is situated deep to this area; important clinically as in uncal herniation.

Upper motor neuron: Neuron located in the motor cortex or other motor areas of the cerebral cortex, giving rise to a descending tract to lower motor neurons in the brainstem (corticobulbar, for cranial nerves) or spinal cord (corticospinal, for body and limbs); may also refer to brainstem neurons projecting to the spinal cord (e.g., reticular formation giving rise to the reticulospinal tract).

Upper motor neuron lesion: A lesion of the brain (cortex, white matter of hemisphere), brainstem, or spinal cord interrupting descending (corticospinal and reticulospinal) motor influences to the lower motor neurons of the spinal cord, characterized by weakness, spasticity and hyperreflexia, and often clonus; accompanied by a Babinski response.

Ventral root: Efferent motor component of a spinal nerve; situated between the spinal cord and the mixed spinal nerve, including its portion traveling in the subarachnoid space within the CSF.

Ventricles: Cerebrospinal (CSF) fluid-filled cavities inside the brain.

Vermis: Unpaired midline portion of the cerebellum, between the hemispheres.

Vertigo: Abnormal sense of spinning, whirling or motion, either of the self or of one's environment.

Visual evoked potential (VEP): A neurophysiological technique using EEG electrodes applied over the occipital cortex to determine the speed of conduction of impulses from the retina through the optic nerve, optic tracts, optic radiations to the occipital cortex; the stimulus may be an alternating checkerboard pattern on a TV monitor or, for young children and comatose patients, a light flash from light-emitting diode (LED) goggles. This test assists in localization by assessing whether there is slowing of transmission of the evoked potential in the various structures through which the signal passes.

White matter: Nervous tissue of the CNS made up of nerve fibers (axons), some of which are myelinated; appears "whitish" after fixation in formalin.

Annotated Bibliography

This is a select list of references with some commentary to help the learner choose additional learning resources about the structure, function and diseases of the human brain.

The perspective is for medical students and residents, and also for non-neurological practitioners, as well as those in related fields in the allied health professions.

The listing includes texts, atlases (some with CDs or publisher Web sites), as well as Web sites and videotapes.

Neuroanatomical Texts and Atlases

The listing includes both neuroanatomical textbooks and atlases, as well as a few neuroanatomical texts with a significant clinical emphasis; recent publications (since the year 2000) have usually been preferentially selected.

Texts (Listed Alphabetically)

Augustine, J.R. *Human Neuroanatomy: An Introduction.* Elsevier, NY: Academic Press, 2008.

The human nervous system is explained in full detail in this book, including mention of injuries to the various systems and other clinical correlations. The illustrations are almost entirely in black and white. There is an extensive reference section at the end.

Conn, P.M., Ed. *Neuroscience in Medicine,* 2nd ed. Totowa, NJ: Humana Press, 2003.

The nervous system is well described from an anatomical and functional perspective in this multi-authored book, with mention of clinical disease states appropriate for each section. Some of the chapters are accompanied by a short section describing clinical disorders (e.g., dementia following the chapter on the cerebral cortex). The illustrations are almost entirely in black and white. The last chapter discusses degeneration, regeneration and plasticity in the nervous system.

Crossman, A.R., and D. Neary. *Neuroanatomy: An Illustrated Colour Text,* 3rd ed. Edinburgh: Elsevier Churchill Livingstone, 2005.

This book format is part of a series of "Illustrated Colour Texts" (the companion *Neurology* text is by Fuller, G. and M. Manford. 2nd ed., 2006). The text is accompanied by many illustrations, including diagrams and photographs of the brain. This book is a useful addendum for students in the health sciences learning about the nervous system for the first time.

FitzGerald, M.J.T., G. Gruener, and E. Mtui. *Clinical Neuroanatomy and Neuroscience*, 5th ed. Edinburgh: Elsevier Saunders, 2007.

The authors have attempted to create an integrated text for medical and allied health professionals, combining the basic neuroscience with clinical material. The neuroanatomical presentation is quite detailed accompanied and richly illustrated, in full color, with large appealing explanatory diagrams, and some MRIs, but there are few actual photographs. The clinical syndromes are in colored panels sometimes accompanied by illustrations. Each chapter ends with a colored panel with core information. A glossary of terms is included at the end of the book.

Gertz, S.D. *Liebman's Neuroanatomy Made Easy and Understandable*, 7th ed. Austin, TX: Pro-ed Publishers, 2007.

The title promises a book that makes neuroanatomy understandable, which it sometimes succeeds in doing by simplifying the subject matter and bringing in some clinical correlations. Making the subject easy is possibly not achievable, certainly not with the rather low quality of illustrations. The book may be usable by students in a nonbiological field of study, but is not nearly sufficient for medical students.

Gilman, S., and S.W. Newman. *Manter and Gatz's Essentials of Clinical Neuroanatomy and Neurophysiology*, 10th ed. Philadelphia, PA: F.A. Davis, 2003.

This slender book attempts to condense a lot of neuroanatomical information with clinical correlations throughout. The illustrations are sketch-like and full of connections. This is possibly a handy review book for those who have taken a full course on the nervous system/neurology, and would not be adequate as a learner's textbook.

Haines, D.E. *Fundamental Neuroscience for Basic and Clinical Applications*, 3rd ed. Philadelphia, PA: Churchill Livingstone, 2006.

This edited, multiauthored large text, with many color illustrations, is an excellent reference book, mainly for neuroanatomical detail.

Kandel, E.R., J.H. Schwartz, and T.M. Jessell. *Principles of Neural Science,* 4th ed. New York: McGraw-Hill, 2000.

This thorough textbook presents a physiological depiction of the nervous system, in some detail, often with experimental details and information from animal studies. It is suitable as a reference book and for graduate students.

Kiernan, J.A. *Barr's The Human Nervous System: An Anatomical Viewpoint,* 9th ed. Baltimore: Lippincott Williams and Wilkins, 2009.

This new edition of Barr's neuroanatomical textbook has additional color, as well as clinical notes (in boxes) and MRIs. It is clearly written and clearly presented, with a glossary. There is no longer an accompanying CD but a publisher's Web site (accessed with the purchase of the book) with some of the illustrations, sample exam questions and some clinical cases.

Kolb, B., and I.Q. Whishaw. *Fundamentals of Human Neuropsychology*, 6th ed. New York: Worth Publishers, 2008.

A classic in the field and highly recommended for a good understanding of the human brain in action. Topics discussed include memory, attention, language and the limbic system.

Martin, J.H. *Neuroanatomy: Text and Atlas*, 3rd ed. New York: McGraw-Hill, 2003.

A very complete text, with a neuroanatomical perspective and accompanied by some fine (two-color) explanatory illustrations, written as the companion to Kandel et al. The material is clearly presented, with explanations of how systems function. A detailed Atlas section is included at the end, as well as a glossary of terms.

Nieuwenhuys, R., J. Voogd, and C. van Huijzen. *The Human Central Nervous System,* 4th ed. Berlin: Springer, 2008.

The authors have transformed a book with unique 3-dimensional drawings of the CNS and its pathways, in half-tones, into a complete textbook of neuroanatomy. Over 200 new diagrams have been added, some with color. The terminology has likewise been altered, from the previous Latin to more contemporary terminology. The text is clearly written and heavily referenced, with some mention of clinical conditions. This book should now be considered as another reference source for information on functional neuroanatomy.

Nolte, J. *Elsevier's Integrated Neuroscience,* Phildelphia, PA: Mosby Elsevier, 2007.

This neuroscience (neuroanatomy) primer is part of a series from Elsevier press integrating the basic sciences – anatomy, histology, biochemistry, physiology, genetics, immunology, pathology, and pharmacology. These books are also linked via the Elsevier 'student consult' Web site (available only to the purchaser of these books). The illustrations are well designed to elucidate the structure and function of the nervous system. The only clinical link is a number of exemplary case studies (and their answers) at the end of the book.

Nolte, J. *The Human Brain: An Introduction to Its Functional Anatomy*, 6th ed. St. Louis, MO: Mosby, 2008.

This is a new edition of an excellent neuroscience text, with anatomical and functional (physiological) information on the nervous system, complemented with clinically relevant material. The textbook includes scores of illustrations in full color, stained brainstem and spinal cord cross-sections, along with three-dimensional brain reconstructions by John Sundsten. There is now added material via a publisher's Web site (accessed with the purchase of the book). The book is probably now too detailed to be considered a textbook for the abbreviated courses of a medical curriculum and may now be considered in the category of a reference book.

Paxinos, G., and J.K. Mai, Eds. *The Human Nervous System,* 2nd ed. Oxford: Elsevier Academic Press, 2004.

Amongst the books reviewed, this is the most extensive in its discussion of the anatomical aspects of the human nervous system – including development, the peripheral and autonomic nervous systems, and the structures of the central nervous system. This would be a reference book to consult for the details.

Rubin, M, and J.E. Safdieh. *Netter's Concise Neuroanatomy.* Philadelphia, PA: Saunders Elsevier, 2007.

This small-sized book with Netter's familiar illustrations has little in the way of text, but contains many tables with a lot of both anatomical and functional information. The illustrations are less useful in their smaller versions.

Steward, O. *Functional Neuroscience.* New York: Springer, 2000.

According to the author, this is a book for medical students that blends the physiological systems approach with the structural aspects. The emphasis is on the "processing" of information, for example, in the visual system. Chapters at the end discuss arousal, attention, consciousness and sleep. Nicely formatted and readable.

Williams, P., and R. Warwick. *Functional Neuroanatomy of Man.* Philadelphia, PA: W.B. Saunders, 1975.

This is the "neuro" section from *Gray's Anatomy.* Although somewhat dated, there is excellent reference material on the central nervous system, as well as the nerves and autonomic parts of the peripheral nervous system. The limbic system and its development are also well described.

Wilson-Pauwels, L., E.J. Akesson, and P.A. Stewart. *Cranial Nerves: Anatomy and Clinical Comments*. Toronto, BC: Decker, 1988.

A handy resource on the cranial nerves, with some very nice illustrations. Relatively complete and easy to follow.

Atlases

DeArmond, S.J., M.M. Fusco, and M.M. Dewey. *Structure of the Human Brain: A Photographic Atlas,* 3rd ed. New York: Oxford University Press, 1989.

An excellent and classic reference to the neuroanatomy of the human CNS. No explanatory text and no color.

England, M.A., and J. Wakely. *Color Atlas of the Brain and Spinal Cord: An Introduction to Normal Neuroanatomy*. Philadelphia, PA: Mosby Elsevier, 2006.

A very well illustrated atlas, with most of the photographs and sections in color. Little in the way of explanatory text.

Felten, D.L., and R.F. Jozefowicz. *Netter's Atlas of Human Neuroscience*. Teterboro, NJ: Icon Learning Systems, 2003.

The familiar illustrations of Netter on the nervous system have been collected into a single atlas, each with limited commentary. Both peripheral and autonomic nervous systems are included. The diagrams are extensively labeled (also see below).

Haines, D. *Neuroanatomy: An Atlas of Structures, Sections and Systems*, 6th ed. Baltimore: Lippincott, Williams and Wilkins, 2004.

A popular atlas that has some excellent photographs of the brain, some color illustrations of the vascular supply, with additional radiologic material, all without explanatory text. The histological section of the brainstem is very detailed. There is a limited presentation of the pathways and functional systems, with text. This edition comes with a CD containing all the illustrations, with some accompanying text.

Hendelman, W.J. *Atlas of Functional Neuroanatomy*, 2nd ed., with CD. Boca Raton, FL: CRC Press, Taylor & Francis, 2006.

This is a learner-oriented atlas, where each illustration has accompanying explanatory text on the opposite page. Each of the sections has a brief introduction (orientation, functional systems, neurological neuroanatomy, and the limbic system). The accompanying CD has all the illustrations, with the added features of "roll-over" (mouse-over) labeling and animation added to the various pathways.

Mai, J.K., G. Paxinos, and T. Voss. *Atlas of the Human Brain,* 3rd ed. Heidelberg: Elsevier Academic Press, 2008.

The human brain is mapped in stereotaxic detail, using a very large-sized format. In part I, there are photographs (above) accompanied by labeled llustrations (below), sometimes accompanied by MRIs. The second part consists of high resolution myelin-stained sections on one side with labeled color diagrams on the opposite page. There are also detailed reconstructions of brain structures. The Atlas includes a DVD of the images plus 3D visualization software.

Netter, F.H. *The CIBA Collection of Medical Illustrations, Vol. 1, Nervous System, Part 1: Anatomy and Physiology*. Summit, NJ: CIBA, 1983.

A classic. Excellent illustrations of the nervous system, as well as the skull, the autonomic and peripheral nervous systems, and embryology. The text is interesting but may be dated.

Nieuwenhuys, R., J. Voogd, and Chr. van Huijzen. *The Human Central Nervous System*. Berlin: Springer, 1981.

Unique three-dimensional drawings of the CNS and its pathways are presented, in tones of gray. These diagrams are extensively labeled, with no explanatory text.

Nolte, J., and J.B. Angevine. *The Human Brain in Photographs and Diagrams*, 3rd ed. Philadelphia, PA: Mosby Elsevier, 2007.

A well-illustrated (color) atlas, with text and illustrations, and neuroradiology. Functional systems are drawn onto the brain sections with the emphasis on the neuroanatomy; the accompanying text is quite detailed. Excellent three-dimensional brain reconstructions by J.W. Sundsten. There is a chapter with clinical imaging. The glossary includes (colored) images. The accompanying CD has all the images (in various formats), with some explanatory text for each.

Woolsey, T.A., J. Hanaway, and M.H. Gado. *The Brain Atlas: A Visual Guide to the Human Central Nervous System*, 3rd ed. Hoboken, NJ: Wiley, 2008.

A complete pictorial atlas of the human brain, with some color illustrations and radiographic material.

Neuroanatomical/Clinical Texts

Afifi, A.K., and R.A. Bergman. *Functional Neuroanatomy: Text and Atlas*, 2nd ed. New York: McGraw-Hill, 2005.

This is a neuroanatomical text with functional information on clinical syndromes. A chapter on the normal is followed by a chapter on clinical syndromes (e.g., spinal cord). The book is richly illustrated (in two colors) using semi-anatomic diagrams and MRIs; there is an atlas of the CNS at the end, but it is not in color. It is a pleasant book visually and quite readable.

Benarroch, E.E., J.R. Daube, K.D. Flemming, and B.F. Westmoreland. *Mayo Clinic Medical Neurosciences: Organized by Neurologic Systems and Levels,* 5th ed. Florence, KY: Informa Healthcare, 2008.

The first part of this book is devoted to the basic sciences, neuroanatomy, neuropathology, neurophysiology and synaptic transmission, underlying the clinical diagnosis of neurologic disorders. The second part consists of a description of the various systems – sensory, motor, consciousness, cerebrospinal fluid and vascular—presented with a clinical perspective. In the third part the nervous system is viewed at various horizontal (axial) levels, starting in the periphery and up to the cortical region.

The book is very nicely illustrated in 4-color, with functional diagrams and other illustrations, including imaging. The illustrations underscore the functional aspects of the nervous system which is being explained, from synapses to pathways. Each chapter includes clinical correlations and clinical problems (in boxes). This full-sized book is printed on high-quality paper and laid out to make it easily readable. It is too extensive for a "required" book, but is perhaps most useful for medical students doing a clinical elective in neurology and also for neurology residents at the beginning of their training.

Blumenfeld, H. *Neuroanatomy through Clinical Cases*. Sunderland, MA: Sinauer Associates, 2002.

This book attempts to bridge the gap between basic neuroanatomy and clinical neurology. Each chapter (e.g., motor pathways, visual system) presents a thorough explanation of the relevant neuroanatomy and is well illustrated, in full color. This is followed by a section presenting key clinical concepts, with examples of signs and symptoms of various clinical disorders relevant to that aspect of the nervous system. The margin includes some illustrations of clinical examinations, some review exercises, plus some mnemonics. Then come the clinical cases, usually several, accompanied by a discussion of the case and its clinical course, and the relevant imaging (unfortunately it is sometimes difficult to keep track of which imaging belongs to which case).

This is a valuable book for the advanced learner, likely a neurology or neurosurgery resident. The book accomplishes what it sets out to do by its title, but perhaps too much so by trying to be too comprehensive in each domain. There is no CD-ROM or Web site with this edition of the book.

Fuller, G., and M. Manford. *Neurology: An Illustrated Colour Text,* 2nd ed. New York: Elsevier Churchill Livingstone, 2006.

The format of this book is part of a series of Illustrated Colour Texts (the companion *Neuroanatomy* text is by Crossman, A.R. and Neary, D., 3rd ed., 2005). This book presents select clinical entities with concise explanations, accompanied by many illustrations (in full color). It is not intended to be a comprehensive textbook. The large format and presentation make this an appealing but limited book.

Marcus, E.M., and S.J. Jacobson. *Integrated Neuroscience: A Clinical Problem Solving Approach.* Boston: Kluwer Academic Publishers, 2003.

This book attempts to bridge the gap between basic neuroanatomy and clinical neurology. The neuroanatomy is presented quite succinctly and accompanied by simplified illustrations, not in color. Diseases relevant to the system being discussed are included within each chapter. Case histories are presented and there are problem-solving chapters for different levels of the nervous system; associated questions invite the problem-solving process, rather briefly. The CD-ROM accompanying the book has a variety of anatomical and MRI images, and some of the problems with additional commentary. Overall, the book does not integrate the anatomy and physiology necessary to resolve the clinical problem and, hence, does not live up to its title, although it may be useful for senior neurology residents preparing for examinations.

Neurologic Examination

Bickley, L.S., Ed. *Bates' Guide to Physical Examination and History Taking,* 10th ed. Philadelphia, PA: Lippincott Williams & Wilkins, 2009.

This is the gold standard for medical students on how to conduct a physical examination, as well as for history-taking. An audio-visual (videotape or CD-ROM) has been prepared for some of the systems.

Campbell, W.W. *DeJong's The Neurologic Examination,* 6th ed. Philadelphia, PA: Lippincott Williams & Wilkins, 2005.

A more advanced and detailed description of the neurologic examination is found in this book, more intended for the resident in neurology and neurosurgery, and perhaps also for rehabilitation physicians (physiatrists). It is appropriately illustrated with some neuroanatomy and with many examples of physical signs associated with diseases of the nervous system.

Clinical Texts

Aminoff, M.J., D.A. Greenberg, and R.P. Simon. *Clinical Neurology*, 6th ed. New York: Lange Medical Books/McGraw-Hill, 2005.

If a student wishes to consult a clinical book for a quick look at a disease or syndrome, then this is a suitable book of the survey type. Clinical findings are given, investigative studies are included, as well as treatment. The illustrations are adequate (in two colors), and there are many tables with classifications and causes.

Asbury, A.K., G.M. McKhann, W.I. McDonald, P.J. Goodsby, and J.C. McArthur. *Diseases of the Nervous System: Clinical Neuroscience and Therapeutic Principles*, 3rd ed. New York: Cambridge University Press, 2002.

A complete neurology text, in two volumes, on all aspects of basic and clinical neurology, and the therapeutic approach to diseases of the nervous system.

Donaghy, M. *Brain's Diseases of the Nervous System*, 12th ed. New York: Oxford University Press, 2008.

A very trusted source of information about clinical diseases and their treatment.

Fauci, A.S., S.L. Hauser, J.L. Jameson, J. Loscalzo, E. Braunwald, D.L. Kasper, and D.L. Longo, Eds. *Harrison's Principles of Internal Medicine*, 17th ed. New York: McGraw-Hill, 2008.

Harrison's is a trusted, authoritative source of information, with few illustrations. Part 2 in Section 3 (Volume I) has chapters on the presentation of disease; Part 15 (Volume II) is on all neurologic disorders of the CNS, nerve and muscle disease, as well as mental disorders. The online version of Harrison's has updates, search capability, practice guidelines and online lectures and reviews, as well as illustrations.

Patten, J. *Neurological Differential Diagnosis*, 2nd ed. New York: Springer, 2004

This comprehensive text systematically reviews clinical disorders affecting each major echelon of the nervous system. Individual chapters are devoted to the clinical assessment of the pupils, visual fields, retina, cranial nerves controlling eye movement, cerebral hemispheres, brainstem, nerve roots (and so on) and the diseases seen in each region. Neuroanatomical information is introduced in each chapter, as required. An important element is the liberal use of brief clinical vignettes to highlight many of the disorders being considered. The anatomical and clinical illustrations are largely the work of the author himself and are in gray-scale. The text is too detailed for medical students but is an excellent resource for trainees in neurology.

Ropper, A.H., and R.H. Brown. *Adams and Victor's Principles of Neurology*, 8th ed. New York: McGraw-Hill, 2005.

A comprehensive neurology text, part devoted to cardinal manifestations of neurologic diseases, and part to major categories of diseases.

Rowland, L.P. *Merritt's Neurology*, 11th ed. Baltimore: Lippincott Williams and Wilkins, 2005.

A well-known, complete and trustworthy neurology textbook, now edited by L.P. Rowland.

Royden-Jones, H. *Netter's Neurology*. Teterboro, NJ: Icon Learning Systems, 2005.

Netter's neurological illustrations have been collected in one textbook, with the addition of Netter-style clinical pictures; these add an interesting dimension to the descriptive text. There is a broad coverage of many disease states, but not in depth, with clinical scenarios in each chapter. The CD accompanying the book contains the complete textbook and its illustrations.

This is definitely a book that both medical students and others in the allied health sciences might consult in the course of their studies and clinical duties.

Schapira, A.H.V., Ed. *Neurology and Clinical Neuroscience*. Philadelphia, PA: Mosby Elsevier, 2007.

This new book, multiauthored, is a compendium of neurological disorders. The subject matter covers the full range, from apraxia to schizophrenia. The chapters include tables and charts, photographs, neuropathology and MRIs, with the judicious use of color. Clearly a reference book.

Weiner, W.J., and C.G. Goetz, Ed. *Neurology for the Non-Neurologist*, 5th ed. Philadelphia, PA: Lippincott Williams & Wilkins, 2004.

This is a thoughtful approach to common neurological symptoms (e.g., vertigo) and diseases (e.g., movement disorders). There are chapters on diagnostic tests and neuroradiology. There are also chapters on principles of neurorehabilitation and medical-legal issues in the care of neurologic patients. The book is sparsely illustrated and not in color.

Pediatric Neurology

Fenichel, G.M. *Clinical Pediatric Neurology: A Signs and Symptoms Approach*, 5th ed. Philadelphia, PA: Elsevier Saunders, 2005.

This book is recommended to medical students and other novices by a highly experienced pediatric neurologist as a basic text with a clinical approach, using signs and symptoms.

Neuropathology

Kumar, V., A.K. Abbas, N. Fausto, S.L. Robbins, and R.S. Cotran, Eds. *Robbins and Cotran Pathologic Basis of Disease*, 7th ed. Philadelphia, PA: Elsevier Saunders, 2005.

A complete source of information for all aspects of pathology for students, including neuropathology. Purchase of the book includes a CD with interactive clinical cases, and access to the Web site.

Kumar, V., R.S. Cotran, and S.L. Robbins, Eds. *Robbins Basic Pathology*, 8th ed. Philadelphia, PA: W.B. Saunders, 2007.

Not as complete as the other text (above).

Web Sites

Web sites should only be recommended *after* they have been critically evaluated by the teaching or clinical faculty. If keeping up with various texts is not difficult enough, a critical evaluation of the various Web resources is an impossible task for any single person. This is indeed a task to be shared with colleagues, and perhaps undertaken by a consortium of teachers and students.

Additional sources of reliable information on diseases are usually available on the disease-specific Web site maintained by an organization, usually with clear explanatory text on the disease often accompanied by excellent illustrations.

The following sites have been visited by the author (WH), and several of them are gateways to other sites—clearly not every one of the links has been viewed. Although some are intended for the lay public, they may contain good illustrations and/or other links.

The usual WWW precaution prevails—look carefully at who created the Web site, and when. A high-speed connection is a must for this exploration!

Society for Neuroscience

http://www.sfn.org/

This is the official Web site for the Society for Neuroscience, a very large and vibrant organization with an annual meeting attended by over 30,000 neuroscientists from all over the world. The Society maintains an active educational branch aimed at the public at large. The following are examples of links to their publications and resources.

Brain Facts

Brain Facts is a primer on the brain and nervous system, published by the Society for Neuroscience. It is a starting point for a lay audience interested in neuroscience. This newly revised edition of *Brain Facts* is available in print and in pdf format. The new edition updates all sections and includes new information on brain development, addiction, neurological and psychiatric illnesses and potential therapies.

Brain Briefings

Information about how basic science discoveries lead to clinical applications is published monthly as a two-page newsletter.

Searching for Answers: Families and Brain Disorders

This four-part DVD shows the human face of degenerative brain diseases. Patients and families describe the powerful physical, emotional, and financial impact of these devastating disorders. Researchers tell how they are working to find treatments and cures for Huntington's disease, Parkinson's disease, amyotrophic lateral sclerosis (ALS), and Alzheimer's disease.

Brain Awareness Week

The Society is responsible for sponsoring a week (in the spring) of special educational activities aimed particularly at students in elementary and high schools. Graduate and medical students, and neuroscientists, are encouraged to go out to the schools to engage young people in exploring the nervous system and its diseases. New resources have just been put on-line at various levels (see http://www.ndgo.net/sfn/nerve/ —Neuroscience Education Resources Virtual Encycloportal—"... gateway to credible information and tools from the World Wide Web for learning about the brain and nervous system").

Digital Anatomis Project

http://www9.biostr.washington.edu/da.html

From: Digital Anatomist Project, Department of Biological Structure, University of Washington, Seattle.

Interactive Brain Atlas: Authored by John W. Sundsten, the material includes 2-D and 3-D views of the brain from cadaver sections, MRI scans, and computer reconstructions.

Neuroanatomy Interactive Syllabus: Authored by John W. Sundsten and Kathleen A. Mulligan, this syllabus uses the images in the Neuroanatomy Atlas above, and many others. It is organized into functional chapters suitable as a laboratory guide, with an instructive caption accompanying each image. It contains 3-D computer graphic reconstructions of brain material; MRI scans; tissue sections, some enhanced with pathways; gross brain specimens and dissections; and summary drawings. Chapters include Topography and Development, Vessels and Ventricles, Spinal Cord, Brainstem and Cranial Nerves, Sensory and Motor Systems, Cerebellum and Basal Ganglia, Eye Movements, Hypothalamus and Limbic System, Cortical Connections, and Forebrain and MRI Scan Serial Sections.

BrainSource

http://www.brainsource.com/

BrainSource.com is an informational Web site aimed at enriching professional, practical, and responsible applications of neuropsychological and neuroscientific knowledge. The Web site is presented by neuropsychologist, Dennis P. Swiercinsky, Ph.D.

The site includes a broad and growing collection of information and resources about normal and injured brains, clinical and forensic neuropsychology, brain injury rehabilitation, creativity, memory and other brain processes, education, brain-body health, and other topics in brain science. BrainSource is also a guide to products, books, continuing education, and Internet resources in neuroscience.

This Web site originated in 1998 for promotion of clinical services and as a portal for dissemination of certain documents useful for attorneys, insurance professionals, students, persons with brain injury and their families, rehabilitation specialists, and others working in the field of brain injury. The Web site is growing to expand content to broader areas of neuropsychological application.

Diseases and Disorders

http://www.mic.stacken.kth.se/Diseases/

This is a "MESH classified" site for the general public and health care professionals of all medical diseases, with links to other resources.

Neuroanatomy and Neuropathology on the Internet

http://www.neuropat.dote.hu

This site has been compiled and designed by Katalin Hegedüs MD, PhD, Department of Neurology, University of Debrecen, Hungary.

It is a source for other sites including neuroanatomy, neuropathology, and neuroradiology, and software (commercial and non-commercial) on the brain, and even including quizzes.

Hardin MD

http://www.lib.uiowa.edu/hardin/md/neuro.html

This site is a service of the Hardin Library for Health Sciences, University of Iowa.

Hardin MD was first launched in 1996, as a source to find the best lists, or directories, of information in health and medicine. Hence, the name Hardin MD comes from Hardin Meta Directory, since the site was conceived as a "directory of directories."

Providing links to high quality directory pages is still an important part of Hardin MD. In recent years, however, other types of links have been added: Just Plain Links pages (http://www.lib.uiowa.edu/hardin/md/jpl.html) have direct links to primary information in circumscribed subjects and many of these pages have links to Medical Pictures.

Medical Neuroscience

http://www.meddean.luc.edu/lumen/MedEd/Neuro/index.htm

Loyola University Medical Education Network, Stritch School of Medicine, Chicago

Whole Brain Atlas

http://www.med.harvard.edu/AANLIB/home.html

The *Whole Brain Atlas* features a Neuroimaging Primer by Keith A. Johnson, M.D., Harvard Medical School.

The Brain from Top to Bottom

http://www.thebrain.mcgill.ca/flash/index_d.html

This site is designed to let users choose the content that matches their level of knowledge. For every topic and subtopic covered on this site, one can choose among three different levels of explanation – beginner, intermediate, advanced.

The major topics include anatomy and function, memory, sensory and motor systems, pain and pleasure, emotion, evolution; other subject areas are under development.

This site focuses on five major levels of organization – social, psychological, neurological, cellular, and molecular. On each page of this site, one can click on a button shown to move among these five levels and learn what role each one plays in the subject under discussion.

The Neurologic Exam: An Anatomical Approach

http://library.med.utah.edu/neurologicexam/html/home_exam.html

This site by Paul D. Larsen, M.D., University of Nebraska School of Medicine, and Suzanne S. Stensaas, Ph.D., University of Utah School of Medicine, includes both an adult and pediatric neurological examination, with video and sound. In addition, there are 4 neurologic cases on this site, with possibly more to come.

The DANA Foundation

http://www.dana.org/

The Dana Foundation is a private philanthropic organization with interests in brain science, immunology, and arts education. It was founded in 1950.

The Dana Alliance is a nonprofit organization of more than 200 pre-eminent scientists dedicated to advancing education about the progress and promise of brain research.

The Brain center is your gateway to the latest research on the human brain. Visit *The BrainWeb & Brain Information* section to access links to validated sites related to more than 25 brain disorders.

Neuroscience for Kids

http://faculty.washington.edu/chudler/neurok.html

Neuroscience for Kids has been created for all students and teachers who would like to learn about the nervous system.

The site contains a wide variety of resources, including images – not only for kids. Sections include exploring the brain, internet neuroscience resources, neuroscience in the news, and reference to books, magazines articles and newspaper articles about the brain.

Neuroscience for Kids is maintained by Eric H. Chudler, Ph.D., and supported by a Science Education Partnership Award (R25 RR12312) from the National Center for Research Resources.

Television Series

http://www.pbs.org/wnet/brain/index.html

The Secret Life of the Brain reveals the fascinating processes involved in brain development across a lifetime.

Episodes include The Baby's Brain, The Child's Brain, The Teenage Brain, The Adult Brain, and The Aging Brain. Also includes History of the Brain, 3-D Brain Anatomy, Mind Illusions, and Scanning the Brain.

The Secret Life of the Brain is a co-production of Thirteen/WNET New York and David Grubin Productions, @2001 Educational Broadcasting Corporation and David Grubin Productions, Inc. All rights reserved.

Videotapes

by W. Hendelman

These edited videotape presentations are on the skull and the brain, as the material would be shown to students in the gross anatomy laboratory. They have been prepared with the same teaching orientation as the *Atlas of Functional Neuroanatomy* by W.J. Hendelman, (2nd ed., 2006) and are particularly useful for self-study or small groups. These videotapes of actual specimens are particularly useful for students who have limited or no access to brain specimens. The videotapes are fully narrated, and each lasts for about 20 to 25 minutes.

Interior of the Skull

This program includes a detailed look at the bones of the skull, the cranial fossa, and the various foramina for the cranial nerves and other structures. Included are views of the meninges and venous sinuses.

The Gross Anatomy of the Human Brain

Part I: The Hemispheres

A presentation on the hemispheres, the functional areas of the cerebral cortex, including the basal ganglia.

Part II: Diencephalon, Brainstem and Cerebellum

A detailed look at the brainstem, with a focus on the cranial nerves, and a functional presentation of the cerebellum.

Part III: Cerebrovascular System and Cerebrospinal Fluid.

A presentation of these two subjects.

Part IV: The Limbic System

A quite detailed presentation on the various aspects of the limbic system, with much explanation and special dissections.

The videotapes are handled by Health Sciences Consortium, a nonprofit publishing cooperative for instructional media. They are now available in DVD format, if so requested.

NOTE: It is suggested that these videotapes be purchased by the library or by an institutional (or departmental) media or instructional resource center.

Information regarding the purchase of these and other videotapes may be obtained from: Health Sciences Consortium, 201 Silver Cedar Ct., Chapel Hill, NC, USA. 27514-1517, Phone: (919) 942-8731. Fax: (919) 942-3689

Index

B

C

D

E

F

O

P

Q

R